# Lecture Notes in Electrical Engineering

Volume 1107

AF811516

The book series *Lecture Notes in Electrical Engineering* (LNEE) publishes the latest developments in Electrical Engineering—quickly, informally and in high quality. While original research reported in proceedings and monographs has traditionally formed the core of LNEE, we also encourage authors to submit books devoted to supporting student education and professional training in the various fields and applications areas of electrical engineering. The series cover classical and emerging topics concerning:

- Communication Engineering, Information Theory and Networks
- Electronics Engineering and Microelectronics
- Signal, Image and Speech Processing
- Wireless and Mobile Communication
- Circuits and Systems
- Energy Systems, Power Electronics and Electrical Machines
- Electro-optical Engineering
- Instrumentation Engineering
- Avionics Engineering
- Control Systems
- Internet-of-Things and Cybersecurity
- Biomedical Devices, MEMS and NEMS

For general information about this book series, comments or suggestions, please contact leontina.dicecco@springer.com.

To submit a proposal or request further information, please contact the Publishing Editor in your country:

**China**

Jasmine Dou, Editor (jasmine.dou@springer.com)

**India, Japan, Rest of Asia**

Swati Meherishi, Editorial Director (Swati.Meherishi@springer.com)

**Southeast Asia, Australia, New Zealand**

Ramesh Nath Premnath, Editor (ramesh.premnath@springernature.com)

**USA, Canada**

Michael Luby, Senior Editor (michael.luby@springer.com)

**All other Countries**

Leontina Di Cecco, Senior Editor (leontina.dicecco@springer.com)

**** This series is indexed by EI Compendex and Scopus databases. ****

Om Hari Gupta · Narayana Prasad Padhy ·
Sukumar Kamalasadan
Editors

# Soft Computing Applications in Modern Power and Energy Systems

## Select Proceedings of EPREC 2023, Volume 4

Springer

*Editors*
Om Hari Gupta
Department of Electrical Engineering
National Institute of Technology
Jamshedpur
Jamshedpur, India

Narayana Prasad Padhy
Department of Electrical Engineering
Indian Institute of Technology Roorkee
Roorkee, Uttarakhand, India

Sukumar Kamalasadan
Department of Electrical and Computer
Engineering
University of North Carolina at Charlotte
Charlotte, NC, USA

ISSN 1876-1100 ISSN 1876-1119 (electronic)
Lecture Notes in Electrical Engineering
ISBN 978-981-99-8006-2 ISBN 978-981-99-8007-9 (eBook)
https://doi.org/10.1007/978-981-99-8007-9



This Springer imprint is published by the registered company Springer Nature Singapore Pte Ltd.
The registered company address is: 152 Beach Road, #21-01/04 Gateway East, Singapore 189721, Singapore

Paper in this product is recyclable.

# Contents

# About the Editors

**Om Hari Gupta** received B.Tech. degree in electrical and electronics engineering from UP Technical University, Lucknow, Uttar Pradesh, India in 2007 and the M.Tech. degree in power electronics and ASIC design from Motilal Nehru National Institute of Technology, Allahabad, Prayagraj, Uttar Pradesh, India, in 2011 and the Ph.D. degree in electrical engineering from Indian Institute of Technology Roorkee, Roorkee Uttarakhand, India, in 2017. He is an awardee of the Canadian Queen Elizabeth II Diamond Jubilee Scholarship for research visiting Canada and from March 2017 to June 2017, he was a visiting Researcher at the Ontario Tech University, Canada. Since 2018, Dr. Gupta has been an Assistant Professor with the Department of Electrical Engineering, National Institute of Technology Jamshedpur, Jharkhand, India. He is one of the founding organizing secretaries of the conference series "Electric Power and Renewable Energy Conference—EPREC". Dr. Gupta is the author of four books and has more than 80 publications including 30 journal papers. He is a senior member of IEEE and a reviewer of many reputed journals like IEEE Transactions on Power Delivery, IET Generation and Transmission, Electric Power Systems Research, Electric Power Components and Systems, and *International Journal of Electrical Power and Energy Systems*. His research interests include power system protection, microgrid, renewable-based distributed generation, and electric power quality.

**Narayana Prasad Padhy** Director, MNIT Jaipur and Director, IIIT Kota, is a Professor (HAG) of Electrical Engineering and has served as Dean of Academic Affairs, IIT Roorkee, Founder Head Mehta Family School of Data Science and Artificial Intelligence, Professor In-Charge Training and Placement, NEEPCO, and Institute Chair Professor at IIT Roorkee. Dr. Padhy is a prolific researcher in the area of Power Systems Engineering and Smart Grid. He has contributed significantly to the domain by way of securing research grants, supervising scholars, publishing research articles, writing books, securing patents, laboratory development and developing online NPTEL courses. Dr. Padhy has held visiting positions at eminent international universities and established research engagements and institutional tie-ups with these universities. He has collaborated with top industries and government agencies for research, laboratory development, and as a knowledge partner. He is well recognised in international and national professional bodies and has actively contributed in various capacities towards their growth. He also represents the country at various international research platforms.

Dr. Padhy is a well awarded and honoured scholarly professional at the national and international levels.

**Sukumar Kamalasadan** is a Duke Energy Distinguished Professor of electric power engineering at the University of North Carolina at Charlotte (UNCC) and the Director of power energy and intelligent systems lab (PEISL) within the Energy Production and Infrastructure Center (EPIC) at UNCC. He received his Ph.D. degree in electrical engineering from the University of Toledo, OH in 2004. His research interests include data-driven approaches to power grid modernization, smart grid, microgrid, power system operation and optimization, power system dynamics, stability and control and renewable-energy-based distributed generation. Professor Kamalasadan is an Associate Editor and paper review chair of the IEEE Transactions on Industry Applications, Editor of IEEE Transactions on Vehicular Technologies, and Editor of Elsevier Electrical Engineering Journal. He is a recipient of several awards, including the National Science Foundation Faculty Early CAREER Award in 2008, best paper awards from IEEE in 2015, 2018, and 2021, outstanding teaching awards in 2004, and Duke Energy Professorship in 2019. He is a Senior Member of the IEEE. Professor Kamalasadan is the past chair of the IEEE Lifelong Learning Subcommittee (LLLSC), and Inaugural Chair of Selection and Quality Control Subcommittee which selects all the webinars and tutorials of IEEE Power and Energy Society (PES). He is a member of PES University Chairs, a group that designs and monitors IEEE PES University. He is also a member of IEEE PES oscillation location taskforce committee, co-chair for subtask under the oscillation location task force, a member of dynamic equivalent systems taskforce, member of working groups in Power System Relaying Committee, Member of IEEE PES distribution subcommittee and Energy storage and DER interconnection Taskforce. Past chair of IEEE IAS Industrial Automation and Control Committee, Chair of IEEE PES Power and Energy Education Committee, and past chair for IEEE PES Charlotte section. He has delivered more than 30 lectures/panels and keynotes in more than 10 countries. He is supervising/has supervised more than 25 M.S. and 30 Ph.D. level research students. He is the author of more than 200 archival type technical publications including a book (monograph type) and four patents.

# Coordinated Design of Damping Controllers for Power System Stability

**Manoj Kumar Kar, Ruchika S. Meshram, Pranali A. Krishna, Shivam V. Patil, and Rohan S. Sonawane**

**Abstract** This paper presents stability improvement of the coordinated design of damping controllers for a single machine infinite bus system. The primary goal of this work is to minimize the rotor speed deviation of the machine. Under various loading circumstances such as nominal loading, light loading and heavy loading, the system's performance has been examined. Power system stabilizer (PSS) and static synchronous series compensator (SSSC) are used as damping controllers. The damping controller parameters are adjusted using a unique sine cosine algorithm (SCA). The responses obtained using SCA are compared with the other three competitive methods to show the superiority of the proposed approach. It was found that the SCA method is capable of effectively suppressing power system oscillations.

**Keywords** FACTS controller · PSS · SCA · SMIB system · SSSC

## 1 Introduction

A power system's capability to continue working steadily in the presence of disturbances is referred to as stability. These disruptions can be large or small [1]. The existing transmission network has been overcrowded due to the rising load demand, leading to a number of stability problems within the power system that required the power system specialists' prompt attention. Balancing load demand and energy generation using the resources already available in the transmission network is the key difficulty in an electric power system. These issues might make controlling power flow difficult for security under post-fault conditions and hence, the fault clearance becomes more difficult for network operators. Furthermore, it is also desirable to deploy a more flexible solution because utilities now must deal with a grid that is

M. K. Kar (✉)
EEE Department, Tolani Maritime Institute, Pune, Maharashtra, India
e-mail: manojkar132@gmail.com

R. S. Meshram · P. A. Krishna · S. V. Patil · R. S. Sonawane
EE Department, AISSMS College of Engineering, Pune, Maharashtra, India

 
O. H. Gupta et al. (eds.), *Soft Computing Applications in Modern Power and Energy Systems*, Lecture Notes in Electrical Engineering 1107,
https://doi.org/10.1007/978-981-99-8007-9_1

extremely dynamic and requires solutions that can accommodate larger changes than in the past. For the last two decades, FACTS devices have drawn attention to solve the above-mentioned issues [2]. Shunt type FACTS devices, like D-STATCOM, are now frequently used to improve the power quality of distribution networks [3, 4]. The FACTS controllers were effectively used for reactive power compensation [5, 6] and to maintain voltage stability [7]. In Gyugyi et al. [8], the performance of the SSSC is compared to that of a TCSC. Wang [9] successfully incorporated SSSC in power system to reduce oscillations in the system. For a sixteen-machine system, PSS and FACTS controllers were designed, and the residue technique, as well as the involvement factor, were used to place the controllers [10]. A GA-based method [11] and a SSA method [12] are presented to tune the SSSC controller to increase the transient stability. In [13], a MGWOA tuned Fractional order PID controller is proposed to increase power system stability of a SMIB system and in [14] a multi-machine power system is considered to suppress the oscillations. The PSO approach [15] is employed to fine-tune the PSS parameters in multi-machine power system. The stability conditions are also met using the damping factor and eigenvalue analysis.

## 2 System Description

A single machine infinite bus (SMIB) system is used as a case study in the current work.

### 2.1 SMIB System

This system comprises of a transformer, a generator, and two transmission lines that provide power to an infinite bus. With SSSC, the SMIB system is illustrated in Fig. 1, where the infinite bus voltage is $V_B$, while $V_A$ is just the generator's terminal voltage.

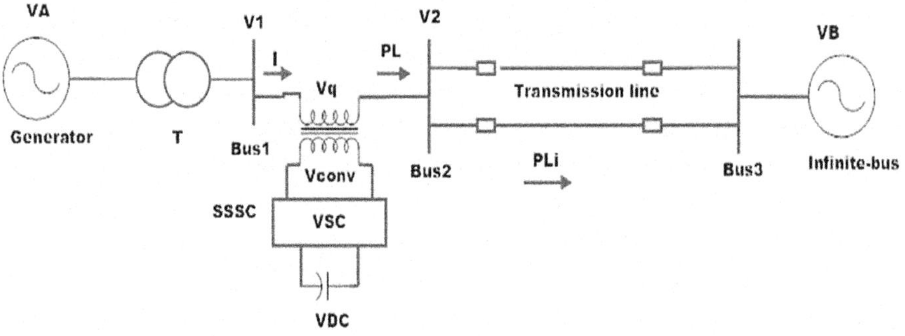


**Fig. 1** Design of SMIB system

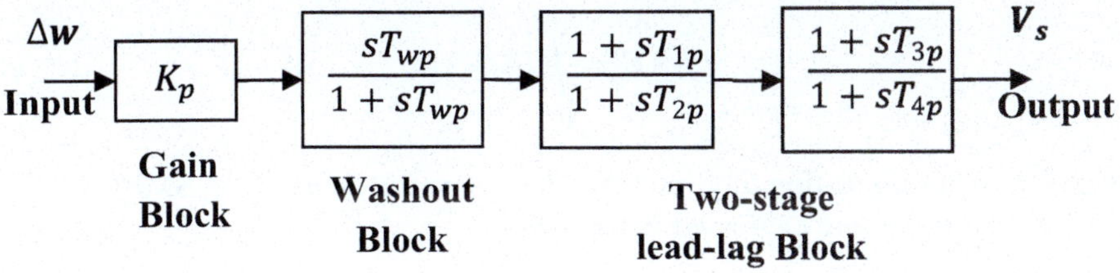


**Fig. 2** Structure of PSS

### 2.2 *Power System Stabilizer (PSS)*

PSSs are Power System Stabilizers, which are widely utilized in modern power systems as an efficient way of improving overall system stability. By varying generator stimulation to reduce ac machines rotor oscillations in respect to one another, the PSS increases the system stability limit. It produces a rotor-mounted electrical torque component that fluctuates in phase with the speed. The framework of PSS is shown in Fig. 2.

### 2.3 *Static Synchronous Series Compensator (SSSC)*

SSSC is a series flexible AC transmission (FACT) device that enables controlled power transmission. It controls power flow on the grid and power lines, both reactive and active. The SSSC uses no reactive power and provides better control as opposed to the phase shift controller. The system incorporates a DC capacitor to supply reactive power. The transmission line's voltage level and power flow are controlled by the SSSC. It has the potential to use the reference using signal strength as a current source and compensator and phase angle regulation. The Direct Voltage Injection Mode is used to power the energy storage. SSSC when connected in series with power transmission lines in a power system, produces a controlled AC voltage source.

## 3 Problem Formulation

A variation in the rotor speed is causing the system to oscillate. The objective of this study is to minimize the speed deviation. The objective function is thereby considered as the ITAE of speed deviations.

The following is the SMIB system's primary objective:

$$ITAE = \int_{0}^{t_s} |\Delta\omega| \cdot tdt \tag{1}$$

where, $t_s$ is the simulation time, and $\Delta\omega$ denotes speed deviation in SMIB.
The constraints are expressed as follows:

$$\text{Minimize } ITAE(K_P, T_{1P}, T_{2P}, T_{3P}, T_{4P}, K_S, T_{1S}, T_{2S}, T_{3S}, T_{4S}) \tag{2}$$

Subjected to

$$\begin{aligned} &K_P^{\min} \leq K_P \leq K_P^{\max} \\ &K_S^{\min} \leq K_S \leq K_S^{\max} \\ &T_{xP}^{\min} \leq T_{xP} \leq T_{xP}^{\max}; x = 1, 2, 3, 4 \\ &T_{yS}^{\min} \leq T_{yS} \leq T_{yS}^{\max}; y = 1, 2, 3, 4 \end{aligned} \tag{3}$$

where, $K_P^{\min}$ and $K_P^{\max}$ are the minimum and maximum values of PSS gain. $K_S^{\min}$ and $K_S^{\max}$ are the lower and upper limits of SSSC gain, $T_{xP}^{\min}$ and $T_{xP}^{\max}$ are the lower and the maximum of PSS's Time Constant, $T_{yS}^{\min}$ and $T_{yS}^{\max}$ are the minimum and maximum limits controller of the SSSC's time constant.

## 4 Proposed Method

A unique meta-heuristic approach based on the sine and cosine trigonometric functions was proposed by Mirjalili [16, 17] in 2016. The SCA technique, in contrast to other meta-heuristic techniques, outperforms them while being simpler, clearer, and easier to implement. It also has less control parameters. The algorithm starts with n number of search agents i.e., with each search agent giving a candidate solution. Each search agent of $t$-th Iteration is a d-dimensional vector with each member corresponding to a variable that is an independent part of the issue. Until the termination requirements are not satisfied shown in eq. 4, the solutions then shift outward or in the direction of the best solution. The optimal solution from the last iteration is utilized to solve the problem after the termination criteria have been satisfied.

$$X_i^{t+1} = \begin{cases} X_i^t + r_1 \times \sin(r_2) \times \left|r_3 \times P_i^t - X_i^t\right|, if r_4 < 0.5, \\ X_i^t + r_1 \times \cos(r_2) \times \left|r_3 \times P_i^t - X_i^t\right|, if r_4 \geq 0.5, \end{cases} \tag{4}$$

where $X_i^t$, $P_i^t$ is used to indicate the ith location of the current solution and the final destination in the current iteration. $X_i^{t+1}$ is the $I$ th solution position in $t + 1$ and $|\blacksquare|$ represents the absolute value. $r_1$, $r_2$, $r_3$ and $r_4$ indicates the algorithm's parameters.

The three parameters $r_1$, $r_2$, and $r_3$ are determined using Eqs. 5, 6 and 7, respectively.

$$r_1 = a \times \left(1 - \frac{t}{T_{max}}\right) \tag{5}$$

$$r_2 = 2 \times pi \times rand\,(0, 1) \tag{6}$$

$$r_3 = 2 \times rand\,(0, 1) \tag{7}$$

where an is just a constant set to 2 under this method and t has been the current iteration number, $T_{max}$ identifies the maximal number of iterations of the algorithm, and rand (0,1) helps to identify a random integer generated from a uniformly distribution between 0 and 1.

## 5 Results and Discussions

Damping controllers are designed and simulated by using the Sim Power System toolkit. A synchronous generator through a 3-phase, 13.8/500 kV step-up transformer, a 2100 MVA, 13.8 kV, 60 Hz output is linked to a 300 km parallel transmission line in the SMIB system with SSSC integrated, which is modelled using MATLAB SIMULINK. The transmission line and the 100 MVA SSSC controller are linked in series. The designed model is put through a simulation, after which the objective function is assessed after a disturbance. Equation 1's fitness value is minimized in order to find the optimum control variables.

### *5.1 Nominal Loading (NL) Condition*

i. Under standard loading conditions (Pe = 0.8 pu) with disturbance, the suggested controller's performance is assessed.
ii. A 3-phase fault is applied at t = 1 s for 5 cycles. The system is restored once the fault is cleared.
iii. Figures 3, 4, 5 and 6 depicts the various responses, which include the SSSC injected voltage (Vq) in pu, tie line power (P) in MW, speed deviation ($\Delta\omega$) in pu, and power angle ($\delta$) in degree.
iv. Based on the responses, it is evident that the suggested strategy is superior to the competition in terms of performance.

**Fig. 3** Speed deviation at NL

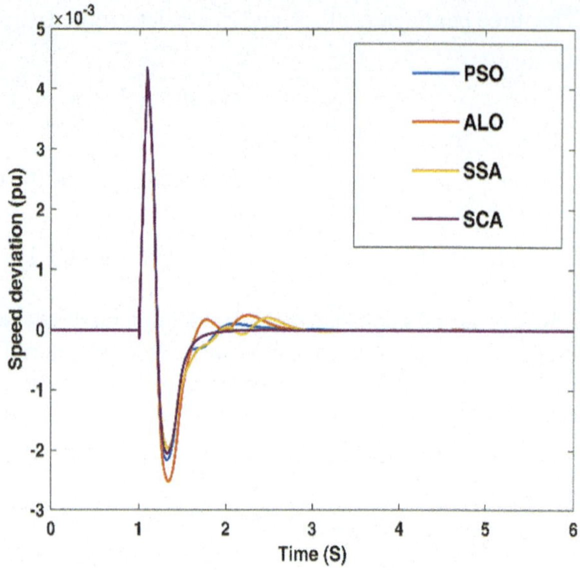


**Fig. 4** Power angle at NL

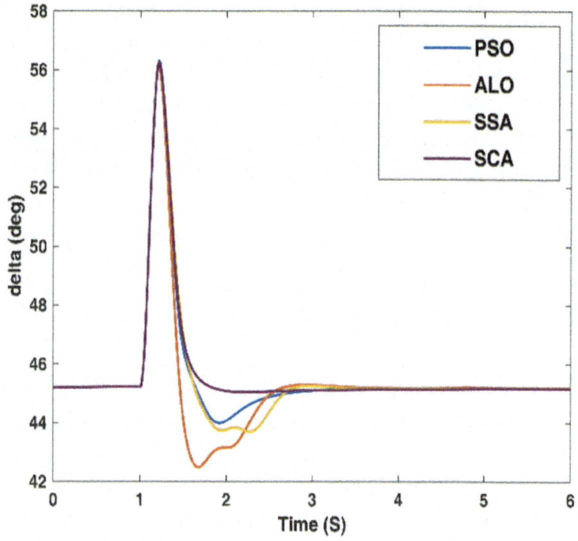


### 5.2 *Light Loading (LL) Condition*

i. Following that, a 3-phase 5 cycle fault with light loading (Pe = 0.5 pu) is considered.
ii. Figures 7, 8, 9 and 10 show the dominance of the SCA approach on various system responses to damp oscillation at nominal loading conditions.

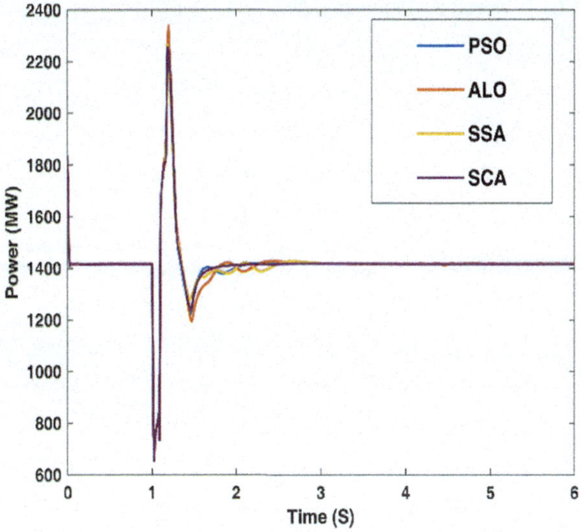


**Fig. 5** Tie line power at NL

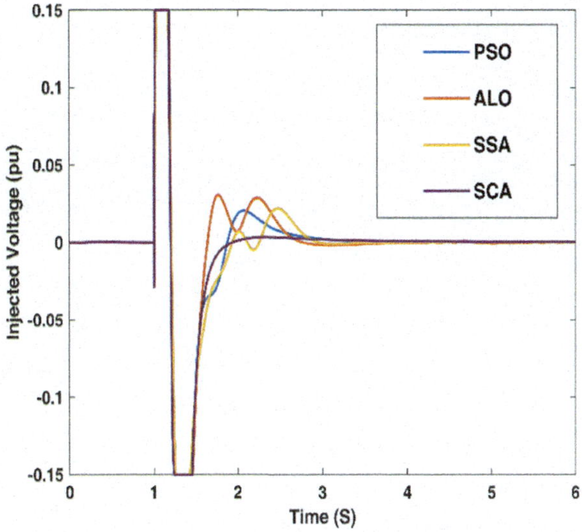


**Fig. 6** SSSC injected voltage at NL

iii. The proposed method's improved damping properties demonstrate why it is preferable to the other approaches taken into account in this study.

**Fig. 7** Speed deviation at LL

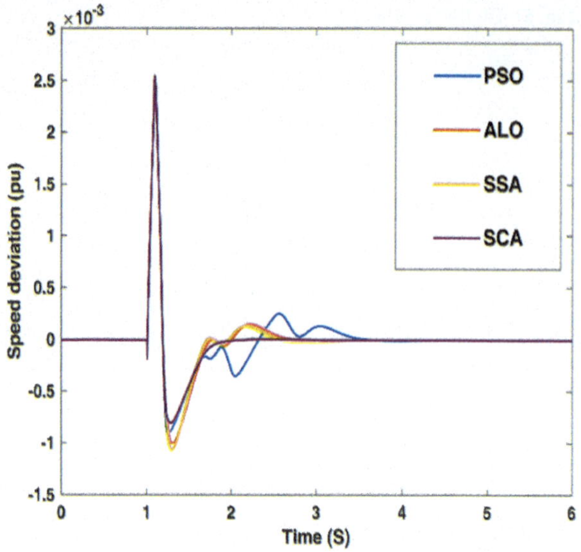


**Fig. 8** Power angle at LL

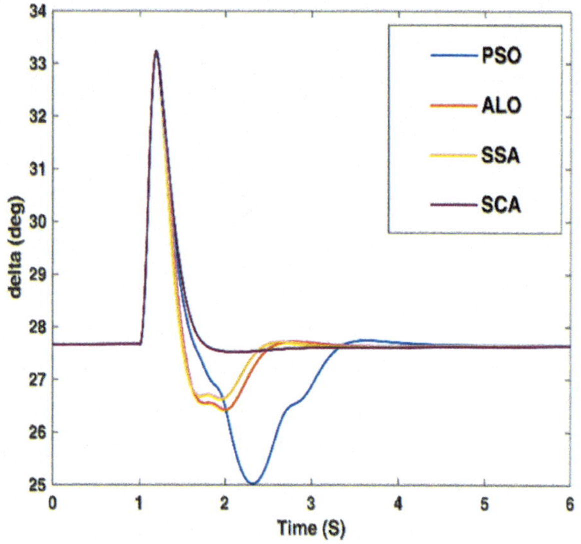


## *5.3 Heavy Loading (HL) Condition*

i. The SCA technique is eventually put to the test by accounting for heavy loading (Pe = 0.95pu).
ii. The results under conditions of heavy loading are seen in Figs. 11, 12, 13 and 14.

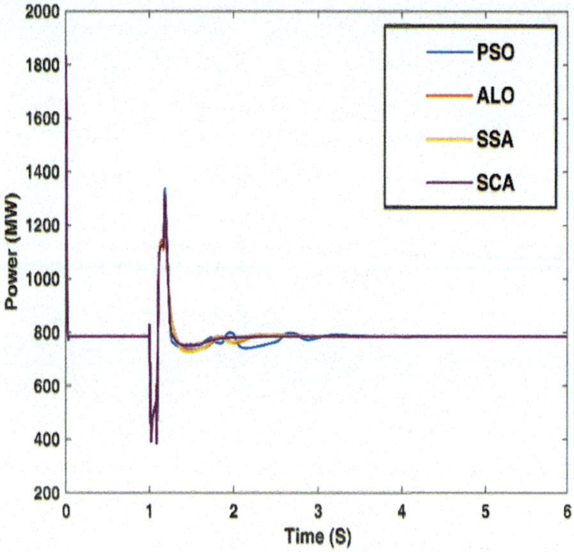


**Fig. 9** Tie line power at LL

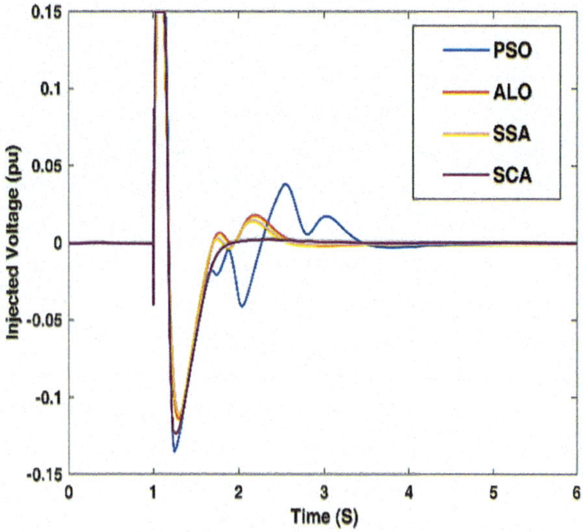


**Fig. 10** SSSC injected voltage at LL

iii. From the figures, it is clear that the SCA outperforms the other techniques taken in this study.

**Fig. 11** Speed deviation at HL

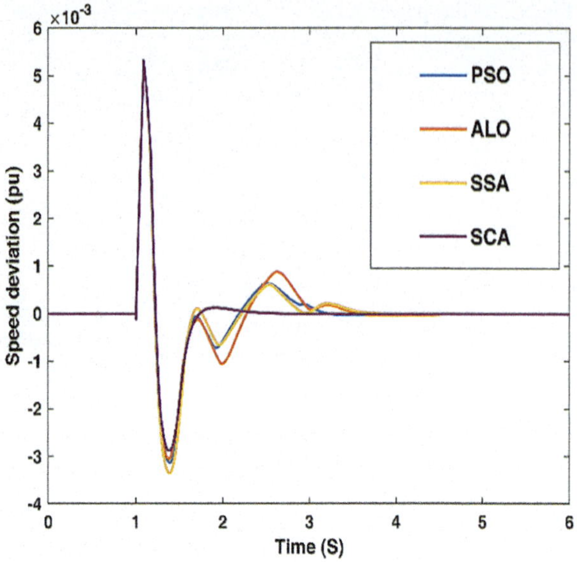


**Fig. 12** Power angle at HL

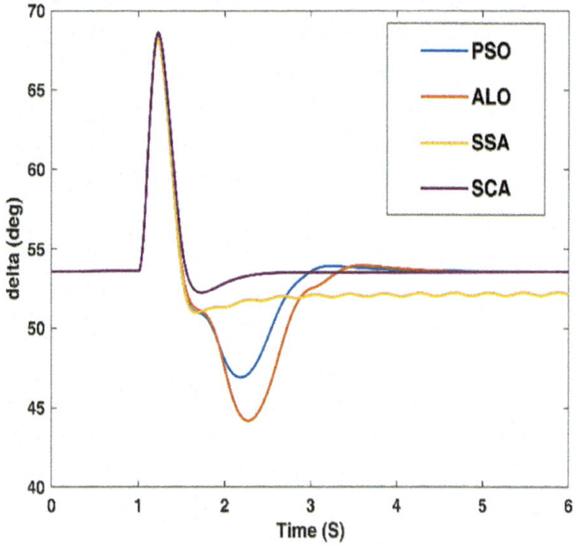

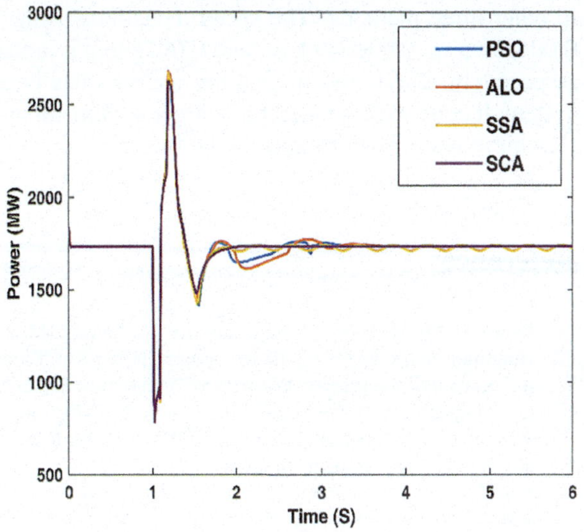


**Fig. 13** Tie line power at HL

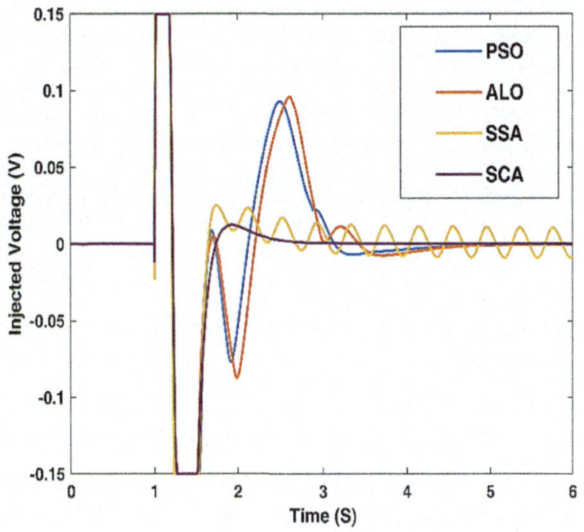


**Fig. 14** SSSC injected voltage at HL

## 6 Conclusion

This paper presents improved power system stability through SCA optimized synchronized structure of SSSC and PSS controller. The robustness of the SCA based damping controllers increases the stability of SMIB system. The effectiveness of the proposed method is measured against other three methods such as PSO, ALO, and DE. It is found that the proposed SCA method gives better results compared to other three methods. The speed deviations using SCA method are found to be 0.001408 p.u., 0.000649 p.u., and 0.001896 p.u. under NL, LL, and HL conditions respectively. The transient stability analysis has been accomplished for an SMIB system. It could be extended for a larger system. Also, some controllers may be used to achieve better performance in the future.

## References

1. Kundur P (1994) Power system stability and control. Mc-Grall Hill
2. Hingoran NG, Gyugyi L (2000) Understanding FACTS concepts and technology of flexible AC transmission systems. Wiley-IEEE Press, New Jersey, NJ, USA
3. Kar MK, Kumar S, Singh AK (2022) Power quality improvement of an interconnected grid system using PWM technique of D-STATCOM. In: Recent advances in power electronics and drives, pp 31–4. Springer, Singapore
4. Kar MK, Rout B, Moharana JK (2014) Improvement of power factor of a grid connected load system using a static compensator. J Found Appl Phys 1(1):5–10
5. Kumar L, Kar MK, Kumar S (2022) Statistical analysis based reactive power optimization using improved differential evolutionary algorithm. Expert Syst e13091. https://doi.org/10.1111/exsy.1309
6. Kumar L, Kar MK, Kumar S (2021) Reactive power management by optimal positioning of FACTS controllers using MFO algorithm. In: 2021 emerging trends in industry 4.0 (ETI 4.0), pp 1–6. https://doi.org/10.1109/ETI4.051663.2021.9619433
7. Kumari S, Kumar L, Kar MK, Kumar S (2023) Application of PSAT for voltage stability improvement using FACTS devices. In: Smart energy and advancement in power technologies, pp 123–129. Springer, Singapore
8. Gyugyi L, Schauder CD, Sen KK (1997) Static synchronous series compensator: a solid-state approach to the series compensation of transmission lines. IEEE Trans Power Deliv 12(1):406–417
9. Wang HF (2000) Static synchronous series compensator to damp power system oscillations. Electr Power Syst Res 54(2):113–119
10. Cai LJ, Erlich I (2005) Simultaneous coordinated tuning of PSS and FACTS damping controllers in large power systems. IEEE Trans Power Syst 20(1):294–300
11. Panda S (2009) Multi-objective evolutionary algorithm for SSSC-based controller design. Electr Power Syst Res 79(6):937–944
12. Kar MK (2023) Stability analysis of multi-machine system using FACTS devices. Int J Syst Assur Eng Manag 1–10
13. Kar MK, Singh AK, Kumar S, Rout B (2023) Application of fractional-order PID controller to mprove stability of a single-machine infinite-bus system. J Inst Eng India Ser B 1–16
14. Kar MK, Kumar S, Singh AK, Panigrahi S, Cherukuri M (2022) Design and analysis of FOPID-based damping controllers using a modified grey wolf optimization algorithm. Int Trans Electr Energy Syst 2022
15. Verdejo H, Pino V, Kliemann W, Becker C, Delpiano J (2020) Implementation of particle swarm optimization (PSO) algorithm for tuning of power system stabilizers in multimachine electric power systems. Energies 13(8):2093
16. Mirjalili S (2016) SCA: a sine cosine algorithm for solving optimization problems. Knowl-Based Syst 96:120–133
17. Kar MK, Kumar S, Singh AK, Panigrahi S (2021) A modified sine cosine algorithm with ensemble search agent updating schemes for small signal stability analysis. Int Trans Electr Energy Syst 31(11):e13058

# The Effect of Electric Vehicle Charging Stations on Distribution Systems While Minimizing the Placement Cost and Maximizing Voltage Stability Index

**Sumeet Kumar and Ashwani Kumar**

**Abstract** Greenhouse gas (GHG) emissions from internal combustion (ICE) vehicles, which are reliant on fossil fuels, are one of the problems in major cities. As a result, the electric vehicle (EV) has become the automotive industry's green alternative. Promoting the worldwide adoption of electric vehicles is viewed as a possible solution to energy security, including energy efficiency, reduced noise, and greenhouse emission reduction. Due to its numerous benefits, including their use of flexible fuels, ease, safe charging, high performance, and cost savings, Plug-In Electric Vehicles (PEVs) will soon replace traditional vehicles as the most affordable option for transportation. Despite the benefits indicated above, improper placement and size of aggregated PEVs cause voltage degradation and loss. Therefore, the best location for charging stations (CS) is crucial for the widespread use of EVs. Thus, in order to allocate fast and slow CS as efficiently as possible, this research suggests two optimization technique that takes into account the cost of installation, operation cost, increased line loss cost, and voltage deviation cost. The CS placement approach of Whale Optimization (WO) and Particle Swarm Optimization (PSO) technique is simulated on IEEE-33 radial distribution systems. The results demonstrate that the recommended technique can choose the best location and size for CS, which will benefit EV owners, EVCS creators, and the power grid.

**Keywords** Distribution system · Electric vehicles · Charging station · Whale optimization · Particle swarm optimization

S. Kumar (✉) · A. Kumar
National Institute of Technology, Kurukshetra, Haryana, India
e-mail: sb786dec@gmail.com

A. Kumar
e-mail: ashwani.k.shrama@nitkkr.ac.in



O. H. Gupta et al. (eds.), *Soft Computing Applications in Modern Power and Energy Systems*, Lecture Notes in Electrical Engineering 1107,
https://doi.org/10.1007/978-981-99-8007-9_2

## 1 Introduction

The market for electric vehicles (EVs) is expanding quickly on a worldwide scale. As per EV volumes, the total number of electric vehicles (including battery electric vehicles [BEVs] and Plug-in hybrid electric vehicles [PHEVs]) on the road increased to 6.75 million in 2021 from 4.2% in 2020. As they aid in lowering pollution and reducing resource depletion, EVs are gaining popularity all over the world. As evidence of how quickly the Indian EV industry is developing, close to 0.32 million vehicles, up 168% YoY, were sold in 2021. The ongoing adoption of electric vehicles in India is supported by the Paris Agreement, which aims to reduce carbon emissions, enhance the quality of the air in metropolitan areas, and decrease oil imports [1].

There is an essential need for widely dispersed, publicly available charging stations (CSs) that provide electrical energy for recharging an EV battery since the number of EVs is growing quickly [2]. Therefore, the construction of the charging infrastructure must be prioritized in order to deploy EVs on a big scale. For the EV business to expand sustainably, coordinated EV CS planning is of utmost importance. In response to the recent demand for PEV supplies, researchers have focused on the optimal layout of charging stations [3]. It is possible to broadly divide CS placement into two types: slow and quick. Due to lower charging costs and accessibility at home or the office, slow charging is the most popular approach. The low charging power used by this approach results in a lengthy charge period for an EV battery. However, in order to drive long distances, EV customers also want an urgent charging option. Therefore, it is essential to have a sufficient number of fast CSs (RCSs) for quick charging. However, the widespread use of EVs places an increased demand on a conventional distribution network, which might have a number of negative effects on the network. With consideration for loss and voltage in distribution networks, charging stations are scaled and constructed for maximum efficiency. The distribution network's properties, such as voltage stability, dependability, power loss, etc., must not be compromised by integrating EV CS into the transport network.

The literature describes many strategies and procedures used by researchers from throughout the world to deploy EV CSs in the best possible positions. A Binary Firefly Algorithm (BFA) approach has been proposed by Islam et al. [2] for the optimal allocation of the rapid charging station in the road distribution network of the township of Bangi, Malaysia. A new hybrid algorithm based on Chicken Swarm Algorithm (CSO) and Teaching Learning Based Optimization Algorithm (TLBO) is proposed in Deb et al. [3] to solve the problem of optimal placement of fast and slow CS. In Ge et al. [4], the authors introduced a unique approach to CS placement using the Grid Partition technique, with the aim function of minimizing user loss on the route to the charging station. To minimize the effects of EVCs in the distribution grid, a novel Genetic Algorithm (GA) based approach using capacitor banks is proposed in Pazouki et al. [5]. The authors of Liu et al. [6] used Adaptive Particle Swarm Optimization (APSO) to report the best sites for EV CSs with a single objective cost function. Parking lots with various levels of charging stations are placed and sized in the best possible way in electric distribution networks is formulated in Mohsenzadeh et al. [7] using Genetic Algorithm (GA). Taking distribution and traffic networks into account [8], the best planning for PEV charging stations and demand response initiatives is done using Genetic Algorithm (GA).

Literature [2–8] describes some of the current research done in the field of EV CSs placement. Additionally, there are a number of restrictions on the objective functions taken into account in the literature, such as a lack of a voltage deviation cost and total line loss cost. In this work, two novel modelling method for the EV CS placement problem is presented that takes into account the superposition of the distribution networks and uses the objective function of installation cost, operating cost, cost of voltage deviation, the extra line lost cost, and Voltage Stability Index (VSI), total line losses. This work proposes a comparison of two metaheuristic algorithms Particle Swarm Optimization (PSO) and Whale Optimization (WO) for the solution of EV CS problem formulation.

The rest of the paper is structured as follows. A brief introduction to EV CS and the work already done is presented in Sect. 1. The concept of the EV CS placement problem is expanded in Sect. 2. Section 3 tells about the optimization technique used in the paper and the flow chart of the optimization technique. The quantitative analysis and simulation result is presented in Sect. 4. The paper is concluded at the end.

## 2 Objective Problem Formulation

The main goal of the objective function is to minimize the overall cost and maximize the VSI. The overall cost is comprised of the installation cost, the cost of operation, the extra cost for voltage variation, and the total line loss cost. Figure 1 provides a diagrammatic depiction of several objective functions. Equations (1)–(18) in the preceding subsections provide further details on the objective function used for the optimization as well as the various constraints.

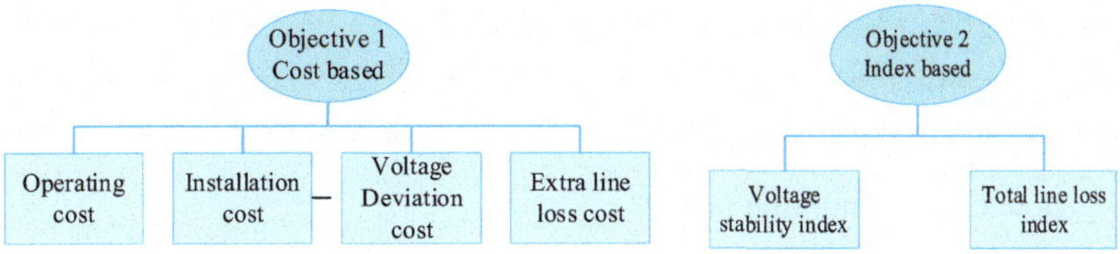


**Fig. 1** Objective function of EV CS

## 2.1 Objective Function 1

The first objective function takes the cost-based approach to find the optimal allocation of EV CS taking the installation cost, operation cost [3], voltage deviation cost [9] and extra line loss cost [10]. The objective function is shown by Eq. 1.

$$\mathrm{MinOF1} = \left(\mathrm{C_{installation}}, \mathrm{C_{\mathbf{operation}}}, \mathrm{C_{deviation}}, \mathrm{C_{extra\,line\,loss}}\right) \tag{1}$$

where,

$C_{installation}$-installation cost of CS

$C_{operation}$-annual operating cost of CS

$C_{deviation}$-cost of per unit voltage deviation

$C_{extralineloss}$-annual cost of extra line loss.

The formulation of different cost function is given below:-

$$\mathrm{C_{installation}} = \mathrm{C_{fast}} + \mathrm{C_{slow}} \tag{2}$$

$$\mathrm{C_{fast}} = \mathrm{n_{fastcs}} * \mathrm{n_{fastcharger}} * \mathrm{n_{fast_{charger_{install}}}} + \mathrm{C_{installationfast}} \tag{3}$$

$$\mathrm{C_{slow}} = \mathrm{n_{slowcs}} * \mathrm{n_{slowcharger}} * \mathrm{C_{slow_{charger_{install}}}} + \mathrm{C_{installationslow}} \tag{4}$$

$$\mathrm{C_{operation}} = \left(\mathrm{n_{fastcs}Cp_{fast}n_{fastcharger}} + \mathrm{n_{slowcs}Cp_{slow}n_{slowcharger}}\right) * \mathrm{P_{electricity}} * \mathrm{T} \tag{5}$$

where,

$n_{fastcs}$-no of fast CS

$n_{slowcs}$-no of slow CS

$C_{installationfast}$-fast CS installation cost

$C_{installationslow}$-slow CS installation cost

$Cp_{fast}$-fast CS consumption power

$Cp_{slow}$-slow CS consumption power

$P_{electricity}$-cost of per unit electricity

T-time period of planning

$$\mathrm{C_{deviation}} = \mathrm{P_{VD}} * \sum_{i=2}^{N} \mathrm{VD}_i^2 \tag{6}$$

$$\mathrm{VD_i = V_i^{base} - V_i^{cs}} \tag{7}$$

where,

$P_{VD}$-cost of per unit voltage deviation

$VD_i$-voltage deviation at $i$th bus

$V_i^{base}$-base case voltage at $i$th bus

$V_i^{cs}$-voltage after placing CS at $i$th bus

$$\mathrm{C_{extra\,line\,loss} = (TPL_{\mathbf{base}} - TPL_{cs}) * \left(k_p + k_e * Lsf * T\right)} \tag{8}$$

$$\mathrm{Lsf = k * Lf + (1 - k) * Lf^2} \tag{9}$$

where,

$\mathrm{TPL_{base}}$-total power loss without CS

$\mathrm{TPL_{cs}}$-total power loss after CS

$\mathrm{k_p}$-annual demand power loss cost

$\mathrm{k_e}$-annual loss of energy cost

Lsf-loss factor

Lf-load factor

## 2.2 *Objective Function 2*

The second objective function takes the index-based approach to find the optimal allocation of EV CS taking the voltage stability index (VSI) [11] and total real power loss of the distribution system. The objective function is shown by Eq. 10.

$$\mathrm{MinOF2} = \alpha_1 * \frac{\mathrm{TPL_{cs}}}{\mathrm{TPL_{base}}} + \alpha_2 * \frac{\Delta \mathrm{VSI_{cs}}}{\Delta \mathrm{VSI_{base}}} \tag{10}$$

$$\Delta \mathrm{VSI} = \max\left(\frac{1 - \mathrm{VSI_i}}{1}\right) \forall i = 2, 3 \ldots\ldots \mathrm{N} \tag{11}$$

$$\mathrm{VSI_{i+1} = V_s^4 - 4V_s^2(R_iP_{Li} + X_iQ_{Li}) - 4(X_iP_{Li} + R_iQ_{Li})^2 \forall i = 2, 3 \ldots\ldots N} \tag{12}$$

where,

$\alpha_1, \alpha_2$-weighting factors

$\Delta VSI_{cs}$-voltage stability index with CS

$\Delta VSI_{base}$-voltage stability index without CS

$N$-number of buses

$VSI_i$-voltage stability index of ith bus

$V_s$-voltage at sending end

$R_i$-resistance of the line

$P_{Li}$-active power load at $i$th bus

$X_i$-inductive reactance of the line

$Q_{Li}$-reactive power load at $i$th bus

## 2.3 *Constraint Used*

Two types of constraints are used in the following proposed objective function.

### 2.3.1 Equality Constraints

Forward and backward sweep method is used for distribution system load flow analysis and is illustrated in Kazmi et al. [12], Martinez and Mahseredjian [13] and Teng et al. [14]. Power balance equation considering EV CS in the distribution system can be defined as follows:

$$\mathrm{P_{Gi}} - \mathrm{P_{CSi}} - \mathrm{P_{Di}} = \sum_{\mathrm{j}=1}^{\mathrm{N}} \mathrm{V_i^2 Y_{ij}^2} \cos\left(\theta_{\mathrm{ij}} + \lambda_{\mathrm{j}} - \lambda_{\mathrm{i}}\right) \forall \mathrm{i} = 1, 2, \ldots\ldots \mathrm{N} \quad (13)$$

$$\mathrm{Q_{Gi}} - \mathrm{Q_{CSi}} - \mathrm{Q_{Di}} = \sum_{\mathrm{j}=1}^{\mathrm{N}} \mathrm{V_i^2 Y_{ij}^2} \sin\left(\theta_{\mathrm{ij}} + \lambda_{\mathrm{j}} - \lambda_{\mathrm{i}}\right) \forall \mathrm{i} = 1, 2 \ldots\ldots \mathrm{N} \quad (14)$$

### 2.3.2 Inequality Constraints

Distribution system should be lime by voltage level at each bus by:

$$\mathrm{V_i^{min}} \leq \mathrm{V_i} \leq \mathrm{V_i^{max}} \quad (15)$$

Maximum and minimum number of fast and slow charging stations can be placed at each bus:

$$0 < n_{fastcs} \leq N_{fastcs\max} \quad (16)$$

$$0 < n_{slowcs} \leq N_{slowcs\max} \quad (17)$$

The total increased load (L) of the network should be less than the maximum load margin ($L_{max}$) of the system:

$$\mathrm{L} \leq \mathrm{L}_{\max} \quad (18)$$

## 3 Optimization Technique Used

Nature-inspired optimization algorithms [15] are metaheuristic algorithms based on biological evolution, swarm behavior patterns, and physical and chemical processes. Nature influenced optimization algorithms are examples of computational intelligence methods that are bioinspired because they contain intelligence. Algorithms influenced by nature are new in their ability to achieve effective solutions with minimal computational resources. Collective intelligence has emerged as a result of biological agents such as ants, bees, crows, bats, cuckoos, and others sharing information and socializing among members of their own species as well as with the environment [16].

In this paper Particle Swarm Optimization (PSO) and Whale Optimization (WO) is used. The social behavior of fish schools and bird flocks served as the basis for the metaheuristic optimization technique known as particle swarm optimization (PSO) [17]. The approach simulates a swarm of particles moving across a search space, each particle standing in for a potential answer to the optimization issue [18]. The flow chart of PSO algorithm is presented in Fig. 2.

Whale Optimization Algorithm (WOA) is a nature-inspired metaheuristic optimization algorithm that was first proposed by Mirjalili et al. in 2016 [19]. It was inspired by the hunting behavior of humpback whales, where the whales work together to encircle their prey and capture it. The algorithm is based on a mathematical model that simulates the behavior of humpback whales, where each whale represents a potential solution to the optimization problem [20]. The flow chart of WO algorithm is presented in Fig. 3.

## 4 Numerical Analysis

In this study, WO and PSO are used to calculate the ideal CS positions. In this part, test system details and the results of the ideal CS setup are provided. IEEE-33 bus radial distribution system is taken as the test system for this analysis. The bus data

**Fig. 2** Flow chart of particle swarm optimization (PSO) algorithm

```
Start
Initialize particles randomly
Set initial velocities randomly
while (stopping criterion is not met) do
  for each particle i do
    Evaluate fitness of particle i
    If fitness of particle i is better than its personal best then
      Update personal best of particle i
    end if
    Identify global best fitness value among all particles
  end for
    for each particle i do
    Update velocity of particle i using formula:
    v_i(t + 1) = ω * v_i(t) + c_1 * r_1 * (p_i − x_i) + c_2 * r_2 * (g − x_i)
    Update position of particle i using formula:
                x_i(t + 1) = x_i(t) + v_i(t + 1)
     end for
  end while
Return global best particle
End
```

**Fig. 3** Flow chart of whale optimization (WO) algorithm

```
START
Generate initial population of random solutions
Evaluate fitness of each solution
Set the global best solution as the one with the highest fitness
WHILE stopping criterion is not met DO
  FOR each solution in population DO
    IF random number < a THEN
      Update solution using equation X_i(t + 1) = X_i(t) − A * D_i(best)
    ELSE IF random number < c THEN
      Update solution using equation X_i(t + 1) = X_i(t) − A * D_i(rand)
    ELSE
      Update solution using equation X_i(t + 1) = X_i(t) − A * D_i(a)
    END IF
    Evaluate fitness of updated solution
    IF updated solution has higher fitness than current solution THEN
      Replace current solution with updated solution
      IF updated solution has higher fitness than global best solution
THEN
        Replace global best solution with updated solution
      END IF
    END IF
  END FOR
END WHILE
RETURN global best solution
```

and line data of IEEE-33 bus radial distribution system is taken from [21] and the one-line diagram of the system is shown in Fig. 4. The total demand on the IEEE-33 bus radial test distribution system with a combined real and reactive load demand of 3.715 MW and 2.3 MVAR. All simulations are carried out using an Intel Core I7 7th Gen CPU with 16 GB of RAM with MATLAB 2022a.

Optimal placement is done based on the two different objective functions using PSO and WO optimization algorithms. Different cases have been formed based on the number of EV fast and slow CS, and also based on the number of charging slots present in that EV CS. Different input parameters have been taken from Islam

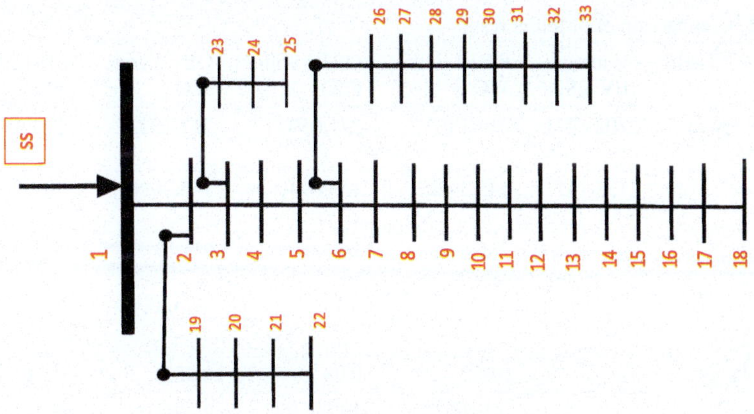


**Fig. 4** One line diagram of IEEE-33 bus radial distribution system

et al. [2], Deb et al. [3] and Pazouki et al. [5] are presented in Table 1. On applying distribution load flow analysis to IEEE-33 bus radial distribution the total active and reactive power losses are 202.6771 KW and 135.141 KVAR. The minimum voltage is 0.91306 p.u at bus no 18 and the maximum voltage is 0.99703 p.u at bus no 2.

Two different objective function is compared with two different Optimization algorithm to find the optimal allocation of EV CS. The comparison results of PSO and WO algorithm are tabulated in Table 2. In Table 2 OBJ 1 is the cost-based objective function formulated by adding installing cost, operation cost, extra line loss cost and voltage deviation cost, the second objective function OBJ 2 contains voltage stability index and the line loss index.

The voltage profile curve of both objectives with PSO and WO is presented in Figs. 5 and 6. The Voltage Stability Index (VSI) of objective function 2 is shown in Fig. 7. The respective active and reactive power loss after placing EV CS with different objective functions is shown in Figs. 8, 9, 10 and 11. The convergence graph of both the objective functions is shown in Fig. 12.

**Table 1** Input parameters

| | | | |
|---|---|---|---|
| $C_{installationfast}$ | 2000 $ | $Cp_{fast}$ | 50 KW |
| $C_{installationslow}$ | 1500 $ | $Cp_{slow}$ | 19.2 KW |
| $C_{fast_{charger_{install}}}$ | 500 $ | $C_{slow_{charger_{install}}}$ | 300 $ |
| $n_{fastcharger}$ | 8 | $n_{slowcharger}$ | 12 |
| $n_{fastcs}$ | 2 | $n_{slowcs}$ | 2 |
| $P_{electricity}$ | 0.065 $/kWhr | T | 8760 |
| $P_{VD}$ | 1000000 $/unit | $k_p/k_e$ | 57.693 $ /0.009611 $ |
| k/Lf | 0.2/0.43 | $\alpha_1, \alpha_2$ | 0.5/0.5 |
| Noofpopulation | 50 | Noofiteration | 200 |

**Table 2** Comparison of different results

| PARAMETERS | Without EV CS | OBJ 1 with WO | OBJ 1 with PSO | OBJ 2 with WO | OBJ 2 with PSO |
|---|---|---|---|---|---|
| Active Power Loss | 202.6771 | 217.1822 | 221.0593 | 217.1822 | 223.9576 |
| Reactive Power Loss | 135.141 | 145.3920 | 149.0940 | 145.3920 | 147.1367 |
| No of fast EV CS/charger | – | 2/8 | 2/8 | 2/8 | 2/8 |
| No of slow EV CS/charger | – | 2/12 | 2/12 | 2/12 | 2/12 |
| Power factor | – | 0.98 | 0.98 | 0.98 | 0.98 |
| Optimal CS Location | – | 2/19/21/20 | 19/20/2/22 | 2/19/21/20 | 2/3/20/19 |
| Optimal CS Size P KW | – | 400/400/ 230.4/230.4 | 400/400/ 230.4/230.4 | 400/400/ 230.4/230.4 | 400/400/ 230.4/230.4 |
| Optimal CS Size Q KVAR | – | 81.22/81.22/ 46.78/46.78 | 81.22/81.22/ 46.78/46.78 | 81.22/81.22/ 46.78/46.78 | 81.22/81.22/ 46.78/46.78 |
| Min/Max voltage location | 2/18 | 2/18 | 2/18 | 2/18 | 2/18 |
| Min/Max voltage | 0.91306/ 0.99703 | 0.91216/ 0.99622 | 0.91216/ 0.99622 | 0.91216/ 0.99622 | 0.91216/ 0.99622 |
| % Increase in line loss | – | 7.156 | 8.8246 | 7.156 | 9.411 |
| Annual extra line loss cost | – | 1167.6 $ | 1439.7 $ | 1167.6 $ | 1535.4 $ |
| Computational time | 0.0641 | 90.5961 | 90.9458 | 91.6607 | 90.8697 |

From Figs. 5 and 6 we can observe that with the placement of EVCS in radial distribution system the voltage profile deviates from the base case. Also, we can see that from Figs. 8,9,10, and 11 that the active and reactive power of the branch and the system increased from the base case. In Fig. 12 whale optimization (WO) algorithm has the better fitness value in both the objective function cases.

The impact on distribution system increases or decreases based on the number of EV CS and number of fast and slow chargers present in the respective CS. The optimal placement of different fast and slow charging stations based on number of chargers presented in Table 3.

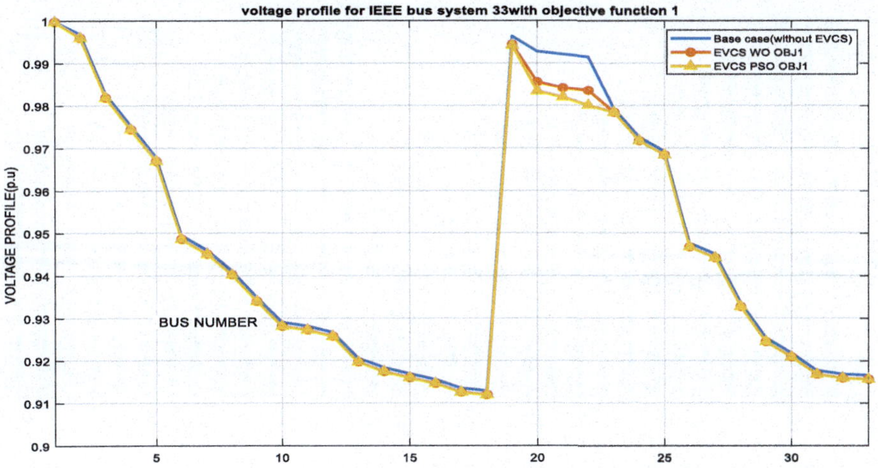


**Fig. 5** Voltage profiles of 33-BUS system after placing EV CS with objective 1

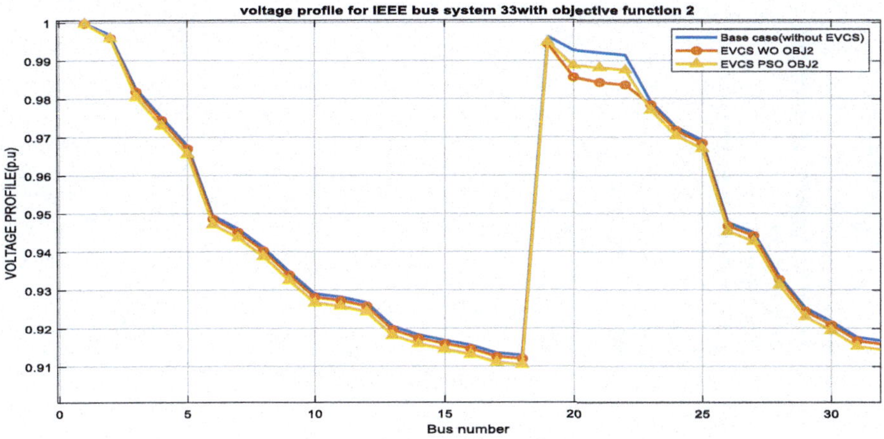


**Fig. 6** Voltage profiles of 33-BUS system after placing EV CS with objective 2

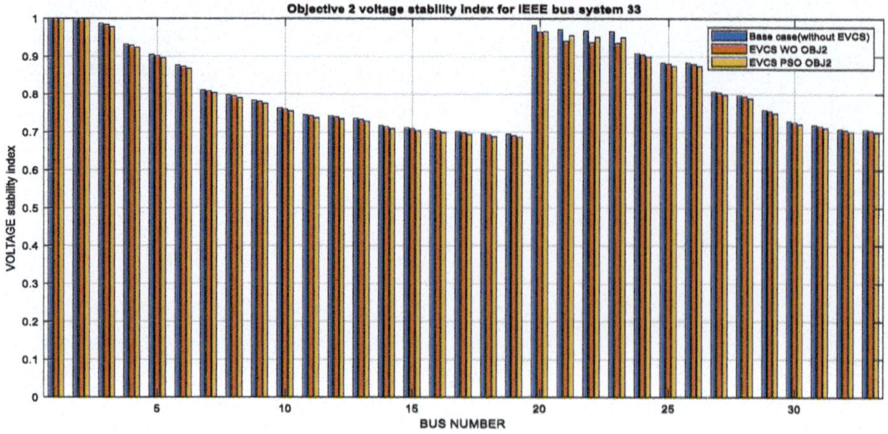


**Fig. 7** Voltage stability index of objective function 2

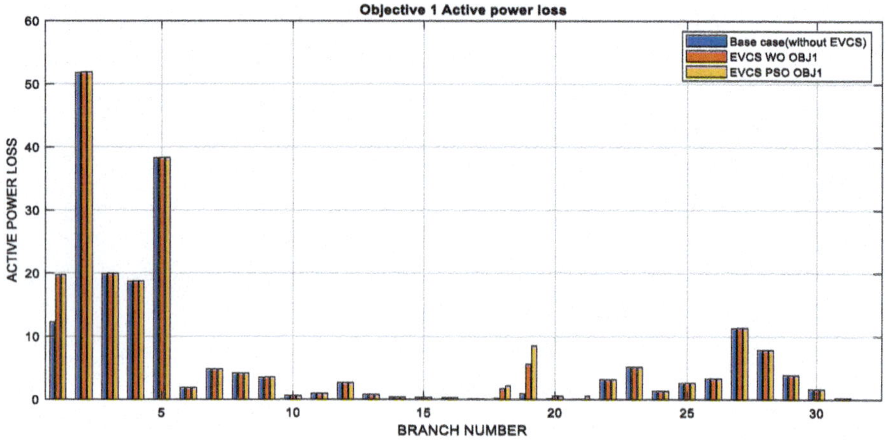


**Fig. 8** Increased active power loss after placing EV CS objective 1

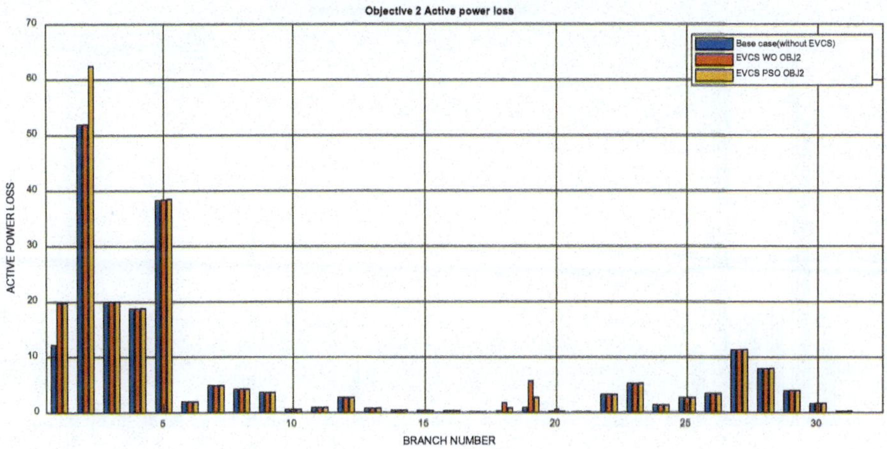


**Fig. 9** Increased active power loss after placing EV CS objective 2

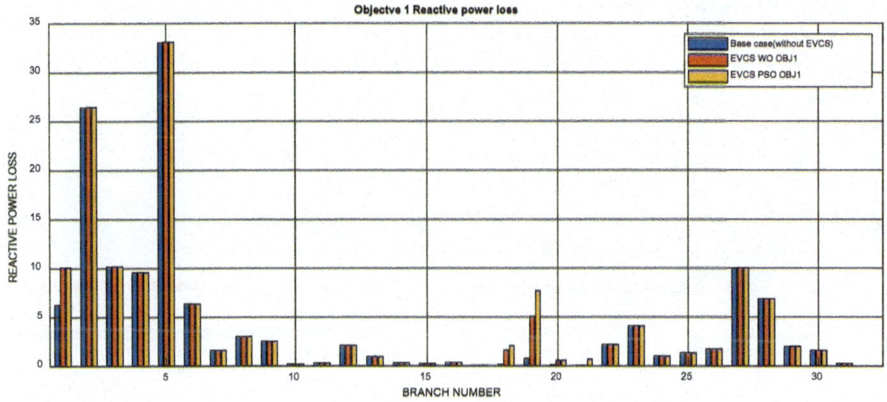


**Fig. 10** Increased reactive power loss after placing EV CS objective 1

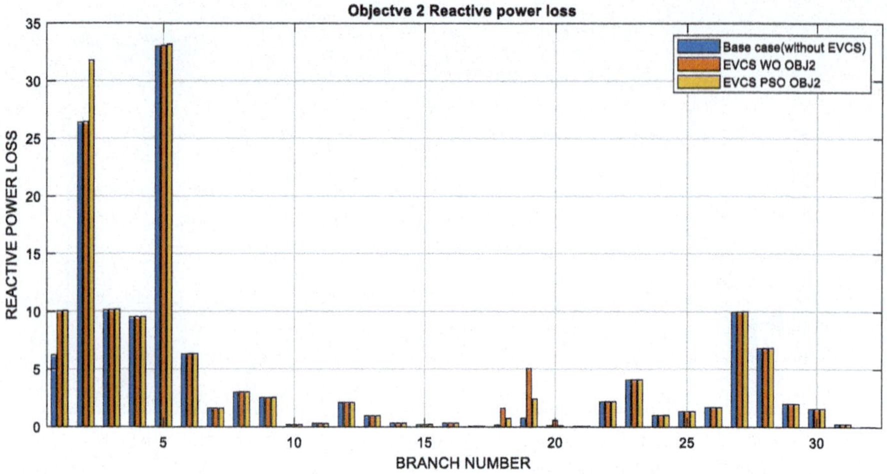


**Fig. 11** Increased reactive power loss after placing EV CS objective 2

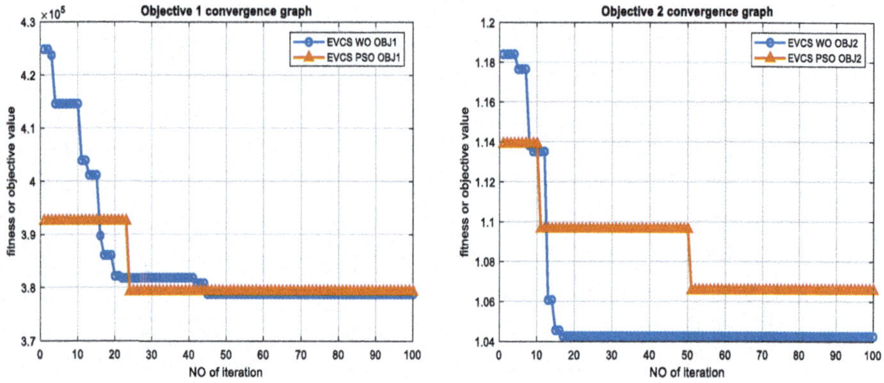


**Fig. 12** Convergence graph of both objective function

**Table 3** Different placement locations for different EVCS numbers

| EV CS | | Optimal placement | | | |
|---|---|---|---|---|---|
| No of fast EV CS/ charger | No of slow EV CS/ charger | OBJ 1 with WO | OBJ 1 with PSO | OBJ 2 with WO | OBJ 2 with PSO |
| 1/8 | 3/15 | 2/20/21/19 | 2/20/21/19 | 2/21/20/19 | 21/22/19/2 |
| 2/8 | 2/12 | 2/19/21/20 | 19/20/2/22 | 2/19/21/20 | 2/3/20/19 |
| 3/8 | 1/10 | 20/19/2/21 | 2/22/20/3 | 2/19/20/21 | 2/19/20/4 |
| 4/8 | 0/0 | 2/20/22/19 | 20/22/19/2 | 2/19/20/21 | 21/20/19/2 |
| 0/0 | 4/15 | 2/19/20/21 | 2/22/21/19 | 2/19/21/20 | 3/2/22/20 |

## 5 Conclusion

For the EV business to flourish quickly, the ideal placement of EV CS. This article provides a unique placement approach for EV CS placement considering the economics and the stability of the distribution system. CS is crucial. The modelling of EV CS contains two objective functions, one focuses on the operation, installation, extra losses and voltage deviation cost and the other objective function aims on the voltage stability index and line loss index. Two nature-based optimization algorithm, Particle Swarm Optimization (PSO) and Whale Optimization (WO) is utilized for the resolution of this difficult positioning issue. In this study, the effectiveness of these two algorithms in handling challenging optimization issues is firmly proven. The best optimal location for EV CS in 33 BUS radial distribution system is 2,19, 20, 21. Whale optimization has given the best result of EV CS allocation out of these two optimization techniques. Further adding DGs to improve the system voltage profile is in the further scope of the paper.

## References

1. IBEF (2022) Electric vehicles market in India. India brand equity foundation, knowledge center blog. https://www.ibef.org/blogs/electric-vehicles-market-in-india
2. Islam MM, Mohamed A, Shareef H (2015) Optimal allocation of rapid charging stations for electric vehicles. In: 2015 IEEE student conference on research and development (SCOReD), Kuala Lumpur, Malaysia, pp 378–383. https://doi.org/10.1109/SCORED.2015.7449360
3. Deb S, Kalita K, Gao X-Z, Tammi K, Mahanta P (2017) Optimal placement of charging stations using CSO-TLBO algorithm. In: 2017 Third international conference on research in computational intelligence and communication networks (ICRCICN), Kolkata, India, pp 84–89. https://doi.org/10.1109/ICRCICN.2017.8234486
4. Ge S, Feng L, Liu H (2011) The planning of electric vehicle charging station based on Grid partition method. In: 2011 International conference on electrical and control engineering, pp 2726–2730. Yichang
5. Pazouki S, Mohsenzadeh A, Haghifam MR, Ardalan S (2015) Simultaneous allocation of charging stations and capacitors in distribution networks improving voltage and power loss. Can J Electr Comput Eng 38:100–105. https://doi.org/10.1109/CJECE.2014.2377653
6. Liu ZF, Zhang W, Ji X, Li K (2012) Optimal Planning of charging station for electric vehicle based on particle swarm optimization. In: IEEE PES innovative smart grid technologies, pp 1–5. Tianjin
7. Mohsenzadeh A, Pazouki S, Ardalan S, Haghifam MR (2017) Optimal placing and sizing of parking lots including different levels of charging stations in electric distribution networks. Int J Ambient Energy 1–8. https://doi.org/10.1080/01430750.2017.1345010
8. Pazouki S, Mohsenzadeh A, Haghifam MR (2013) Optimal planning of plug-in electric vehicles (PEVs) charging stations and demand response programs considering distribution and traffic networks, pp 90–95. https://doi.org/10.1109/SGC.2013.6733806
9. Bilil H, Ellaia R, Maaroufi M (2012) A new multi-objective particle swarm optimization for reactive power dispatch, pp 1119–1124. https://doi.org/10.1109/ICMCS.2012.6320157
10. Wu A, Ni B (2015) Line loss analysis and calculation of electric power systems. https://doi.org/10.1002/9781118867273
11. Jamian JJ, Musa H, Mustafa MW, Mokhlis H, Adamu SS (2011) Combined voltage stability index for charging station effect on distribution network. Int Rev Electr Eng 6:3175–3184

12. Kazmi SA, Shahzad M, Shin D (2017) Voltage stability index for distribution network connected in loop configuration. IETE J Res 63:1–13. https://doi.org/10.1080/03772063.2016.1257376
13. Martinez JA, Mahseredjian J (2011) Load flow calculations in distribution systems with distributed resources. A review. In: 2011 IEEE power and energy society general meeting, Detroit, MI, USA, pp 1–8. https://doi.org/10.1109/PES.2011.6039172
14. Teng J-H (2003) A direct approach for distribution system load flow solutions. IEEE Trans Power Deliv 18:882–887. https://doi.org/10.1109/TPWRD.2003.813818
15. Doagou-Mojarrad H, Gharehpetian GB, Rastegar H, Olamaei J (2013) Optimal placement and sizing of DG (distributed generation) units in distribution networks by novel hybrid evolutionary algorithm. Energy 54:129–138. https://doi.org/10.1016/j.energy.2013.01.043
16. Nguyen T, Truong V, Tuấn P (2016) A novel method based on adaptive cuckoo search for optimal network reconfiguration and distributed generation allocation in distribution network. Int J Electr Power Energy Syst 78:801–815. https://doi.org/10.1016/j.ijepes.2015.12.030
17. Azli NA, Nayan NM, Ayob SM (2013) Particle swarm optimisation (PSO) and its applications in power converter systems. Int J Artif Intell Soft Comput 3. https://doi.org/10.1504/IJAISC.2013.056848
18. Aje OF, Josephat AA (2020) The particle swarm optimization (PSO) algorithm application–a review. Glob J Eng Technol Adv 3:001–006. https://doi.org/10.30574/gjeta.2020.3.3.0033
19. Gharehchopogh FS, Gholizadeh Hojjat (2019) A comprehensive survey: whale optimization algorithm and its applications. Swarm Evol Comput 48:1–24. https://doi.org/10.1016/j.swevo.2019.03.004
20. Mirjalili SZ, Mirjalili S, Saremi S, Mirjalili S (2020) Whale optimization algorithm: theory, literature review, and application in designing photonic crystal filters: methods and applications. https://doi.org/10.1007/978-3-030-12127-3_13
21. Bhadra J, Chattopadhyay TK (2015) Analysis of distribution network by reliability indices. In: 2015 International conference on energy, power and environment: towards sustainable growth (ICEPE), Shillong, pp 1–5

# SFO Based Economic Load Dispatch with FACTS Devices for DC Link Placement Problem

**B. Suresh Babu**

**Abstract** This Paper Presents a SFO based Economic Load Dispatch with FACTS devices for DC link Placement Problem. While most of the existing DC links are designed for point to point transmission, the multi-terminal DC system operation has become a reality and its usage is expected to increase in the future with a view of making the operation more flexible, secure and economical. The realization of multi-terminal DC systems cannot be done at once but can be executed by replacing the existing AC transmission lines with DC links over a period of time. The other development in power electronics pioneered FACTS devices that make the operation of the power system more flexible, secure and stable and have become popular in the recent years. The FACTS placement problem is another important optimization problem in power systems. New solution strategies involving FO have been suggested for the developed OPF with DC link and FACTS placement problem with a view of obtaining the global best solution in this paper. It presents simulation results of IEEE57 bus test systems with a view of demonstrating the effectiveness of the developed strategies.

**Keywords** Firefly optimization (FO) · AC/DC power flow · FACTS devices · Valve point effect

## Nomenclature

| | |
|---|---|
| $\alpha_j \ \beta_j \ \chi_j$ | Fuel cost coefficients |
| $d_j \ e_j$ | Coefficients of valve point effects of the generator |
| TFC | Total fuel cost |
| TRPL | Total Real power loss |
| NVSI | Net Voltage stability index |

B. S. Babu (✉)
Electrical and Electronics Engineering, Shri Vishnu Engineering College for Women, Mahiravani, Trimbak Road, Nashik, Maharashtra 422213, India
e-mail: sbeee77@yahoo.com

 
O. H. Gupta et al. (eds.), *Soft Computing Applications in Modern Power and Energy Systems*, Lecture Notes in Electrical Engineering 1107,
https://doi.org/10.1007/978-981-99-8007-9_3

| | |
|---|---|
| LVSI | Largest Voltage stability index |
| FACTS | Flexible AC transmission systems |
| $Type_f$ | Integer number in the range of (1–3) denoting the type of $f$-th FACTS device, 1 represents SVC, 2 denote TCSC and 3 indicates UPFC |

## 1 Introduction

Over the years, numerous mathematical programming techniques such as gradient method by Carpentier 1962 [1], a new innovative technique for OPF validation has been described. The Load balancing and loss reduction are achieved by reconfiguring the network. The global search technique is rendered owing to the discontinuous nature of the problem [2–5]. A modified strategy for OPF along with radial distribution system. This approach explores an optimal location of shunt capacitors thereby enhancing voltage stability and power factor [6–8]. A new imperative technique for OPF has been proposed by modern stochastic algorithms such as Particle Swarm Optimization (PSO), Evolutionary Programming (EP) and Harmony Search Optimization (HSO) [9–14], a hierarchical approach for optimal location of FACTS devices through OPF model [15]. A complementary approach for solving a variety of ELD problems [16–19] and found to yield satisfactory results. Self Adaptive FO with a view of obtaining the global best solution through simulation results on IEEE57 bus test system.

## 2 Problem Statement with FACTS Devices and Solution Process

CASE 1: TFC

$$\text{Minimize } \Phi_1(x,u) = \sum_{j\in\Pi} \alpha_j P_j^{g^2} + \beta_j P_j^g + \chi_j + \left| d_j \sin\left(e_j\left(P_j^g(\min) - P_j^g\right)\right)\right| \tag{1}$$

$$\text{FACTS device constraints } -0.8 \le \eta_z \le 0.2 \text{ for TCSC and UPFC} \tag{2}$$

$$-100m\text{var} \le Q_i^f \le +100m\text{var for SVC and UPFC} \tag{3}$$

CASE 2: TRPL

CASE 3: LVSI

CASE 4: Combination of CASE 1 and CASE 2

CASE 5: Combination of CASE 1 and CASE 3

CASE 6: Combination of CASE 2 and CASE 3

CASE 7: Combination of CASE 1, CASE 2 and CASE 3.

Evaluate the AOF, highest $f$ in the population as the optimal solution

$$\text{Maximize} \quad f = \frac{1}{1 + AOF} \tag{4}$$

## 2.1 *Representation of Decision Variables*

The control variable in vector representation in FO method is as follows

$$f = [Type_f,\ Loc_f,\ \eta_f,\ Q_f^F,\ P_k^G,\ V_j^G,\ T_v,\ L_p^{dc},\ I_p^{dc}] \tag{5}$$

- *Read the problem data*
- *Choose FOPM parameters such as number of fireflies, maximum number of iteration and convergence check.*
- *Set iter = 0*
- *Compare the FO of all the firefly in the population for each firefly*
- *Repair the firefly and Replace the transmission lines by DC links, place FACTS devices and set the control parameters according to the firefly values*
- *Evaluate the values of firefly parameters*
- *Carryout the AC/DC load flow*
- *Evaluate TFC, TRPL and LVSI, light intensity function and objective function*
- *Rank the fireflies and highest light intensity in the population as the optimal solution.*

# 3 Results and Discussions

IEEE 57 bus test system by using fix the Table 1 FO parameter (Number of firefly and Maximum number of iteration) and fix the Table 2 Number of devices to be placed (Number of DC link and Number of FACTS devices) [20].

It is observed from Table 3 tested results from IEEE57 bus system that FOPM in CASE 1, CASE 2 and CASE 3 is the reduction of the TFC, TRPL and LVSI. It is seen from the results of CASE 4, CASE 5, CASE 6 and CASE 7 that the PM as

**Table 1** FO parameter

| Parameter | Value |
|---|---|
| Number of firefly | 30 |
| Maximum number of iteration | 300 |

**Table 2** Number of devices to be placed

| Test system | IEEE 57 |
|---|---|
| Number of DC links | 3 |
| Number of FACTS devices | 5 |

well as the HSO and PSO offer a compromised solution, which lies in between the respective best and worst objective function values obtained in CASE 1, CASE 2 and CASE 3 in IEEE57 bus system. It is observed from Tables 4, 5 and 6 that the Optimal solution of FOPM, HSO and PSO. The %TFC savings, %TRPL savings and %LVSI for IEEE57 are in Figs. 1, 2 and 3.The lower and upper load bus voltages of all the cases of the FOPM are graphically displayed in Fig. 4 respectively for IEEE57 bus system.

## 4 Summary and Future Research

In this paper, FO solution technique for OPFDC with FACTS placement are developed and tested on three IEEE 57 test systems. The algorithms use sequential AC/DC load flow involving NR technique for computing the objective function during search and are able to offer the global best solution. It can also be observed that the proposed methods perform better with FACTS devices. Hybrid algorithms involving classical and evolutionary algorithms and the classical can be used to solve the OPFDC with FACTS placement problem in order to explore new search regions with the possibility of landing at a better solution point.

**Table 3** Performance Comparison of IEEE57 bus system

| CASE | | Before placement | IEEE 57 | | |
|---|---|---|---|---|---|
| | | | FOPM | HSO | PSO |
| 1 | TFC | 4556.5930 | 3796.1226 | 3800.4563 | 3801.2526 |
| | TRPL | 28.8037 | 29.8226 | 30.6259 | 30.7040 |
| | NVSI | 5.7914 | 4.8494 | 5.4457 | 5.4407 |
| | LVSI | 0.2887 | 0.2316 | 0.2440 | 0.2437 |
| 2 | TFC | 4556.5930 | 5336.0222 | 5324.7986 | 5325.3278 |
| | TRPL | 28.8037 | 12.0663 | 12.2299 | 12.2348 |
| | NVSI | 5.7914 | 5.2935 | 4.8339 | 4.8351 |
| | LVSI | 0.2887 | 0.2657 | 0.2438 | 0.2461 |
| 3 | TFC | 4556.5930 | 5365.7795 | 5065.1341 | 5157.2746 |
| | TRPL | 28.8037 | 17.0111 | 16.2796 | 15.3382 |
| | NVSI | 5.7914 | 3.7220 | 3.7745 | 4.1688 |
| | LVSI | 0.2887 | 0.1895 | 0.1990 | 0.2069 |
| 4 | TFC | | 5231.6348 | 5100.1084 | 4738.6463 |
| | TRPL | | 25.7949 | 26.9970 | 21.3307 |
| | NVSI | | 5.6759 | 6.1861 | 6.0059 |
| | LVSI | | 0.2607 | 0.2843 | 0.2997 |
| 5 | TFC | | 3796.7276 | 3800.7042 | 3802.8248 |
| | TRPL | | 29.7640 | 30.5586 | 30.5075 |
| | NVSI | | 4.2749 | 4.8859 | 4.8829 |
| | LVSI | | 0.2100 | 0.2226 | 0.2216 |
| 6 | TFC | | 5343.4440 | 5336.5989 | 5323.0015 |
| | TRPL | | 12.1275 | 12.4479 | 12.2558 |
| | NVSI | | 4.9801 | 4.5069 | 4.8364 |
| | LVSI | | 0.2500 | 0.2265 | 0.2440 |
| 7 | TFC | | 5043.1763 | 4388.2177 | 4696.4122 |
| | TRPL | | 24.4751 | 35.0490 | 31.4009 |
| | NVSI | | 6.0508 | 6.4466 | 6.2837 |
| | LVSI | | 0.3029 | 0.2853 | 0.3328 |

**Table 4** Optimal solution of FOPM for IEEE57 bus system

| | Before Placement | Case-1 FC | Case-2 Loss | Case-3 VS | Case-4 FC + Loss | Case-5 FC + VS | Case-6 Loss + VS | Case-7 FC + Loss + VS |
|---|---|---|---|---|---|---|---|---|
| $P^G$ | 359.604 | 462.617646 | 181.462598 | 83.233323 | 251.083683 | 462.671541 | 179.782401 | 350.215180 |
| | 35.000 | 10.000216 | 23.074543 | 19.911095 | 32.283245 | 10.000011 | 22.658904 | 66.809328 |
| | 40.000 | 20.000028 | 137.227388 | 127.581494 | 119.125354 | 20.067936 | 138.349395 | 86.202688 |
| | 50.000 | 10.000498 | 71.434464 | 70.230099 | 79.102837 | 10.029188 | 70.901078 | 33.155198 |
| | 450.000 | 512.857139 | 397.129814 | 521.331321 | 391.552414 | 512.702591 | 397.842674 | 405.368769 |
| | 35.000 | 10.000059 | 42.537516 | 39.269849 | 31.677943 | 10.060605 | 43.393073 | 58.731220 |
| | 310.000 | 255.146992 | 410.000000 | 406.253950 | 371.769429 | 255.032113 | 410.000000 | 274.792717 |
| $V^G$ | 1.040 | 1.024861 | 1.060493 | 1.053420 | 1.063891 | 1.024746 | 1.060338 | 1.049412 |
| | 1.010 | 1.001172 | 1.047607 | 1.043578 | 1.039963 | 1.001133 | 1.048917 | 1.027369 |
| | 0.985 | 0.973865 | 1.050133 | 1.052315 | 1.009913 | 0.973982 | 1.049264 | 1.004695 |
| | 0.980 | 0.949210 | 1.020176 | 1.050313 | 0.978106 | 0.949245 | 1.021557 | 0.970000 |
| | 1.005 | 0.951456 | 1.053694 | 1.056456 | 0.957631 | 0.951414 | 1.052667 | 0.990616 |
| | 0.980 | 0.940000 | 1.035665 | 1.021059 | 0.950622 | 0.940000 | 1.034548 | 0.962065 |
| | 1.015 | 0.969682 | 1.041373 | 1.037053 | 0.947697 | 0.969814 | 1.040161 | 0.963040 |

(continued)

**Table 4** (continued)

| | Before Placement | Case-1 FC | Case-2 Loss | Case-3 VS | Case-4 FC + Loss | Case-5 FC + VS | Case-6 Loss + VS | Case-7 FC + Loss + VS |
|---|---|---|---|---|---|---|---|---|
| $T$ | 0.970 | 0.900791 | 1.045080 | 1.025172 | 1.082997 | 0.900934 | 1.044109 | 0.917207 |
| | 0.978 | 0.938516 | 0.971183 | 0.995455 | 1.060580 | 0.938586 | 0.971389 | 1.021423 |
| | 0.967 | 1.094362 | 1.061663 | 1.020132 | 1.060143 | 1.094285 | 1.062275 | 0.970165 |
| | 0.940 | 0.968448 | 1.034135 | 1.013482 | 0.968371 | 0.968584 | 1.033208 | 1.068509 |
| | 0.930 | 1.063404 | 0.988566 | 0.990920 | 1.070236 | 1.063403 | 0.987028 | 0.953288 |
| | 0.955 | 1.041107 | 1.044604 | 1.040796 | 1.077040 | 1.041144 | 1.045321 | 1.023742 |
| | 0.958 | 0.965675 | 0.943548 | 0.931453 | 0.949622 | 0.965522 | 0.942518 | 0.940430 |
| | 0.895 | 0.944795 | 1.015784 | 0.995345 | 0.957638 | 0.944747 | 1.016365 | 1.055602 |
| | 0.900 | 0.923083 | 0.900000 | 0.910062 | 0.958811 | 0.923148 | 0.900211 | 1.067993 |
| | 0.955 | 0.977203 | 1.040957 | 1.015597 | 0.940213 | 0.977212 | 1.041466 | 1.055175 |
| | 1.043 | 0.942371 | 0.990961 | 0.984296 | 1.040755 | 0.942473 | 0.991404 | 1.004839 |
| | 1.000 | 0.971777 | 0.950661 | 0.946771 | 1.050580 | 0.971853 | 0.952131 | 1.026834 |
| | 1.000 | 0.904493 | 0.941213 | 0.930236 | 0.942708 | 0.904343 | 0.942044 | 0.928273 |
| | 1.043 | 0.902387 | 1.029155 | 1.006633 | 0.949441 | 0.902537 | 1.029173 | 0.958606 |
| | 0.975 | 1.099783 | 0.925550 | 0.900000 | 0.957768 | 1.099635 | 0.925445 | 0.907315 |
| | 0.980 | 0.924334 | 0.975958 | 0.964894 | 0.933217 | 0.924329 | 0.974657 | 1.018789 |
| | 0.958 | 0.945002 | 1.042022 | 1.011908 | 0.944943 | 0.945154 | 1.041207 | 0.933183 |
| $L_p^{dc}$ | | 72 | 64 | 64 | 17 | 72 | 64 | 57 |
| | | 16 | 58 | 57 | 9 | 16 | 57 | 24 |
| | | 10 | 22 | 10 | 4 | 10 | 22 | 7 |
| $I_p^{dc}$ | | 0.394364 | 0.132061 | 0.149387 | 0.206635 | 0.393644 | 0.130698 | 0.100000 |
| | | 0.721434 | 0.226453 | 0.269190 | 0.360441 | 0.720758 | 0.227122 | 0.168781 |
| | | 0.623835 | 0.636398 | 0.621154 | 0.231266 | 0.624614 | 0.634546 | 0.492025 |
| $\alpha$ | | 0.000000 | 0.087894 | 0.113588 | 0.000002 | 0.000000 | 0.090660 | 0.561667 |
| $\beta_o$ | | 0.487171 | 0.712100 | 0.752049 | 0.066466 | 0.486560 | 0.705684 | 0.000000 |
| $\gamma$ | | 0.941813 | 0.367858 | 0.475204 | 0.637109 | 0.941158 | 0.369257 | 0.447836 |

(continued)

**Table 4** (continued)

| | Before Placement | Case-1 FC | Case-2 Loss | Case-3 VS | Case-4 FC + Loss | Case-5 FC + VS | Case-6 Loss + VS | Case-7 FC + Loss + VS |
|---|---|---|---|---|---|---|---|---|
| Facts type | | 2<br>3<br>2<br>3<br>2 | 2<br>2<br>2<br>2<br>1 | 2<br>2<br>2<br>2<br>1 | 3<br>2<br>2<br>3<br>3 | 2<br>3<br>2<br>3<br>2 | 2<br>2<br>2<br>2<br>1 | 2<br>1<br>2<br>2<br>3 |
| Facts Location | | 38<br>34<br>63<br>52<br>67 | 50<br>24<br>76<br>43<br>79 | 48<br>66<br>35<br>50<br>78 | 61<br>38<br>50<br>39<br>71 | 38<br>34<br>63<br>52<br>66 | 51<br>24<br>76<br>43<br>79 | 67<br>59<br>54<br>47<br>45 |
| VAR support | | –<br>73.454<br>–<br>30.167<br>– | –<br>–<br>–<br>–<br>74.799 | –<br>–<br>–<br>–<br>86.282 | 28.893<br>–<br>–<br>74.300<br>82.285 | –<br>73.410<br>–<br>30.130<br>– | –<br>–<br>–<br>–<br>74.700 | —<br>75.746<br>—<br>—<br>40.884 |
| Line com fact | | −0.438<br>−0.522<br>−0.148<br>− 0.429<br>−0.800 | 0.198<br>−0.800<br>−0.389<br>−0.059<br>– | 0.133<br>−0.800<br>−0.323<br>−0.329<br>– | −0.378<br>−0.736<br>−0.269<br>−0.463<br>−0.462 | −0.439<br>−0.522<br>−0.149<br>−0.428<br>−0.800 | 0.198<br>−0.799<br>−0.396<br>−0.050<br>– | −0.594<br>–<br>−0.262<br>−0.640<br>−0.066 |

**Table 5** Optimal solution of HSO for IEEE 57 bus system

| | Before placement | Case-1 FC | Case-2 Loss | Case-3 VS | Case-4 FC + Loss | Case-5 FC + VS | Case-6 Loss + VS | Case-7 FC + Loss + VS |
|---|---|---|---|---|---|---|---|---|
| $P^G$ | 359.604 | 458.749051 | 191.540202 | 192.280681 | 361.003303 | 458.196356 | 184.685493 | 309.799420 |
| | 35.000 | 10.000000 | 21.237587 | 19.492152 | 65.719155 | 10.000000 | 21.252586 | 42.727071 |
| | 40.000 | 20.000000 | 134.310812 | 120.199662 | 62.374994 | 20.072869 | 135.425907 | 53.144791 |
| | 50.000 | 10.000000 | 76.556256 | 67.756690 | 41.701145 | 10.000000 | 76.672846 | 27.338433 |
| | 450.000 | 516.652693 | 390.817510 | 455.701792 | 321.568575 | 517.111135 | 398.426724 | 550.000000 |
| | 35.000 | 10.000000 | 40.917647 | 29.526207 | 63.937772 | 10.002054 | 40.876397 | 10.000000 |
| | 310.000 | 256.024148 | 407.649937 | 382.122412 | 361.492101 | 255.976223 | 405.907913 | 292.839243 |
| $V^G$ | 1.040 | 1.025648 | 1.057447 | 1.072134 | 1.096505 | 1.025547 | 1.056565 | 1.025908 |
| | 1.010 | 1.000973 | 1.044933 | 1.059960 | 1.074612 | 1.000546 | 1.042679 | 1.010049 |
| | 0.985 | 0.971684 | 1.042585 | 1.054608 | 1.029689 | 0.971186 | 1.043565 | 0.981356 |
| | 0.980 | 0.950045 | 1.043960 | 1.048589 | 0.980029 | 0.950008 | 1.044880 | 0.943083 |
| | 1.005 | 0.951397 | 1.051868 | 1.056020 | 0.964365 | 0.951183 | 1.053717 | 1.007568 |
| | 0.980 | 0.940000 | 1.035183 | 1.017499 | 0.940000 | 0.940000 | 1.038479 | 0.980000 |
| | 1.015 | 0.970893 | 1.047025 | 1.035970 | 0.959283 | 0.970203 | 1.049850 | 1.012209 |

(continued)

Table 5 (continued)

| | Before placement | Case-1 FC | Case-2 Loss | Case-3 VS | Case-4 FC + Loss | Case-5 FC + VS | Case-6 Loss + VS | Case-7 FC + Loss + VS |
|---|---|---|---|---|---|---|---|---|
| $T$ | 0.970 | 0.900156 | 1.045530 | 1.023242 | 0.975823 | 0.900012 | 1.048973 | 0.992339 |
| | 0.978 | 0.936852 | 0.958937 | 0.990536 | 1.076706 | 0.937584 | 0.962894 | 0.944658 |
| | 0.967 | 1.093223 | 1.066948 | 1.049559 | 1.010248 | 1.092712 | 1.066446 | 1.056918 |
| | 0.940 | 0.967468 | 1.043704 | 1.027968 | 1.021186 | 0.966695 | 1.047031 | 1.055952 |
| | 0.930 | 1.062674 | 0.987728 | 1.014592 | 0.938395 | 1.063262 | 0.989500 | 0.938053 |
| | 0.955 | 1.041559 | 1.041904 | 1.046034 | 1.063521 | 1.040995 | 1.037994 | 1.007675 |
| | 0.958 | 0.965530 | 0.938656 | 0.930498 | 1.004560 | 0.966022 | 0.941399 | 1.037655 |
| | 0.895 | 0.944611 | 1.014698 | 0.994809 | 1.006734 | 0.945014 | 1.014368 | 0.922681 |
| | 0.900 | 0.922717 | 0.900572 | 0.901871 | 0.926789 | 0.921791 | 0.900000 | 1.100000 |
| | 0.955 | 0.979537 | 1.041342 | 1.025341 | 1.050855 | 0.979907 | 1.041346 | 0.931155 |
| | 1.043 | 0.943430 | 0.991051 | 0.996817 | 1.055747 | 0.943500 | 0.992368 | 0.947504 |
| | 1.000 | 0.971621 | 0.950513 | 0.965070 | 1.027211 | 0.971698 | 0.949621 | 0.965858 |
| | 1.000 | 0.905113 | 0.939528 | 0.942107 | 1.022409 | 0.905181 | 0.937367 | 0.951411 |
| | 1.043 | 0.902282 | 1.038625 | 1.021435 | 0.917750 | 0.902891 | 1.038483 | 0.975357 |
| | 0.975 | 1.098738 | 0.931701 | 0.934289 | 0.923554 | 1.099189 | 0.932424 | 1.027763 |
| | 0.980 | 0.925334 | 0.976539 | 0.982060 | 1.074519 | 0.925998 | 0.973422 | 0.935815 |
| | 0.958 | 0.942780 | 1.039893 | 1.014203 | 0.951583 | 0.942776 | 1.040810 | 0.984881 |
| $L_p^{dc}$ | | 72 | 62 | 64 | 60 | 72 | 61 | 21 |
| | | 16 | 58 | 57 | 27 | 16 | 58 | 17 |
| | | 9 | 18 | 10 | 8 | 9 | 18 | 14 |
| $I_p^{dc}$ | | 0.389317 | 0.149973 | 0.117741 | 0.294783 | 0.388535 | 0.130331 | 0.822265 |
| | | 0.721960 | 0.251302 | 0.264139 | 0.132183 | 0.719397 | 0.249869 | 0.628072 |
| | | 0.637236 | 0.623659 | 0.727385 | 0.941559 | 0.636149 | 0.611109 | 0.109369 |

(continued)

**Table 5** (continued)

| | Before placement | Case-1 FC | Case-2 Loss | Case-3 VS | Case-4 FC + Loss | Case-5 FC + VS | Case-6 Loss + VS | Case-7 FC + Loss + VS |
|---|---|---|---|---|---|---|---|---|
| Facts type | | 2 | 2 | 2 | 3 | 2 | 2 | 2 |
| | | 3 | 2 | 2 | 2 | 3 | 2 | 2 |
| | | 2 | 2 | 2 | 2 | 2 | 2 | 2 |
| | | 3 | 2 | 2 | 2 | 3 | 2 | 2 |
| | | 2 | 1 | 1 | 3 | 2 | 1 | 2 |
| Facts location | | 38 | 51 | 48 | 34 | 38 | 51 | 69 |
| | | 34 | 19 | 66 | 61 | 34 | 20 | 40 |
| | | 63 | 43 | 31 | 69 | 63 | 41 | 44 |
| | | 52 | 45 | 47 | 45 | 52 | 46 | 60 |
| | | 67 | 78 | 79 | 51 | 66 | 78 | 46 |
| VAR support | | – | – | – | 79.300 | – | – | – |
| | | 72.822 | – | – | – | 73.001 | – | – |
| | | – | – | – | – | – | – | – |
| | | 30.275 | – | – | – | 30.063 | – | – |
| | | – | 73.287 | 97.650 | 34.502 | – | 73.294 | – |
| Line com fact | | −0.435 | 0.187 | 0.028 | − 0.416 | −0.433 | 0.200 | −0.577 |
| | | −0.525 | −0.800 | −0.744 | 0.014 | −0.526 | −0.800 | 0.200 |
| | | −0.152 | −0.426 | −0.388 | 0.013 | −0.151 | −0.431 | −0.148 |
| | | −0.431 | −0.055 | −0.163 | −0.194 | −0.434 | −0.038 | −0.021 |
| | | −0.796 | – | – | −0.561 | −0.797 | – | −0.674 |

**Table 6** Optimal solution of PSO for IEEE57 bus system

| | Before placement | Case-1 FC | Case-2 Loss | Case-3 VS | Case-4 FC + Loss | Case-5 FC + VS | Case-6 Loss + VS | Case-7 FC + Loss + VS |
|---|---|---|---|---|---|---|---|---|
| $P^G$ | 359.604 | 457.642435 | 192.472994 | 168.149830 | 356.541872 | 455.148703 | 193.260030 | 371.099700 |
| | 35.000 | 10.009901 | 21.235018 | 22.973653 | 10.000000 | 10.000000 | 21.127468 | 41.291106 |
| | 40.000 | 20.007200 | 134.371368 | 122.119310 | 84.335719 | 20.000000 | 134.332012 | 89.266755 |
| | 50.000 | 10.000000 | 76.476374 | 64.161023 | 38.853538 | 10.217802 | 76.517272 | 32.714853 |
| | 450.000 | 517.792720 | 389.977631 | 460.961332 | 334.414249 | 519.368920 | 389.682379 | 497.164826 |
| | 35.000 | 10.000000 | 41.180872 | 37.181741 | 59.377206 | 10.000000 | 41.079506 | 36.547424 |
| | 310.000 | 256.051763 | 407.320582 | 390.591292 | 388.608137 | 256.572043 | 407.057102 | 214.116204 |
| $V^G$ | 1.040 | 1.026355 | 1.058005 | 1.066527 | 1.027153 | 1.025931 | 1.057965 | 1.062788 |
| | 1.010 | 1.000310 | 1.044613 | 1.055470 | 0.985563 | 1.000330 | 1.044879 | 0.961858 |
| | 0.985 | 0.972076 | 1.042573 | 1.060232 | 0.990769 | 0.971671 | 1.042735 | 0.993286 |
| | 0.980 | 0.950379 | 1.043585 | 1.054229 | 1.016093 | 0.950128 | 1.043072 | 1.009223 |
| | 1.005 | 0.950717 | 1.051730 | 1.061244 | 1.064833 | 0.950175 | 1.051471 | 1.027935 |
| | 0.980 | 0.940000 | 1.035905 | 1.024926 | 1.016490 | 0.940000 | 1.036069 | 0.991753 |
| | 1.015 | 0.970023 | 1.046466 | 1.038861 | 1.008339 | 0.970186 | 1.046636 | 1.006558 |

(continued)

**Table 6** (continued)

| | Before placement | Case-1 FC | Case-2 Loss | Case-3 VS | Case-4 FC + Loss | Case-5 FC + VS | Case-6 Loss + VS | Case-7 FC + Loss + VS |
|---|---|---|---|---|---|---|---|---|
| $T$ | 0.970 | 0.900495 | 1.045764 | 1.028358 | 0.900000 | 0.900000 | 1.045484 | 1.038726 |
| | 0.978 | 0.937643 | 0.959461 | 0.981766 | 0.993318 | 0.937130 | 0.958981 | 1.060923 |
| | 0.967 | 1.093048 | 1.067497 | 1.049749 | 1.065053 | 1.093006 | 1.068012 | 1.057364 |
| | 0.940 | 0.967390 | 1.043021 | 1.027908 | 0.900000 | 0.967548 | 1.043365 | 0.977552 |
| | 0.930 | 1.062804 | 0.988993 | 0.995810 | 1.048910 | 1.063306 | 0.989081 | 0.963966 |
| | 0.955 | 1.041430 | 1.041872 | 1.038969 | 0.920769 | 1.041359 | 1.041389 | 1.017417 |
| | 0.958 | 0.964856 | 0.939463 | 0.933229 | 1.023365 | 0.965972 | 0.939408 | 1.042187 |
| | 0.895 | 0.944172 | 1.015176 | 0.994710 | 0.917875 | 0.944855 | 1.014740 | 1.100000 |
| | 0.900 | 0.922545 | 0.900256 | 0.911401 | 1.079476 | 0.922187 | 0.900000 | 1.016696 |
| | 0.955 | 0.979686 | 1.042137 | 1.020718 | 0.900000 | 0.979683 | 1.042673 | 0.961096 |
| | 1.043 | 0.942531 | 0.990661 | 0.984990 | 0.980770 | 0.942861 | 0.990936 | 0.900000 |
| | 1.000 | 0.971429 | 0.951004 | 0.966005 | 1.096300 | 0.971399 | 0.950847 | 1.092813 |
| | 1.000 | 0.905018 | 0.938707 | 0.937198 | 0.952308 | 0.905047 | 0.938632 | 0.976852 |
| | 1.043 | 0.901794 | 1.037877 | 1.018054 | 0.900000 | 0.901424 | 1.037959 | 1.015214 |
| | 0.975 | 1.098943 | 0.931352 | 0.921157 | 0.921016 | 1.099431 | 0.930927 | 1.001390 |
| | 0.980 | 0.925656 | 0.977287 | 0.981776 | 1.007667 | 0.925813 | 0.976562 | 0.900000 |
| | 0.958 | 0.943358 | 1.039930 | 1.020837 | 1.074962 | 0.943498 | 1.040016 | 1.009429 |
| $L_p^{dc}$ | | 72 | 62 | 64 | 16 | 72 | 62 | 28 |
| | | 16 | 58 | 56 | 11 | 16 | 58 | 16 |
| | | 9 | 18 | 10 | 2 | 9 | 18 | 1 |
| $I_p^{dc}$ | | 0.389913 | 0.152233 | 0.100668 | 0.478137 | 0.389165 | 0.151369 | 0.305651 |
| | | 0.717052 | 0.247949 | 0.199367 | 0.139523 | 0.716298 | 0.249467 | 0.678300 |
| | | 0.638314 | 0.627080 | 0.642785 | 1.000000 | 0.640415 | 0.625663 | 0.486425 |

(continued)

**Table 6** (continued)

| | Before placement | Case-1 FC | Case-2 Loss | Case-3 VS | Case-4 FC + Loss | Case-5 FC + VS | Case-6 Loss + VS | Case-7 FC + Loss + VS |
|---|---|---|---|---|---|---|---|---|
| Facts type | | 2<br>3<br>2<br>3<br>2 | 2<br>2<br>2<br>2<br>1 | 2<br>2<br>2<br>2<br>1 | 2<br>2<br>2<br>2<br>2 | 2<br>3<br>2<br>3<br>2 | 2<br>2<br>2<br>2<br>1 | 2<br>2<br>2<br>2<br>2 |
| Facts location | | 38<br>34<br>63<br>52<br>67 | 51<br>19<br>42<br>45<br>78 | 48<br>66<br>34<br>46<br>79 | 74<br>31<br>75<br>44<br>4 | 37<br>34<br>63<br>52<br>66 | 51<br>19<br>43<br>45<br>78 | 63<br>68<br>44<br>49<br>58 |
| VAR support | | –<br>73.714<br>–<br>29.360<br>– | –<br>–<br>–<br>–<br>73.100 | –<br>–<br>–<br>–<br>90.640 | –<br>–<br>–<br>–<br>– | –<br>73.458<br>–<br>29.851<br>– | –<br>–<br>–<br>–<br>73.246 | –<br>–<br>–<br>–<br>– |
| Line com fact | | −0.437<br>−0.527<br>−0.153<br>−0.433<br>−0.795 | 0.189<br>−0.799<br>−0.427<br>−0.054<br>– | 0.100<br>−0.752<br>−0.353<br>−0.195<br>– | −0.305<br>−0.486<br>−0.684<br>−0.507<br>−0.376 | −0.436<br>−0.523<br>−0.152<br>−0.429<br>−0.797 | 0.188<br>−0.800<br>−0.426<br>−0.053<br>– | −0.696<br>−0.052<br>−0.414<br>−0.288<br>−0.739 |

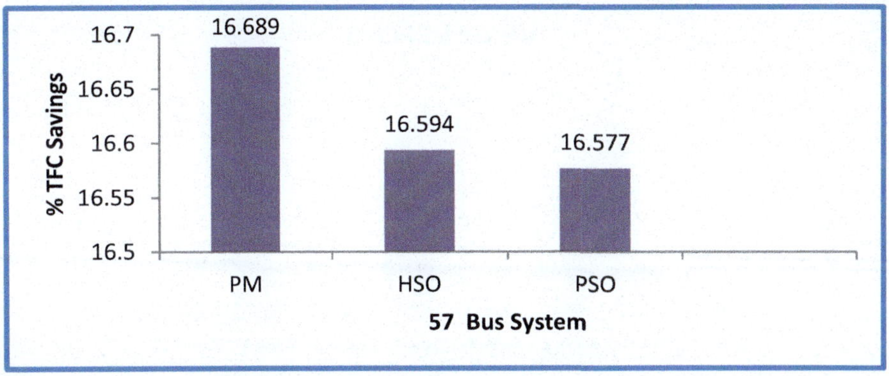


**Fig. 1** Comparison % savings of TFC for IEEE 57

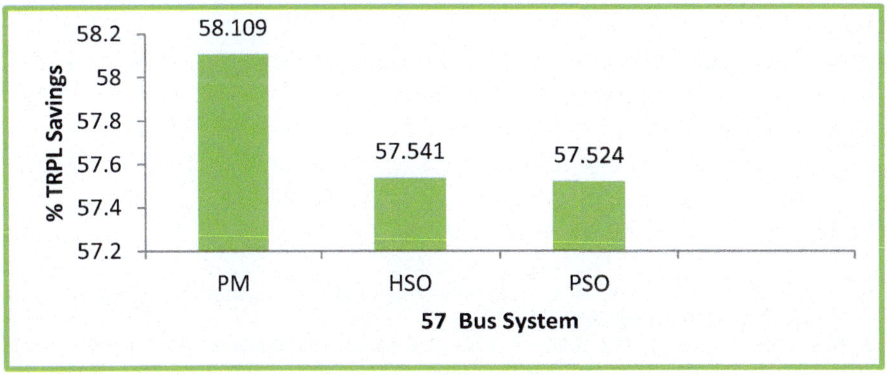


**Fig. 2** Comparison % savings TRPL for IEEE 57

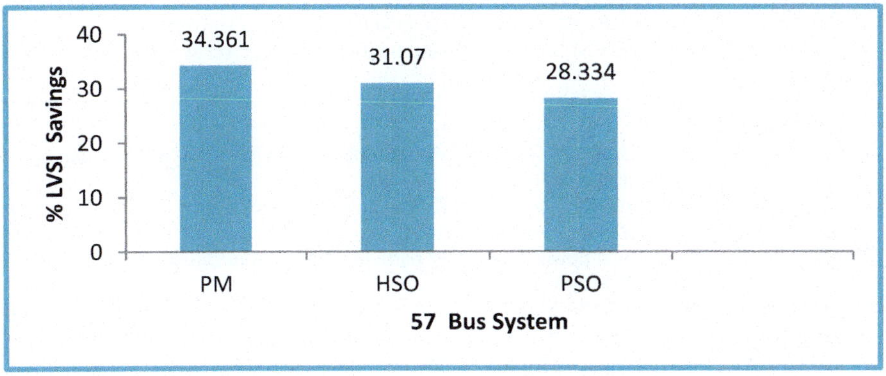


**Fig. 3** Comparison % savings LVSI for IEEE 57

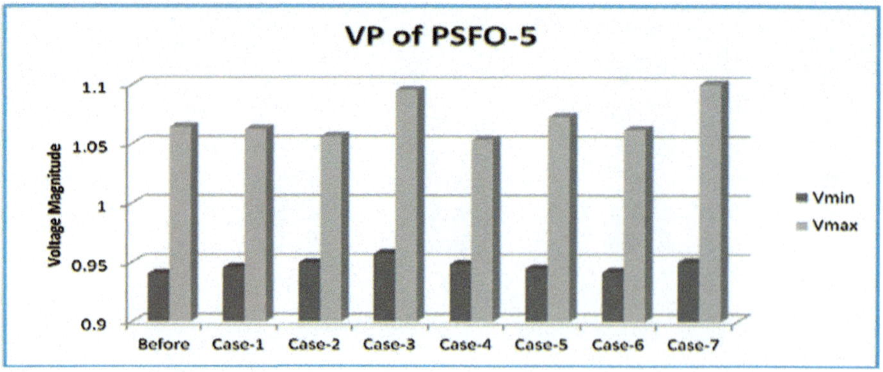


**Fig. 4** FOPM of lower and upper VMs for IEEE57

**Acknowledgements** The author gratefully acknowledges the authorities of Sandip Institute of Technology and Research Centre for their continued support, encouragement and the facilities provided to carry out this work.

## References

1. Carpentier J (1962) Contribution a l'Etude du Dispatching Economique'. Bulletin de la Societe Francaise des Electriciens 3:431–474
2. Momoh JA, El-Hawary ME, Adapa R (1999) A review of selected optimal power flow literature to 1993 Part I: nonlinear and quadratic programming approaches. IEEE Trans Power Syst 14:96–104
3. Yuryevich J, Wong KP (1999) Evolutionary programming based optimal power flow. IEEE Trans Power Syst 144:1245–1257
4. Nagsarkar TK, Sukhija MS (2007) Power system analysis. Oxford University Press
5. Arrillaga J, Watson NR (2013) Computer modelling of Electrical power systems. Wiley & Sons Ltd.
6. Arrillaga J, Arnold CR, Harker BJ (1983) Computer modelling of Electrical power systems. Wiley & Sons Ltd.
7. Lee KY, Park YM, Oritz JL (1984) Fuel-cost minimization for both real and reactive power dispatches. IEE Proc C 131(3):85–93
8. O.Alsac,and B. Scott. (1974). Optimal load flow with steady state security, IEEE Transactions on power Apparatus and systems. PAS: 745–751.
9. Abido MA (2002) Optimal power flow using particle swarm optimization'. Proc Int J Electr Power Energy Syst 24(7):563–571
10. Abido MA, Al-Ali NA (2009) Multi-objective differential evolution for optimal power flow. In: International conference on power engineering, energy and electrical drives, POWERENG 2009, Lisbon, Portugal, 18–20 March 2009, pp 101–106
11. Abou El Ela AA, Abido MA (2010) Optimal power flow using differential evolution algorithm. Electr Power Syst Res 80(7):878–885
12. Naresh Babu AV, Ramana T, Sivanagaraju S (2014) Analysis of optimal power flow problem based on two stage initialization algorithm. Int J Electr Power Energy Syst 55:91–99
13. Niknam T, Narimani MR, Jabbari M, Malekpour AR (2011) A modified shuffle frog leaping algorithm for multi-objective optimal power flow. Energy 36:6420–6432

14. Sivasubramani S, Swarup KS (2011) Multi-objective harmony search algorithm for optimal power flow problem. Int J Electr Power Energy Syst 33(3):745–752
15. Hingorani NG, Gyugy L (2001) Understanding FACTS concepts and technology of flexible AC transmission systems. Standard Publishers Distributors, New Delhi
16. Vinod Kumar K, Lakshmi Phani G (2011) Combined economic emission dispatch-pareto optimal front approach. Int J Comput Appl 30(12):16–21
17. Sulaiman MH, Mustafa MW, Zakaria ZN, Aliman O, Abdul Rahim SR (2012) Firefly algorithm technique for solving economic dispatch problem. In: 2012 IEEE international power engineering and optimization conference (PEOCO2012), Melaka, Malaysia, 6–7 June 2012
18. Chandrasekaran K, Simon SP (2012) Tuned fuzzy adapted firefly lambda algorithm for solving unit commitment problem. J Electr Syst 8(2):132–150
19. Yang X-S (2013) Multiobjective firefly algorithm for continuous optimization. Eng Comput 18(2):175–184
20. Test archive. Systems. http://www.ee.washington.edu/research/pstca/. Last Accessed Dec 2012

# Selection of Batteries for Electric Vehicle Applications

**Gaurang Swarup Sharma, G. Agam Swarup, and Subho Upadhyay**

**Abstract** Storage of energy is considered one of the most important factors for electric utilities and, electric vehicles. A battery is the primary source of energy for an electric vehicle and a properly selected battery will directly affect the reliability of the E-Vehicle. This necessitates understanding the behavior of the batteries on a basic level in ideal as well as practical conditions. The main goal of this work is to increase the reliability of the electric vehicle through a better selection of batteries. This work mainly focuses on simulating the charging and discharging characteristics of different types of batteries using MATLAB/Simulink and compares them to propose suitable battery types for different applications.

**Keywords** Batteries · Simulation · MATLAB/Simulink · Charging · Discharging · Electric vehicles

## 1 Introduction

Global Warming is a recognized concern for the entire world and fuel-consuming technologies are one of the main reasons. To mitigate its effects the world today is ever-changing and switching towards the use of renewable energy. Sustainable forms of energy are being recognized and preferred over conventional fossil fuels [1].

In India, green energy technologies are being adopted in a big way. Government policy viz. According to the Government of India guidelines for Tariff Based Competitive Bidding Process for the Procurement of Power from Grid Connected Solar PV Power Projects (3 August 2017, MNRE), The National Electric Mobility Mission Plan (NEMMP), The FAME India Scheme (Faster Adoption and Manufacture of (Hybrid and Electric Vehicles))are highly supportive in achieving the target of shifting to Electric Vehicle.

G. S. Sharma · G. A. Swarup · S. Upadhyay (✉)
Department of Electrical Engineering, Dayalbagh Educational Institute, Agra, India
e-mail: subhome2k10@gmail.com



O. H. Gupta et al. (eds.), *Soft Computing Applications in Modern Power and Energy Systems*, Lecture Notes in Electrical Engineering 1107,
https://doi.org/10.1007/978-981-99-8007-9_4

Electric energy, owing to features like clean transportable, and storable, is the most prominent form of energy in India. However, the issues mentioned below are of concern to the researchers.

1. Renewable Energy sources: The utilization of renewable energy in vehicle applications is one of the major points of discussion and research today. Work is being done to overcome the challenges of hybrid and electric vehicles [2, 3]. These are plagued with a lack of inertia which reduces their acceptance as a credible energy source at the utility-scale. Energy storage is being explored as a provider of inertia to overcome this problem. Energy storage also provides reliability to island renewable energy sources.
2. Electric vehicles: To reduce pollution levels, the Government of India is emphasizing the use of E-Vehicles for personal public and goods transportation. The range of power and other dynamics of the vehicles are directly dependent on the electrical energy available on board. Research on the batteries and different energy-storing systems in an electric vehicle [4, 5] have provided deeper insights into the internal working of a battery system. These techniques provide a deeper idea of the batteries used in EVs. There has been work done in the Overview of batteries and battery management for electric vehicles [6]. There has been work on the optimal charging and discharging techniques of grid-connected EVs [7, 8].Claude Ziad El- Bayeh et al. have researched the Charging and Discharging Strategies of battery-operated vehicles [9].

The study of electric vehicles their architecture, and the challenges faced in their use is an important aspect of choosing a battery for its use [10]. To understand the compatibility of converters and different components in a vehicle with different batteries a literature review is done on its architecture, with an understanding of converter topologies and control techniques is depicted in Vidhya and Allirani [11]. To study the recent developments in EVs, a futuristic approach is included in the article [12].

An electric vehicle's battery should have a high-speed charging rate and a considerably low discharging rate compared to general-purpose batteries. This project is a fundamental model to compare battery specifications such as State of Charge (SoC) and Current using MATLAB/Simulink. There has been plenty of advancement studies relating to different batteries. Lithium-ion and its state-of-the-art design in electric vehicles is a popular and advancing field of research [13]. Similar studies have been made for lead-acid [14], nickel–cadmium [15], and nickel metal hydride batteries [16]. The focus was made on the characteristics of the lithium-ion battery in a theoretical aspect by studying its use in an electric vehicle [17–19]. The study of a state of the art charge estimator, management system, and power capability of a lithium-ion battery was focused on Hannan et al. [20] and Xiong et al. [21]. Research has been done on an overview of the aging effect of lithium-ion battery [22]. A data-driven state of charge estimator for lithium-ion batteries in an electric vehicle [23] can be done to succeed in this work. A graphical study is done for the characteristics of 4 batteries: lithium-ion, lead-acid, nickel cadmium, and nickel metal hydride batteries. It is considered that how much time the battery takes to charge to full SOC from the lowest considerable state (20%) and for discharging, the time taken for a battery to discharge from a certain level of SOC (95%) is observed. So, the simulation of the batteries for their performance is done based on the following parameters.

State Of Charge (SOC): The state of charge of an electric battery in relation to its capacity is called the state of charge. SOC is the capacity that remains in a battery and is an important parameter that determines the control strategy. SOC shows the characteristics of the battery during charging and discharging and depends on its level or range. When the voltage is high, the battery is fully charged, and when it reaches its lowest point, it is empty.

Current: For discharging the battery is connected to a load of 5 ohms and the load current characteristics are observed. As for the charging, considering Constant Voltage Charging, voltage is constant hence the current characteristics are observed in order to check the better performance.

Power to Weight Ratio: It is a crucial parameter for automobiles. It provides a measure of the performance of the power source. The performance of power is calculated by the ratio of power output and the curb weight of the power source.

Modeling and Simulation of Battery: Modeling and simulation of the battery are required to know the mathematical relationship between battery input and output parameters and the simulator. Here a controlled voltage source is dependent on the actual state of charge (SOC). As it is concluded from the above the battery plays an important role in the performance of the electric vehicle so its optimum selection is a must and this can be achieved by modeling and simulation. So, the battery model is required to monitor battery parameters such as SOC, voltage, and current.

## 2 Methodology

The methodology followed for selection of a battery model is shown in Fig. 1.

### *2.1 For Charging*

- Grid charging is preferred as per the current situation. Here, an external DC source is applied to the battery in a parallel combination i.e. positive terminal of the source to the anode of the battery and the negative terminal of the source to the cathode of the battery as shown in Fig. 2.
- As the current leaves the voltage source to charge a battery, the current is denoted with a negative sign along with its magnitude.

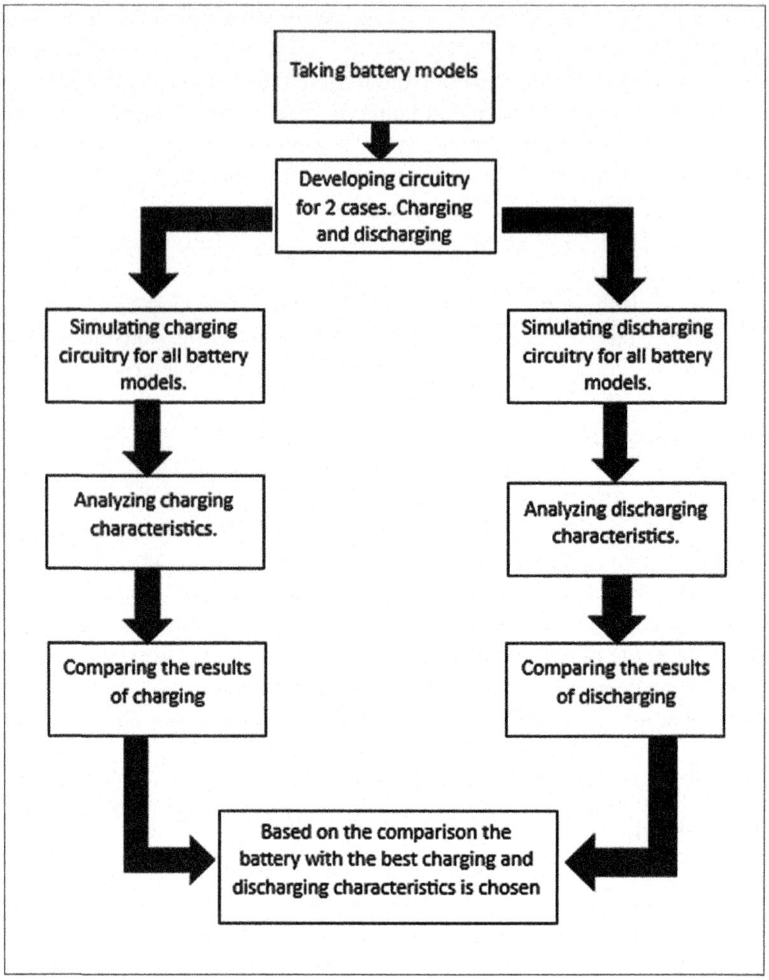


**Fig. 1** Flowchart of methodology

## *2.2 For Discharging*

- A series Resistive type load is connected to the battery for determining discharging characteristics of the battery.
- For controlled discharging a MOSFET is connected. In order to use the MOSFET as a switch the gate of MOSFET it is connected to the comparator which will decide the on and off the MOSFET at a specific value of SOC.
- Another end of the battery is connected to the Bus Selector through which the current, voltage and state of charge of the battery are observed. This SOC is

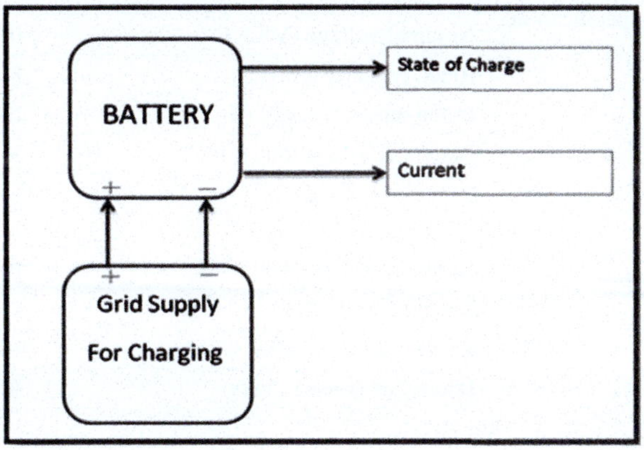


**Fig. 2** Block diagram for charging

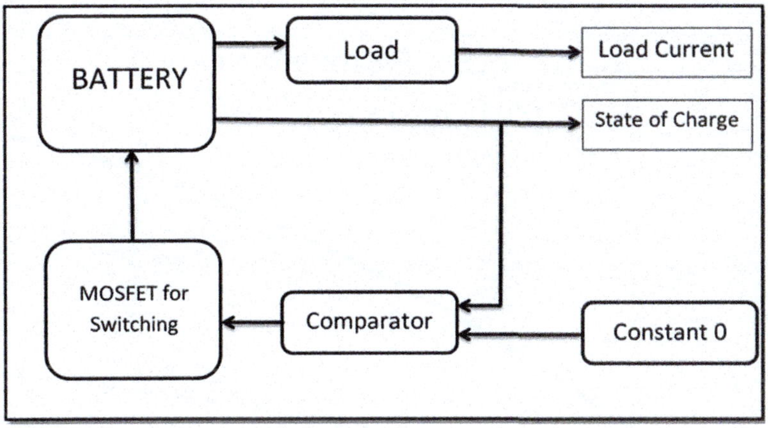


**Fig. 3** Block diagram for discharging

provided to the comparator for the controlled switching. The block diagram for discharging is shown in Fig. 3.

## 3 Battery Specifications

The batteries of 10 V and 7AH ratings as depicted in Tables 1 and 2 which were selected for simulating the charging and discharging effects. Here the simulation begins by charging a 20% initial SOC of batteries to 95%. The time required to charge to 95% is determined for each type of battery from Fig. 4, while Fig. 5 shows the current drawn for charging in different types of batteries. Similarly Fig. 6 shows

**Table 1** Battery specification for charging

| | |
|---|---|
| Nominal voltage (V) | 10 |
| Rated capacity (Ah) | 7 |
| Initial-state of charge (%) | 20 |
| Battery response time (s) | 30 |

**Table 2** Battery specification for discharging

| | |
|---|---|
| Nominal voltage (V) | 10 |
| Rated capacity (Ah) | 7 |
| Initial-state of charge (%) | 95 |
| Battery response time (s) | 30 |

the current discharges from each battery during discharging operation. Finally Fig. 7 shows the SOC discharging period for the battery.

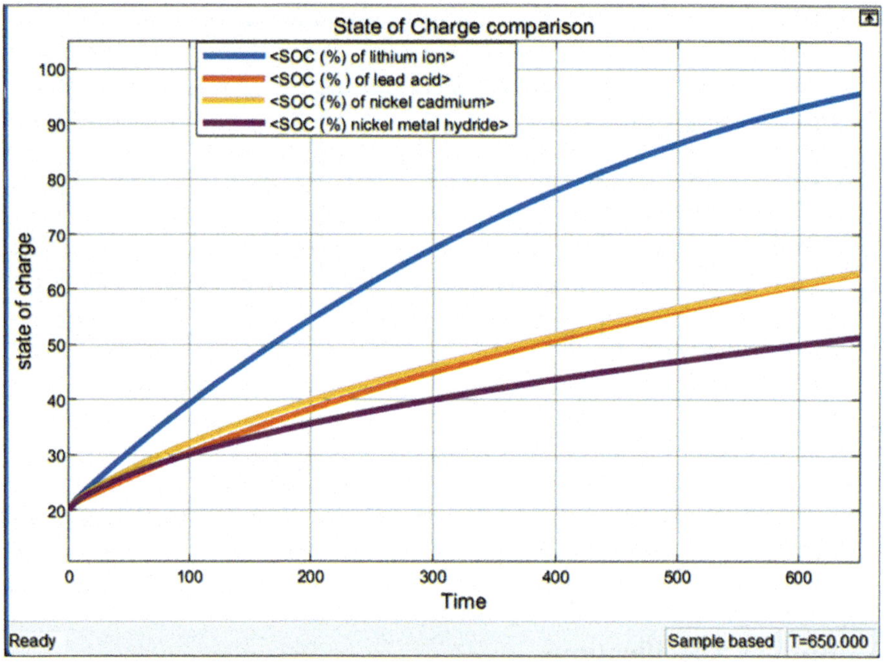


**Fig. 4** SoC charging characteristics of different batteries

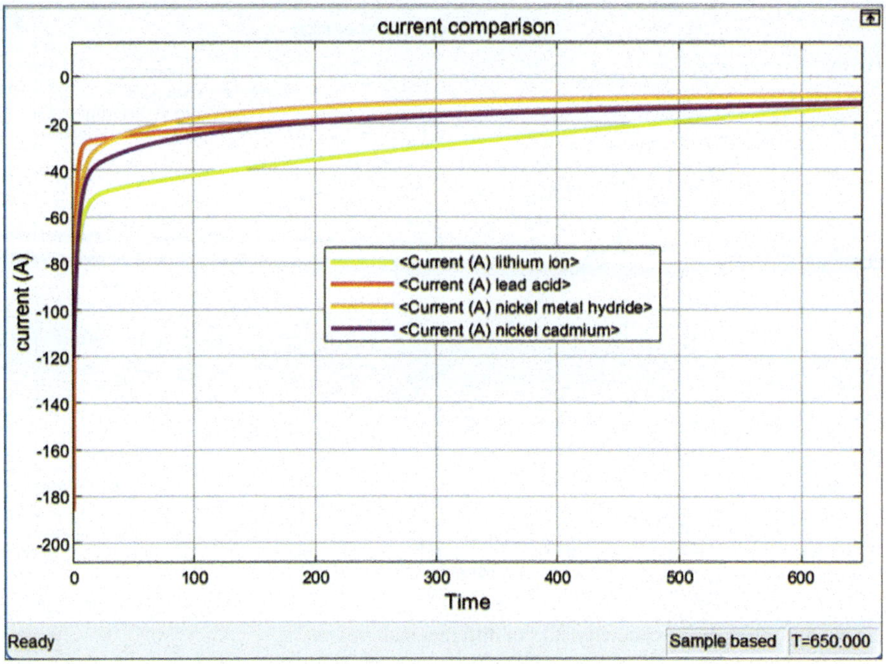


Fig. 5 Current charging characteristics of different batteries

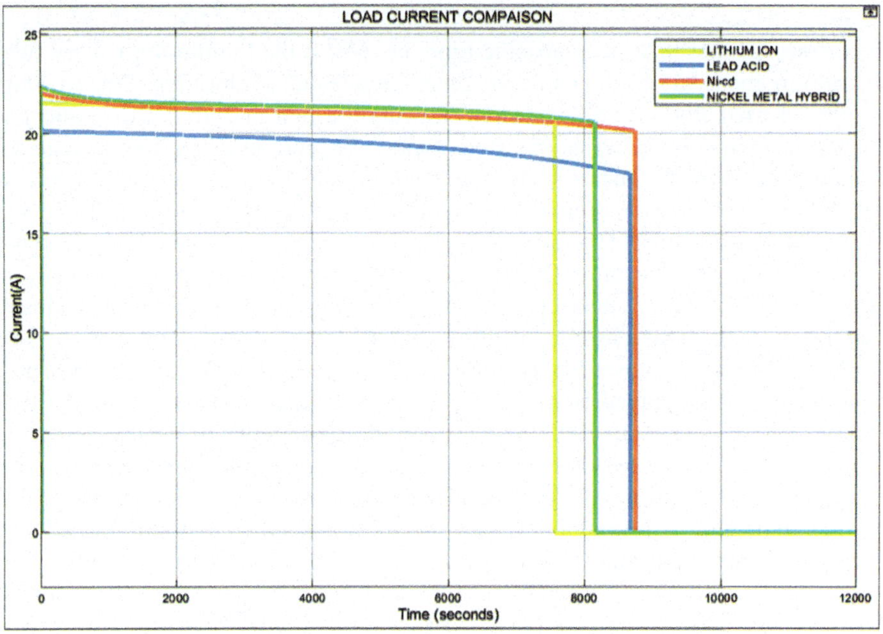


Fig. 6 Current discharging characteristics of different batteries

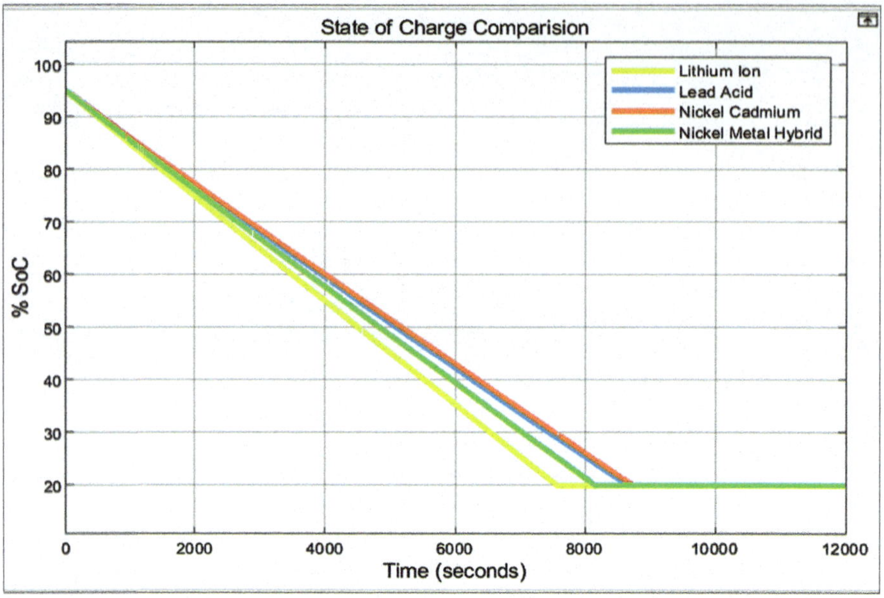


**Fig. 7** SoC discharging characteristics of different batteries

# 4 Results and Discussion

It is observed that in a given specific time, i.e. 650 s, the lithium-ion battery fully charges to about 95% SoC, while the other batteries take a lot more time to attain a fully charged state. Also, the current flowing through the lithium-ion battery is higher than the other batteries (negative because the current flows from an external source to the battery). The time taken by the lead-acid battery to attain 95% SoC is 1979s which is thrice as compared to the lithium-ion battery as shown in Table 3.

The lithium-ion type of battery would discharge sooner as compared to other types of battery. The SOC becomes 95–20% at 11556 s, which specifies why lithium-ion batteries are not used in houses or for other storage purposes. Whereas lead-acid batteries have the largest discharging time i.e., at 13689 s, hence, these they are used for higher storage purposes like in-house or charge storing from solar power plants.

Another point of comparison is the Power to Weight Ratio which is the highest for lithium-ion batteries (160–170 W/Kg) i.e. lithium-ion battery will generate a specific amount of power with least weight amongst all batteries. But from the perspective of current, the lithium-ion battery provides a higher and less varying current which is used in the field of Automobiles. This statement is justified by the simulated data presented in the above table.

**Table 3** Observed comparison between batteries

| Battery | Power to weight ratio [24] | Charging time (simulation) | Discharging time (simulation) | Load current variation $\Delta I(\%) = \left(\frac{I_{95}-I_{20}}{I_{95}} \times 100\right)$ |
|---|---|---|---|---|
| Lithium ion | 160–170 W/kg | 650 s/10.83 min | 11,556 s/ 150.667 min | 3.5 |
| Lead acid | 20–30 W/kg | 1979s/ 32.98 min | 13,689 s/ 171.45 min | 11 |
| Nickel cadmium | 45–55 W/kg | 1943s/ 32.38 min | 13,509 s/173.5 min | 8.62 |
| Nickel metal hybrid | 65–80 W/kg | 3900 s/65 min | 12,690 s/ 161.53 min | 7.95 |

## 5 Conclusion

Through the above simulation results, it can be concluded that both lithium-ion and lead-acid batteries were the best-performing batteries out of the four but are meant for different uses in our daily lives. Nickel–cadmium battery is also a good choice in terms of charging time as it takes lesser time to charge and less time to discharge charge as compared to lead-acid battery but the major disadvantage of this battery is the "Memory Effect" [25].

The lithium-ion battery takes less time to charge to full SoC (650 s) but also takes the least time to discharge i.e. 11556 s (simulation time) which is considerable.

The Power to weight ratio of the lithium-ion battery is the highest (160–170 W/kg) due to which it has the best performance amongst the other batteries.

A lead-acid battery takes thrice as long as a lithium-ion battery to attain a fully charged state but has a higher discharge time of 13689 s (simulation time) than the lithium-ion battery of 11556 s (simulation time). These factors make lithium-ion battery preferable for E-Vehicles and lead-acid batteries for use in household power supply.

## References

1. Miguel CB, Killian L, Pedro N, Filipe S (2014) Sustainable energy systems in an imaginary island. Renew Sustain Energy Rev 37:229242
2. Hannan MA, Azidin FA, Mohamed A (2014) Hybrid electric vehicles and their challenges: a review. Renew Sustain Energy Rev 29:135150
3. Azidin FA, Hannan MA, Mohamed A (2013) Renewable energy technologies and hybrid electric vehicle challenges. Przeglad Elektrotechniczny 89(8):150156

4. Nanditta RV, Nagulash B, Nidish D, Arvind KJ, Ajaykumar A (2021) Comprehensive review on various batteries and energy storage systems for electric vehicles. Turk Online J Qual Inq 12:6506–6520
5. Hannan MA, Hoque MM, Mohamed A, Ayob A (2017) Review of energy storage systems for electric vehicle applications: issues and challenges. Renew Sustain Energy Rev 69:771789
6. Liu W, Placke T, Chau KT (2022) Overview of batteries and battery management for electric vehicles. Energy Rep 8:4058–4084
7. Solanke TU, Ramachandaramurthy VK, Yong JY, Pasupuleti J, Kasinathan P, Rajagopalan A (2020) A review of strategic charging–discharging control of grid-connected electric vehicles. J Energy Storage 28:101193
8. Sortomme E, El-Sharkawi MA (2010) Optimal charging strategies for unidirectional vehicle-to-grid. IEEE Trans Smart Grid 2(1):131–138
9. El-Bayeh CZ, Alzaareer K, Aldaoudeyeh A-MI, Brahmi B, Zellagui M (2021) Charging and discharging strategies of electric vehicles: a survey. World Electr Veh J 12(1)
10. Kempton W, Letendre SE (1997) Electric vehicles as a new power source for electric utilities. Transp Res Part D Transp Environ 2(3):157–175
11. Vidhya H, Allirani S (2021) A literature review on electric vehicles: architecture, electrical machines for power train, converter topologies and control techniques, pp 565–575
12. Cheng KWE (2009) Recent development on electric vehicles. In: 3rd International conference on power electronics systems and applications, pp 1–5. PESA
13. State-of-the-art and energy management system of lithium-ion batteries in electric vehicle applications: issues and recommendations-scientific figure on research gate. https://www.researchgate.net/figure/Charging-and-discharging-phenomenon-of-Li-ion-battery_fig2_323922170
14. Meena N, Baharwani V, Sharma D, Sharma A, Choudhary B, Parmar P, Stephen R (2014) Charging and discharging characteristics of lead acid and Li-ion batteries, pp 1–3
15. Sperandio G, Nascimento Jr C, Adabo G (2011) Modeling and simulation of nickel-cadmium batteries during discharge, pp 1–9
16. Mebarki B, Draoui B, Rahmani L, Allaoua B (2013) Electric automobile Ni-MH battery investigation in diverse situations. Energy Procedia 36:130–14. ISSN 1876-6102
17. Lithium-ion battery (2022). Wikipedia. https://en.wikipedia.org/wiki/Lithium-ion_battery
18. Kennedy B, Patterson D, Camilleri S (2000) Use of lithium batteries in electric vehicles. J Power Sour 90:156162
19. Omar N et al (2010) Evaluation of performance characteristics of various lithium-ion batteries for use in BEV application. In: IEEE vehicle power and propulsion conference, p 16. Lille, France
20. Hannan MA, Lipu MSH, Hussain A, Mohamed A (2017) A review of lithium-ion battery state of charge estimation and management system in electric vehicle applications: challenges and recommendations. Renew Sustain Energy Rev 78:834854
21. Xiong R, He H, Sun F, Liu X, Liu Z (2012) Model-based state of charge and peak power capability joint estimation of lithium-ion battery in plug-in hybrid electric vehicles. J Power Sour 229:159169
22. Marongiu A, Nußbaum FG, Waag W, Garmendia M, Sauer DU (2016) Comprehensive study of the influence of aging on the hysteresis behavior of a lithium iron phosphate cathode-based lithium ion battery–an experimental investigation of the hysteresis. Appl Energy, Elsevier 171(C):629–645
23. Xiong R, Sun F, Gong X, Gao C (2014) A data-driven based adaptive state of charge estimator of lithium-ion polymer battery used in electric vehicles. Appl Energy 113:14211433
24. Wagner R, Preschitschek N, Passerini S, Leker J, Winter M (2013) Current research trends and prospects among the various materials and designs used in lithium-based batteries. J Appl Electrochem 43. https://doi.org/10.1007/s10800-013-0533-6
25. Jeyaseelan C, Jain A, Khurana P, Kumar D, Thatai S (2020) Ni-Cd batteries. In: Inamuddin RB, Pothu R, Asiri AM (eds) Rechargeable batteries

# Issues and Solutions for Optimum Overcurrent Relays Co-Ordination in Medium Voltage Radial Distribution System

**Shanker D. Godwal and Kartik S. Pandya**

**Abstract** In medium voltage distribution systems, generally, electrical power is supplied through the radial feeder and ring main distribution systems. Further to protect such feeder's overcurrent relay coordination is preferred. The overcurrent relay protection is very much efficient for these feeders. Still, there are some issues related to the overcurrent relay coordination scheme. These issues are the maloperation of relays under normal operating conditions, backup relays operation before primary response, speed of primary and backup relays, underreach and overreach problems, and coordination of overcurrent relays with fuses. In this paper, two main issues are discussed. These are backup relays responding without failure of primary relay and speed of primary and backup relays. The solution to such issues is presented. Further, one new constraint in the overcurrent relay coordination scheme has been considered along with one objective function reported in the literature. The reported objective function, after essential modification, is considered for problem formulation of over-current relay coordination. Considering the new objective function and constraint, the optimum relay setting of two systems (radial and ring main distribution systems) is achieved. Results show the effectiveness of the modified objective function and new constraint over the conventional method. For optimum overcurrent relay coordination, a genetic algorithm is used.

**Keywords** Overcurrent relays · Coordination · Plug setting time setting · Plug setting multiplier

S. D. Godwal (✉)
CSPIT CHARUSAT, Changa & IT-NU, Ahmedabad, India
e-mail: shankergodwal@gmail.com; shanker.godwal@nirmauni.ac.in

K. S. Pandya
Ahmedabad, India



O. H. Gupta et al. (eds.), *Soft Computing Applications in Modern Power and Energy Systems*, Lecture Notes in Electrical Engineering 1107,
https://doi.org/10.1007/978-981-99-8007-9_5

# 1 Introduction

These days electricity demand has been intensifying as compared to a decade ago. The power demand has inflated tremendously compared to decades ago and so more and more power suppliers came forward to install more power plants with maximum power handling capacity and installed capacity, and these have distended their capability to fulfill the energy demand. To supply reliable power to consumers efficiently with maintaining power quality, the role of interconnected power systems and power system protection is increased [1]. Huge and complex interconnected power systems required an efficient and robust power system protection scheme. Many power system protection schemes protect the entire power system in an efficient manner and out of these schemes inverse overcurrent relay protection scheme is widely used for the distribution system and backup protection of transmission lines [2]. Overcurrent relays coordination is a very crucial task to protect the entire distribution system properly and back up the transmission line. High speed and proper selection of overcurrent relays also add complexity to relay coordination i.e., parameters associated with inverse overcurrent relays must be optimum otherwise the relay may mal-operate and miscoordination may take place [3]. If coordination is not proper among overcurrent relays the protection scheme may fail. Recently many researchers have proposed a large number of philosophies regarding over-current relay coordination schemes to achieve optimum parameters like plug setting (PS), time multiplier setting (TMS), and operating time of relays [4]. The optimum overcurrent relay coordination scheme states that all primary relay operating time must be minimized. The objective function to achieve minimum or optimum parameters of overcurrent relays is associated with some constraints which are a constraint on plug setting, constraints on TMS constraint on operating time, and constraint on coordination criteria [5].

*The contribution of this paper is "To add one more constraint in overcurrent relay coordination formulation to achieve optimum and improved results". The plug setting constraints which are added in the relay coordination problem will improve the results and will lead to results toward accuracy. In the previously available literature, only coordination time interval and plug setting of individual relays have been added which sometimes leads to mal-operation under overload conditions. This paper has overcome the said problem of miscoordination between primary and secondary relays during overload conditions.*

**Problem Statement**

The requirement of the said constraint can be understood by following the radial system. Let the radial feeder contain three zones A, B, and C. Each zone is protected by three IDMT overcurrent relays. As per the protective scheme relay A must reach for fault beyond B and up to C point. Similarly, relay B must reach up to D as these two relays will provide the backup protection for the R2 and R3. Further, the relay R1, R2, and R3 are IDMT Relays and as per standard relays may operate between 1.05 to 1.3 value of its pickup. Now let the feeder rated current be 200 amp and fault

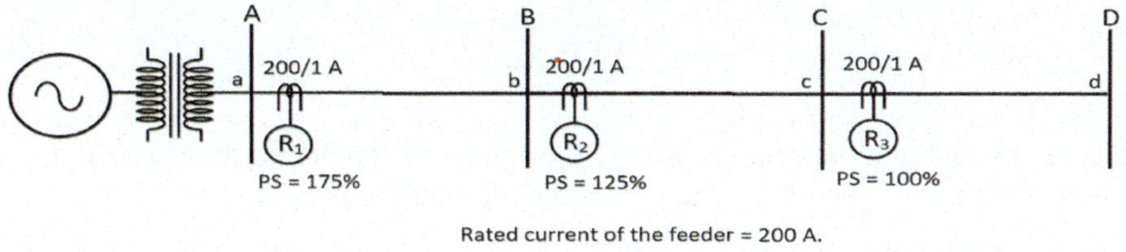


**Fig. 1** Simple radial network with three sections [8]

has been taken place at D bus bar. In that case, as per coordination if R3 will work as the primary and R2 will work as a backup relay. Now the current seen by the relay R3 and R2 is 200 amp and if pickup of R3 is 1.3 and pickup of R2 is 1.05 then these are kept at 100% plug setting considering that 20% overloading is allowed. In such case, under normal conditions, the feeder will carry 240 amp current. Further for 100% the plug setting relay R3 will pick up at 260 amp whereas R2 will operate at 210 amp. In this way, both relays will mal-operate. So for Proper operation of both relays, it is required that relay R3 pickup must be 1.05 times and relay R2 pickup must be 1.3 times otherwise both relays will mal-operate. Hence it is required to add a new constraint in overcurrent relays coordination considering the pickup range of IDMT relays (Fig. 1).

Further, the speed of overcurrent relays is supposed to be high which means the action of the relay must be very fast. In this regard, it is required to achieve plug setting and time setting in such a way that the operating time of the relay will become very fast. Apart from this relays constrains should not get violated. It means the primary relay should get the first chance to operate and if it fails to operate then backup relays must operate. Failing to do so, will lead to miscoordination. To avoid such a problem the problem formulation part of overcurrent relays coordination must be robust. In this paper, one reported objective function is considered and after essential modification, the modified objective function is proposed.

## 2 Problem Formulation

The following mathematical structure can be used to optimize the plug setting and time setting of the overcurrent relay coordination problem for radial distribution systems. As per the overcurrent relay coordination philosophy, the sum of all relays' operating time must be minimum when these relays are working as primary relays at their individual allocated zones [1]. The goal of overcurrent relay coordination is to reduce the operation time of all primary relays. In this continuation, the Objective function of the overcurrent relay can be written as per Eq. (1) [6–14]

$$(\mathrm{OF}) = \sum_{i=1}^{m} t_{i,k} \tag{1}$$

where m is the total no of overcurrent relays in the radial distribution system, $t_{i,k}$ is the time taken by individual relays to operate for fault at k.

**Reported Objective Function (ROF)**

The reported objective function (ROF) is considered from Rajput et al. [1]. Where m is the total no. of primary relays, ti is the operating time of the ith primary relay, $\Delta t_{pbk}$ is the discrimination time between kth P/B relay pair. The total number of P/B relay pairs is defined by n, and k represents each P/B relay pair. The operating time of the primary and backup relay is denoted as $t_{pk}$ and $t_{bk}$ respectively for the kth pair. $\alpha_1$ and $\alpha_2$ are the constants and considered as unity for simplicity. The effectiveness of ROF is presented in Rajput et al. [1]. The major drawback of ROF is, its first term consists of t2, and as per the overcurrent relay coordination state the operating time of all primary relays should be minimum. Here it can be seen that the operating time is square. So for the proper comparison, it is required to replace the first term of the square of the sum of operating time with the sum of operating time of all primary relays.

$$\mathrm{ROF} = \alpha_1 \sum_{i=1}^{m} t_i^2 + \alpha_2 \sum_{k=1}^{n} \left( |\Delta t_{pbk} - |\Delta t_{pbk}|| \cdot \frac{t_{pk}^2}{t_{bk}^2} + (\Delta t_{pbk} + |\Delta t_{pbk}|) \cdot t_{bk}^2 \right) \tag{2}$$

**Modified Objective Function (MOF)**

Modification in the reported objective function. The modification required in ROF is implemented in MOF

$$\mathrm{MOF} = \alpha_1 \sum_{i=1}^{m} t_i + \alpha_2 \sum_{k=1}^{n} \left( |\Delta t_{pbk} - |\Delta t_{pbk}|| \cdot \frac{t_{pk}^2}{t_{bk}^2} + (\Delta t_{pbk} + |\Delta t_{pbk}|) \cdot t_{bk}^2 \right) \tag{3}$$

In MOF it can be seen that sum of all primary relays in POF is minimum as compared to OF. To reduce or minimize the sum of all relays operating time five basic constraints are considered in the past literature which are as follows.

## *2.1 Constraint Set I—Coordination Criteria*

As per coordination criteria, there should be a specific difference between backup relay operating time and primary relay operating time which is known by the coordination time interval CTI [7, 8]

$$t_{i,k} - t_{j,k} \geq CTI \tag{4}$$

where $t_{i,k}$, $t_{j,k}$ are operating times of the primary relay and backup relays respectively.

### 2.2 Constraint Set II—Bounds on Relay Operating Time

Each relay will take a certain time to operate at the same time the relay should not take a long time to operate. This constraint specifies the maximum and minimum time taken by the relay [9–11]

$$t_{i,min} \leq t_i \leq t_{i,max} \tag{5}$$

where $t_{i,max}$, $t_{i,min}$ are the maximum and minimum time taken by relays for a fault on i point which is the maximum operating time of the relay at i location.

### 2.3 Constraint Set III—Bounds on the TMS of Relays

For the coordination of overcurrent relays, it is required to create some intentional time delay so this constraint can be represented as per Eq. (4) [12–14]

$$TMS_{i,min.} \leq TMS \leq TMS_{i,max.} \tag{6}$$

where $TMS_{i,min.}$, $TMS_{i,max.}$ Maximum and minimum time delay for relay $R_i$.

### 2.4 Constraint Set IV—Bounds on the Plug Setting (PS) of Relays

The Plug setting of the relay may vary from maximum to minimum so the bound can be defined as per Eq. (5) [4, 13].

The bounds on relays can be stated as

$$PS_{i,min} \leq PS \leq PS_{i,max} \tag{7}$$

$PS_{i,min}$, $PS_{i,max.}$ Minimum maximum plug setting of relay $R_i$; respectively.

### 2.5 *Constraint Set V—Relay Characteristic*

As per this criteria, it is required to specify the operating characteristic of the over current relay for proper coordination. In many research articles, Eq. (6) is considered for relay coordination.

$$\text{Relay operating time} = \frac{\beta}{(\text{PSM})^{\alpha} - 1} \times \text{TMS} \tag{8}$$

where β and α constant decide the stiffness of overcurrent relay characteristics.

**Constraint Set VI: Constraint Considering Plug Setting of Primary and Backup Relay**

As in per this constraint, the relay may be operated in between 1.05 and 1.3 times of the plug setting value. As discussed earlier, this constraint will not allow the backup relay to operate before the primary relay operation

$$\text{PS of RB} > (1.3/1.05) \times \text{PS of RP} \tag{9}$$

where $R_P$ is the primary relay and $R_B$ is the backup relay associated with $R_P$.

## 3 Distribution Network

Figure 2 shows the distribution system having two generators with equal ratings. These generators are feeding power to a 66 kV bus and two more feeders. Relays R1–R7 are kept for the protection of the entire feeder. Further, $R_1$ & R2 relays are backup for the R3 relay similarly R1 and R2 relays are also backup for the R4 relay. As in said feeder, power is being fed by two parallel lines hence backup of the R5 relay will be R4 and the backup of relay R6 will be R3. As R5 and R6 are directional overcurrent relays so back of relay 7 will be R3 and R4. Further, for the sake of simplicity fault locations are considered mid of the line in case of fault in sections B and C with fault level 2600 MVA. Also, the fault level at bus C is considered 1000 MVA. Further fault current data are given in Table 1.

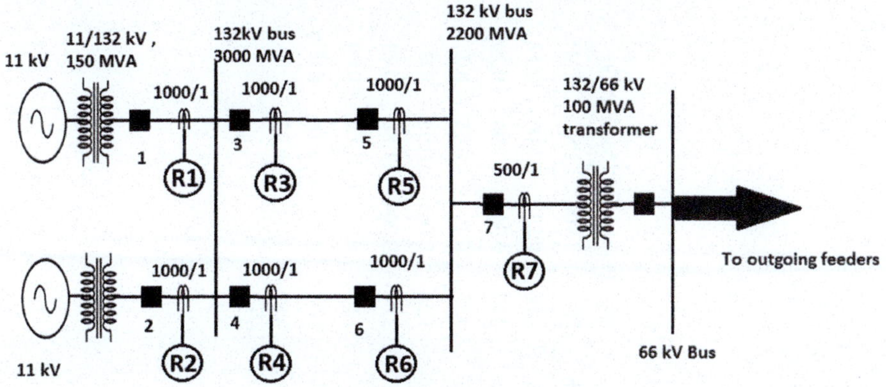


**Fig. 2** Medium voltage level distribution system [15]

**Table 1** Relay fault currents for distribution network

| S. No | Primary relay | Primary relay CT ratio | Back up relay | Backup relay CT ratio | Primary relay fault current (amp) | Backup relay fault current |
|---|---|---|---|---|---|---|
| 1 | R1 | 1000:1 | – | – | 6560 | – |
| 2 | R2 | 1000:1 | – | – | 6560 | – |
| 3 | R3 | 1000:1 | R1 | 1000:1 | 8529 | 2843 amp |
| 4 | R3 | 1000:1 | R1 | 1000:1 | 8529 | 8529 amp |
| 5 | R4 | 1000:1 | R2 | 1000:1 | 8529 | 8529 amp |
| 6 | R4 | 1000:1 | R1 | 1000:1 | 8529 | 2843 amp |
| 7 | R5 | 1000:1 | R4 | 1000:1 | 2843 | 2843 amp |
| 8 | R6 | 1000:1 | R3 | 1000:1 | 2843 | 2843 amp |
| 9 | R7 | 500:1 | R3 | 1000:1 | 4373 | 2186 amp |
| 10 | R7 | 500:1 | R4 | 1000:1 | 4373 | 2186 amp |

## 4 Results and Discussions

To prove the need for MOF and constraint on plug setting, the three soft computing techniques are used. To achieve optimum results of the operating time of each relay, three soft computing techniques are considered. These techniques are genetic algorithm (GA), Harmonic Search Algorithm (HSA), and Particle swarm optimization. It is observed that HSA performs more optimum as compared to the rest of the two algorithms. Further, Table 2 depicts the plug setting and time setting achieved from various cases i.e. for the conventional objective function, reported objective function, and modified objective function. Figure 3 depicts the sum of the operating time of primary and backup relays for the distribution system shown in Fig. 2.

**Table 2** Plug setting and time multiplier setting obtained for different constraints

| Primary relay | OF | | ROF GA | | MOF GA | | ROF HAS | | MOF HAS | | ROF PSO | | MOF PSO | |
|---|---|---|---|---|---|---|---|---|---|---|---|---|---|---|
| | PS [15] | TMS [15] | PS | TMS | PS | TMS | PS | TMS | PS | TMS | PS | TMS | PS | TMS |
| R1 | 2 | 0.15 | 2.03 | 0.09 | 1.45 | 0.14 | 2.4 | 0.056 | 1.79 | 0.31 | 2.06 | 0.23 | 1.06 | 0.12 |
| R2 | 2 | 0.15 | 2.34 | 0.19 | 2.09 | 0.281 | 1.32 | 0.25 | 2.16 | 0.29 | 2.12 | 0.34 | 1.96 | 0.29 |
| R3 | 1.5 | 0.15 | 1.563 | 0.09 | 0.79 | 0.1 | 0.63 | 0.047 | 0.89 | 0.06 | 0.85 | 0.1 | 0.93 | 0.2 |
| R4 | 1.5 | 0.15 | 1.687 | 0.12 | 0.72 | 0.17 | 2.07 | 0.08 | 1.36 | 0.09 | 1.54 | 0.061 | 0.92 | 0.08 |
| R5 | 1 | 0.1 | 1.04 | 0.082 | 1.38 | 0.06 | 1.48 | 0.055 | 1.67 | 0.1 | 2.04 | 0.064 | 1.86 | 0.13 |
| R6 | 1 | 0.1 | 1.56 | 0.067 | 1.42 | 0.05 | 1.41 | 0.089 | 1.92 | 0.1 | 1.08 | 0.032 | 1.25 | 0.09 |
| R7 | 1 | 0.25 | 1.34 | 0.074 | 1.98 | 0.081 | 1.34 | 0.054 | 1.2 | 0.12 | 1.67 | 0.05 | 1.78 | 0.096 |
| OTp | 4.51 s | | 2.96 s | | 4.19 s | | 2.55 s | | 3.76 s | | 2.89 s | | 3.93 s | |
| OTb | 2.38 s | | 1.73 s | | 3.85 s | | 1.78 s | | 2.90 s | | 2.76 s | | 3.46 s | |

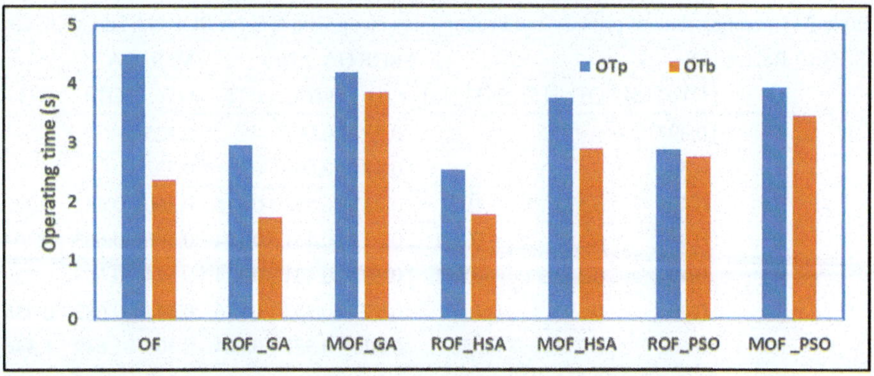


**Fig. 3** Comparative analysis of various techniques OF with ROF and MOF

It can be seen from Fig. 2 that the operating time in the case of the conventional objective function is quite higher as compared to the rest of the two objective functions. Table 3 presents the operating time of primary and backup relays. It can be noted that there are major miscoordination has been observed in the case of the conventional objective function. Whereas in the case of reported objective function and modified objective functions no miscoordination has been observed. Table 4 depicts the operating time of primary and backup relays for ROF and MOF. It can be noted that even though the reported objective function gives optimum relay setting as compared to rest techniques but there is a major issue associated with ROF. So it is required to adopt MOF over ROF. The result achieved from MOF however quite higher as can be seen from Table 2 but as compared to OF the results are more optimized. Figures 4 and 5 present a comparative analysis of the primary relay's operating time for ROF with OF and MOF with OF respectively. Similarly Figs. 6 and 7 present a comparative analysis of the backup relay's operating time for ROF with OF and MOF with OF respectively.

**Table 3** Operating time of primary and backup relay along with CTI for ROF GA and MOF GA

| S. No. | Relays | OF | | | ROF GA | | | MOF GA | | |
|---|---|---|---|---|---|---|---|---|---|---|
| | | OTp [15] | OTb [15] | CTI [15] | OTp | OTb | CTI | OTp | OTb | CTI |
| | R1 | 0.397 | – | – | 0.237 | 0.000 | – | 0.425 | | |
| | R2 | 0.397 | – | – | 0.474 | 0.000 | – | 0.732 | | |
| | R3 | 0.401 | 0.593 | 0.192 | 0.237 | 0.353 | 0.116 | 0.360 | 0.682 | 0.322 |
| | R3 | 0.401 | 0.359 | −0.041 | 0.237 | 0.431 | 0.194 | 0.360 | 0.663 | 0.303 |
| | R4 | 0.401 | 0.359 | −0.041 | 0.307 | 0.431 | 0.124 | 0.644 | 0.663 | 0.020 |
| | R4 | 0.401 | 0.593 | 0.192 | 0.307 | 0.353 | 0.046 | 0.644 | 0.682 | 0.038 |
| | R5 | 0.662 | 0.713 | 0.050 | 0.524 | 0.527 | 0.004 | 0.303 | 1.649 | 1.346 |
| | R6 | 0.662 | 0.713 | 0.050 | 0.310 | 0.416 | 0.106 | 0.247 | 0.858 | 0.611 |
| | R7 | 0.789 | 0.873 | 0.084 | 0.205 | 0.506 | 0.301 | 0.193 | 1.274 | 1.081 |
| | R7 | 0.789 | 0.873 | 0.084 | 0.205 | 0.635 | 0.430 | 0.193 | 2.612 | 2.419 |

**Table 4** Operating time of primary and backup relay along with CTI for ROF HAS, MOF HAS, ROF PSO, and MOF PSO

| Relays | ROF HAS | | | MOF HAS | | | ROF PSO | | | MOF PSO | | |
|---|---|---|---|---|---|---|---|---|---|---|---|---|
| | OTp | $OT_b$ | CTI | OTp | OTb | CTI | OTp | OTb | CTI | OTp | OTb | CTI |
| R1 | 0.138 | | | 0.859 | | | 0.602 | | | 0.425 | | |
| R2 | 0.793 | | | 0.746 | | | 0.881 | | | 0.775 | | |
| R3 | 0.192 | 0.200 | 0.008 | 0.203 | 1.312 | 1.109 | 0.346 | 0.895 | 0.548 | 0.662 | 0.753 | 0.091 |
| R3 | 0.192 | 0.705 | 0.513 | 0.203 | 0.677 | 0.474 | 0.346 | 0.799 | 0.452 | 0.662 | 0.701 | 0.039 |
| R4 | 0.190 | 0.705 | 0.516 | 0.251 | 0.677 | 0.426 | 0.162 | 0.799 | 0.637 | 0.266 | 0.701 | 0.434 |
| R4 | 0.190 | 0.200 | 0.011 | 0.251 | 1.312 | 1.061 | 0.162 | 0.895 | 0.733 | 0.266 | 0.753 | 0.487 |
| R5 | 0.264 | 0.310 | 0.046 | 0.442 | 0.460 | 0.017 | 0.250 | 0.285 | 0.035 | 0.537 | 0.577 | 0.040 |
| R6 | 0.443 | 0.561 | 0.119 | 0.405 | 0.448 | 0.043 | 0.197 | 0.786 | 0.589 | 0.491 | 1.426 | 0.935 |
| R7 | 0.150 | 1.025 | 0.875 | 0.349 | 0.627 | 0.278 | 0.127 | 1.123 | 0.996 | 0.238 | 1.959 | 1.721 |
| R7 | 0.150 | 0.365 | 0.216 | 0.349 | 0.572 | 0.223 | 0.127 | 0.348 | 0.220 | 0.238 | 0.796 | 0.558 |

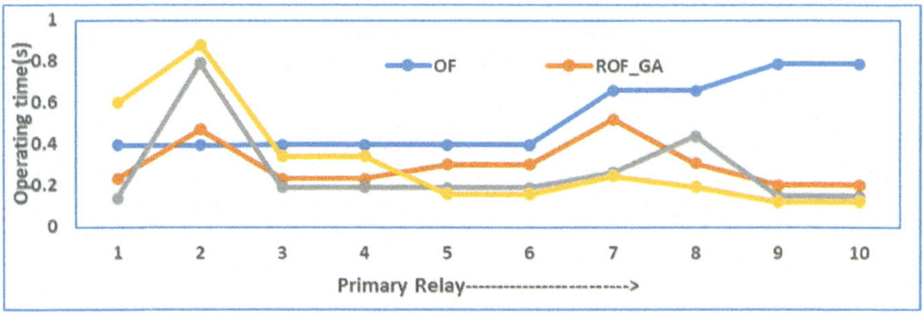


**Fig. 4** Comparative analysis of primary relay's operating time for ROF with OF

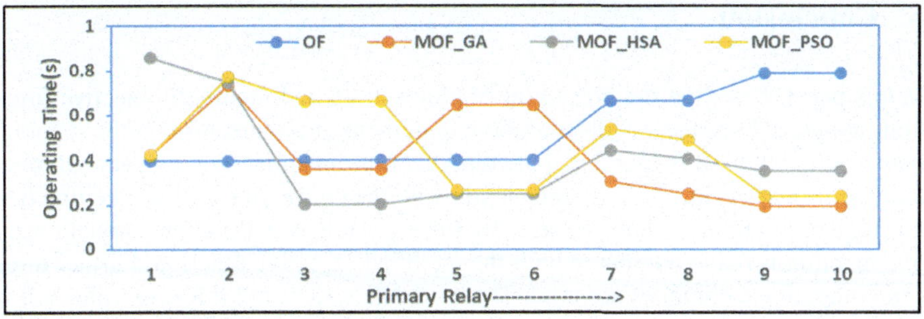


**Fig. 5** Comparative analysis of primary relay's operating time for MOF with OF

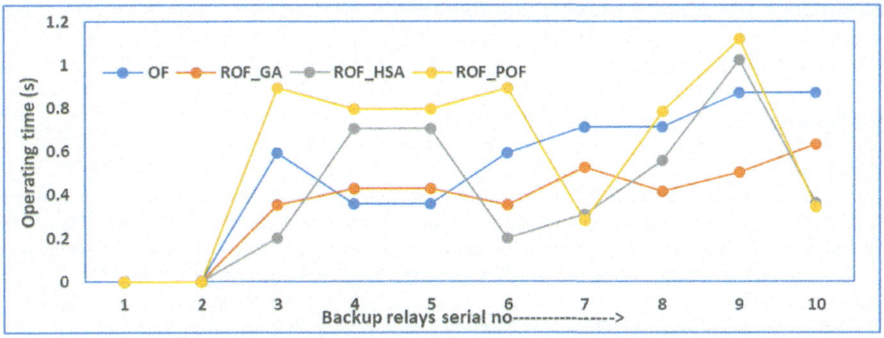


**Fig. 6** Comparative analysis of backup relay's operating time for ROF with OF

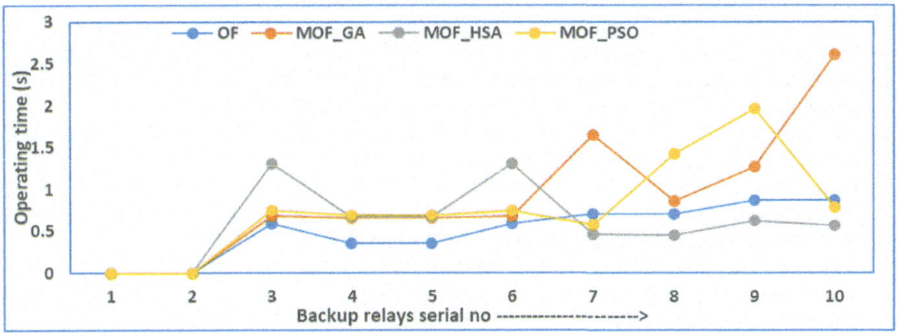


**Fig. 7** Comparative analysis of backup relay's operating time for MOF with OF

## 5 Conclusion

In this paper, based on the constraint of plug setting and modified objective functions are used. Further, it is disused the requirement of modification in the reported objective function. It is found that the reported objective function delivers optimum relay setting but to have a fair comparison with the conventional objective function it is required to have a common base. To fulfill this criterion the conventional objective function is being compared with the modified objective function. Further three soft computing techniques are used. These are HSA, GA, and PSO algorithms. It is found that harmonic search algorithms work more effectively. Further to examine the effectiveness of constraint on plug setting and modify objective function redial distribution system has been taken and obtained results are compared with the conventional formulation of overcurrent relay coordination.

## References

1. Rajput VN, Adelnia F, Pandya KS (2018) Optimal coordination of directional overcurrent relays using improved mathematical formulation. IET Gener Transm Distrib 12(9):2086–2094
2. Rajput VN, Pandya KS (2016) A novel approach for optimum coordination of directional overcurrent relays including far-end faults in inter connected power systems. Int J Eng Technol (IJET) 8(6)
3. Birla D, Maheshwari RP, Gupta H (2006) A new nonlinear directional overcurrent relay coordination technique, and banes and boons of near-end faults based approach. IEEE Trans Power Deliv 21(3):1176–1182
4. Amraee T (2012) Coordination of directional overcurrent relays using seeker algorithm. IEEE Trans Power Deliv 27(3)
5. Birla D, Maheshwari RP, Gupta H (2007) An approach to tackle the threat of sympathy trips in directional overcurrent relay coordination. IEEE Trans Power Deliv 22(2):851–858
6. Bedekar PP, Bhide SR (2011) Optimum coordination of directional overcurrent relays using the hybrid GA-NLP approach. IEEE Trans Power Deliv 26(1):109–119
7. Rajput VN, Pandya KS (2017) Coordination of directional overcurrent relays in the interconnected power systems using effective tuning of harmony search algorithm. Sustain Comput Inform Syst 15:1–15
8. Oza BA, Nair NC, Mehta RP, Makwana VH (2013) Power system protection and switchgear. McGraw Hill education(India) Private Limited, New Delhi
9. Kiliçkiran HC, Şengör İ, Akdemir H, Kekezoğlu B, Erdinç O, Paterakis NG (2018) Power system protection with digital overcurrent relays: a review of nonstandard characteristics. Electr Power Syst Res 164:89–102
10. Srivastava A, Tripathi JM, Mohanty SR et al (2016) Optimal over-current relay coordination with distributed generation using hybrid particle swarm optimization–gravitational search algorithm. Electr Power Compon Syst 44(5):506–517
11. Coordination using MRFO algorithm: a case study of adaptive protection of the distribution network of the Hatay province of Turkey (2021). Electr Power Syst Res 192;106998
12. Godwal SD, Pandya KS, Rajput VN, Vora SC (2020) A review on approaches employed for solving directional overcurrent relays' coordination problem, pp 35–51. https://doi.org/10.1007/978-981-15-0206-4_4

13. Draz A, Elkholy MM, El-Fergany AA (2021) Slime mould algorithm constrained by the relay operating time for optimal coordination of directional overcurrent relays using multiple standardized tripping curves. Neural Comput Applic 33:11875–11887. https://doi.org/10.1007/s00521-021-05879-x
14. El-Fergany AA, Hassanein HM (2019) Water cycle algorithm for optimal overcurrent relays coordination in electric power systems. Soft Comput 23:12761–12778. https://doi.org/10.1007/s00500-019-03826-6

# Powering Electric Vehicles with Solar Panels with Both the G2V and V2G Charging Modes

**Pranjali Kumari and Sanjay Kumar**

**Abstract** Solar PV based charging of Electric vehicle (EV) charging circuit is presented in this research. To obtain the optimal performance of the solar photovoltaic system under Typical Test Circumstances, incremental conductance MPPT algorithm is employed. With a boost converter with a 400 V output, a 100 AH battery may be charged by a solar PV array. Then, the voltage is lowered to meet the needs of buck operation and a 220 V battery. Short-term storage capacity is shown to completely charge rapidly. The objective is accomplished by the appropriate application of the calculated parameters of the passive components (filter elements on the converters and inverters of the system). When power from solar photovoltaic panels is not available, electric cars are powered from the AC source. We use a PI (proportional plus integral) regulator with a 10 rad/s corner frequency. An H-bridge rectifier provides a DC–DC bidirectional converter with a 400 V dc output voltage. One thing to note is how fast the SOC of the battery is reached. The battery current and voltage are shown during charging and draining modes. During charging, it is evident that the current and voltage of grid are synchronized. It is believed that they are out of phase during discharge, signifying a opposite power movement. Power semiconducting components are controlled with switching frequency of 10 kHz PWM pulse. Electric vehicle on-board chargers are available for use in parking lots and residences. Less grid-based power needs and emission-free, quiet driving are hallmarks of solar PV connected electric vehicles (EVs).

**Keywords** PV system · Grid connected mode · V2G · Electric vehicle · G2V · PI control · Incremental conductance MPPT

P. Kumari (✉) · S. Kumar
Department of Electrical Engineering, National Institute of Technology, Jamshedpur, Jharkhand, India
e-mail: pranjali2408@gmail.com

S. Kumar
e-mail: skumar.ee@nitjsr.ac.in

 
O. H. Gupta et al. (eds.), *Soft Computing Applications in Modern Power and Energy Systems*, Lecture Notes in Electrical Engineering 1107,
https://doi.org/10.1007/978-981-99-8007-9_6

## 1 Introduction

Use of solar energy to charge electric vehicle (EV) batteries is feasible in urban areas. Thus, the grid's burden is reduced when EVs are linked directly to solar charging stations [1]. So, in this study, we propose using solar energy to charge electric vehicles. The transportation industry is a major contributor to emissions of greenhouse gases [2]. This problem is made worse by the EVs' unpredictable charging habits [3, 4] and the rising popularity of quick charging [5, 6], both of which draw significant amounts of electricity from the grid for very little periods of time. Figure 1 illustrates the charging of electric vehicles using solar photovoltaics (PV), the grid, and PV.

Battery SOC (state of charge) must be maintained within safe limits for a long lifetime in electric cars. Nickel–cadmium batteries, lead-acid batteries, and lithium-ion batteries are the most common types found in electric vehicles. Lithium-ion batteries are favoured for EVs due to their high specific energy, powerful ability to discharge, and extended lifetime [6]. A typical PV-fed dc bus charging station consists of PV arrays, EV chargers, ESUs, and multiple DC–DC power converters that connect all of these components to a common dc bus. A DC–AC converter is necessary when the bus is operating in grid-connected mode, otherwise it is operating in the off-grid or islanded mode [7].

Several elements of EV charging utilising the grid-only utility supply have been thoroughly discussed in literature for many years [8]. Moreover, a number of review articles in this field have been published [9]. The bidirectional dc charger regulates the voltage and current to fit the recharging EV or ESU. Power may go in either way. Uncertainty the energy from the bus of dc side has to be supplied into the AC side, and this device acts as a converter that converts DC to AC energy and it is known as the inversion mode of operation. In contrast, if the grid power is required to charge the dc bus, a converter with property of AC–DC modulation must be installed and it is known as the rectification mode. Since it might alter battery life and maintenance needs, an efficient charger is crucial [10].

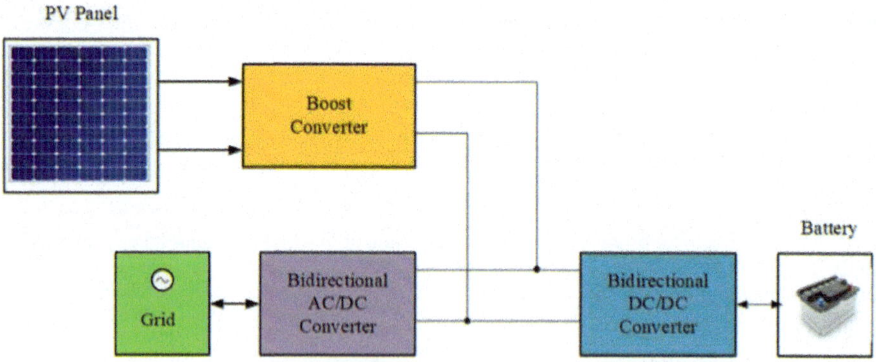


**Fig. 1** EV with solar charging configuration

Lithium batteries are now the standard due to their superior performance in almost every category [11, 12]. In addition, it has a low self-discharge rate, a long life cycle, no memory effect when charging, a wide operating temperature range, and fast charging. A broad variety of chemical compounds are used in lithium-based batteries; for example, Lithium ferro phosphate ($LiFePO_4$) is safer and less likely to explode if the battery is unintentionally short-circuited or overcharged, making it more convenient to use. Lithium-titanate, the most up-to-date version, offers a broader working temperature range, recharges faster, and takes bigger recharge rates (>10 C) [13]. A better version of the lithium-based battery now in use in EVs is in the works.

The key benefit in the use of solar energy to power electric vehicles (EVs) has the potential to drastically reduce carbon emissions and boost energy independence. EVs may be charged with solar energy throughout the day utilising the grid-to-vehicle (G2V) and vehicle-to-grid (V2G) charging modes, and they can provide extra energy back to the grid at times of peak demand. Solar panels may be placed at home or at public charging stations to power EVs in the G2V mode. This enables EV users to lessen their dependency on fossil fuels by charging their cars with clean, renewable energy. Additionally, utilising solar power throughout the day to charge EVs might ease the load on the grid during times of high demand. EVs may provide extra energy to the grid in the V2G mode during times of high demand. This may lessen the demand for peaker plants that use fossil fuels and increase grid stability. Additionally, EV owners may make money by reselling extra energy to the grid, which reduces the cost of owning an EV. EVs may run on solar energy and provide beneficial grid services by combining the G2V and V2G charging modes. But there are certain issues that need to be resolved, such as creating standardised communication protocols for EVs and the grid and making sure that V2G charging does not shorten EV batteries' lives. Overall, utilising solar panels to power electric vehicles while employing the G2V and V2G charging modes has the potential to provide a viable and affordable option for lowering carbon emissions and enhancing grid stability.

## 2 History of Electric Vehicle Technologies

### *2.1 EV's Developing History*

A vehicle powered by one or more electric motors is often referred to as an electric vehicle (EV). It can include a motorbike or scooter, a truck or a bus. Nevertheless, Hybrid electric vehicles (HEVs), plug-in hybrid electric vehicles (PHEVs), and pure battery electric cars (BEVs) are the only types of EVs considered in this research. Further benefits of the EV include less vibration, stench, noise, or difficulty changing gears as compared to a gasoline-powered car. With respect to developments in power electronics, battery, control, converters, and microelectronics, the EV is expected to have a substantial influence on the automobile industry.

## 2.2 Battery

It's not only electric cars that rely on electricity; it's crucial to modern living. The battery is essential to the operation of an electric vehicle because it stores electricity for later use as motive force (EV). Rechargeable batteries have surpassed all other forms of energy storage for usage in electric vehicle propulsion systems due to its convenience, safety, and high energy density. Ultra-capacitors, flywheels, and hydrogen fuel cells are a few examples of alternative storage components that have been developed, but none of them have been able to compete with the low cost of batteries. So far, a battery has numerous significant disadvantages, including: (1) a small amount of energy; (2) a lengthy charging process; and (3) restricted charging cycles. These elements lead to a reduced driving range and pricey replacements. A battery's performance is determined by its particular energy, efficiency, maintenance requirements, cost, and environmental effect.

## 2.3 PV System

In order to generate electricity, PV modules absorb sunlight. These days, most solar panels use modules made with either polycrystalline or monocrystalline silicon. Yet lately, thin films have gained popularity because they are simpler to make, more affordable, and function better at greater temperatures. While other, more advanced technologies, including multijunction modules and heterojunction with intrinsic thin layers, are more effective, they are seldom employed for broad applications because of their significantly huge prices. The open circuit voltage and power of a typical commercial module range from around 20 to 30 V dc and 200 to 300 W, respectively. In order to generate the required voltage for operation, the modules are linked in series strings; This ranges from around 200 V dc to about 500 V dc, and it's used in electric vehicles. When the quantity of parallel threads rises, the power of the array rises as well, thanks to the series–parallel link. One may learn about the performance of a PV scheme under different temperature and irradiance ($G$) circumstances by analysing its current–voltage ($I$–$V$) and power–voltage ($P$–$V$) characteristics ($T$). Changing values of $G$ and $T$ cause a nonlinear $I$–$V$ curve. Power is always greatest at a constant operating point called the maximum power point (MPP). When $G$ or $T$ changes, the MPP swings continually. Thus, a maximum power point tracker (MPPT) is necessary to guarantee that the modules are always drawing the maximum feasible power, independent of the values of $T$ and $G$. Several MPPT strategies are utilized, including as the incremental conductance, perturb and observe, and hill climbing approaches. particle swarm optimization (PSO), Artificial neural networks (ANN), fuzzy logic, and other advanced soft computing approaches have been developed in recent years.

### 2.3.1 P&O MPPT

Due to its low cost and ease of installation, the perturbation and observation approach has been widely employed in solar PV systems. Using the voltage and current information the PV module has detected, this method calculates the PV power. By sampling both the PV Array current and voltage, the MPPT algorithm used in the P&O technique determines the PV output power and the power change. The voltage of the solar array is regularly increased or decreased to power the tracker [4]. A second perturbation is created in the same (opposite) direction if a particular perturbation causes the output power of the PV to grow (reduce). Up until the maximum power point has been achieved, the duty cycle of the DC chopper is changed. In actuality, the MPP is where the system oscillates. The oscillation may be minimised by reducing the step size of the perturbation. Small step sizes, however, cause the MPPT to lag. The PV array would show multiple characteristic curves for various irradiance and cell temperature levels. Each curve has a peak in its power. The converter is then given the equivalent maximum voltage at this moment. The P&O Algorithm flowchart is shown in Fig. 2.

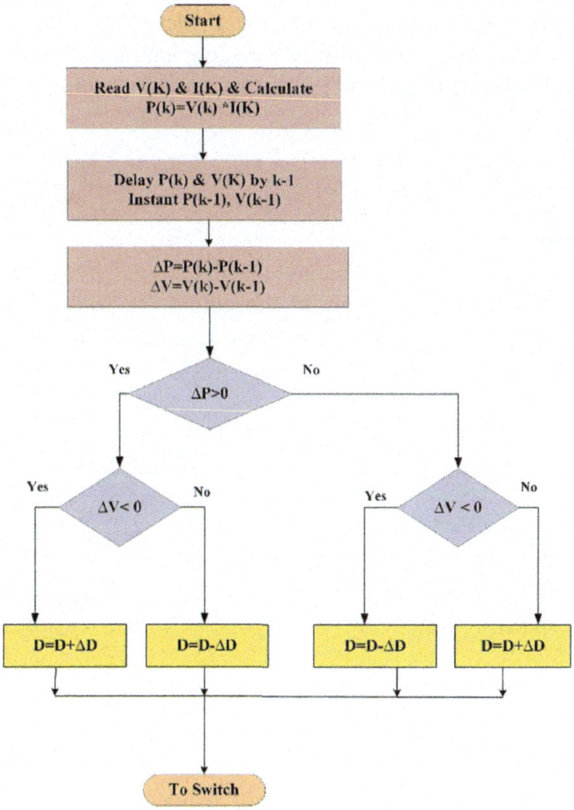


**Fig. 2** Flowchart of pertub and observe algorithm

## 3 EV Circuit Configuration and Design

For the EV to run as efficiently as possible, the passive filters utilised in the circuit must be designed well. This article presents the design of passive components. A high value DC capacitor is needed on the DC side of the inverter of the single phase grid because the DC output voltage contains harmonics with second order terms. Between the input and output, there is an energy storage device called a DC capacitor. The amount of energy storage will determine the capacitor value. (i) If the input power factor is one, the input power is supplied.

$$P_{in} = v_{in} \times i_{in} = \frac{VI}{2} - \frac{VI}{2}\cos 2\omega t \tag{1}$$

Power of the inductor at input side is given by (2)

$$P_L = \frac{\partial\left(\frac{1}{2}L(I\sin\omega t)^2\right)}{\partial t} = \omega L I^2 \sin\omega t \cos\omega t \tag{2}$$

The H-bridge converter receives energy from the input inductor and uses it to charge the output dc capacitor. If there are no losses in power in the devices, the output capacitor will store the same amount of power as the transformation among the input and inductor. Using this, one can determine the power flow via the capacitor (3)

$$P_c = P_{in} - P_L = \left(\frac{VI}{2} - \frac{VI}{2}\cos 2\omega t\right) - \left(\omega L I^2 \sin\omega t \cos\omega t\right) \tag{3}$$

The capacitor is charged and discharged by second order components in Eq. (3), which results in ripple in the output DC voltage. By simplifying and taking the integration to get the instantaneous power for a haf cycle, where (4) is the ripple energy

$$E_c = \int_0^{T/2} \sqrt{\frac{V^2I^2}{4} + \frac{\omega^2L^2I^2}{4}}\sin 2\omega t dt = \sqrt{\frac{V^2I^2}{4} + \frac{\omega^2L^2I^2}{4}}/\omega \tag{4}$$

One may determine the relationship between the voltage of the DC side, capacitor at the DC side, and inductor at the input side as follows from the swell energy in the capacitor: (5)

$$C_{dc} = \frac{\sqrt{\frac{V^2I^2}{4} + \frac{\omega^2L^2I^2}{4}}}{2 \times V_{dc} \times \Delta V_{dc} \times \omega} \tag{5}$$

$C_{dc}$ is determined to be 2 mF using (5).

(ii) The current ripples passing over the filter inductor informs the design of the input side filter ($L_s$ and $C_s$) ($L_s$).

$$L_s = \frac{V_{dc} \times \frac{V \sin \omega t}{Vdc} \times \left(1 - \frac{V \sin \omega t}{Vdc}\right)}{2\Delta I f_{sw}} \tag{6}$$

The instant rate of the AC input voltage is represented through $V$ sin$t$, while $Vdc$ is the DC voltage with fluctuations. The current ripple $I$ is considered to be 10% of the current at the input side and $f_{sw}$ is the switching frequency. The input LC filter capacitor is used to dampen the output voltage's harmonics, which are represented by (7)

$$C_s = \left(\frac{1}{2\pi f_{sw}}\right)^2 \times \frac{1}{L_s} \tag{7}$$

The 20 F and 0.75 mH values for the filter capacitor $C_s$ and filter inductor $L_s$ are attained. (iii) ($L_o$ and $C_o$) output filter using (8) and (9) separately, the filter at output i.e. capacitor $C_o$ and inductor $L_o$ are determined.

$$L_o = \frac{\left(D(1 - D^2)r\right)}{2 f_{sw}} \tag{8}$$

$$C_o = \frac{(1 - D)}{8L_o\left(\frac{\Delta V_o}{V_o}\right) f_{sw}^2} \tag{9}$$

To keep the output voltage constant, a very large output side capacitor is needed. $D$ is the duty cycle, resistance on the internal part denoted by $r$, buck/boost converter $f_{sw}$ switching frequency, output voltage is $V_o$, and voltage ripple, which is five percent of voltage at the output end, is $V_o$. It is determined that $L_o$ and $C_o$ are 41 H and 600 F. In this study, a single phase 120 V rms system is taken into account.

## 4 Simulation Results and Discussions

This Simulink model consists of 2000 W solar PV linked to the DC bus through a step up converter, the 240 V EV battery is connected to the Dc bus through DC DC to buck boost converter and the grid is connected to the DC bus via a single phase inverter (Fig. 3).

This control logic consists of the incremental conductance extreme power chasing method (MPPT) for the PV scheme. Incremental conductance MPPT receives the PV voltage and current and generates the duty cycle and duty processed via the PWM generator for generating the pulse for the Boost converter (Fig. 4). Bidirectional converter of the battery is controlled by the voltage control method. The voltage of

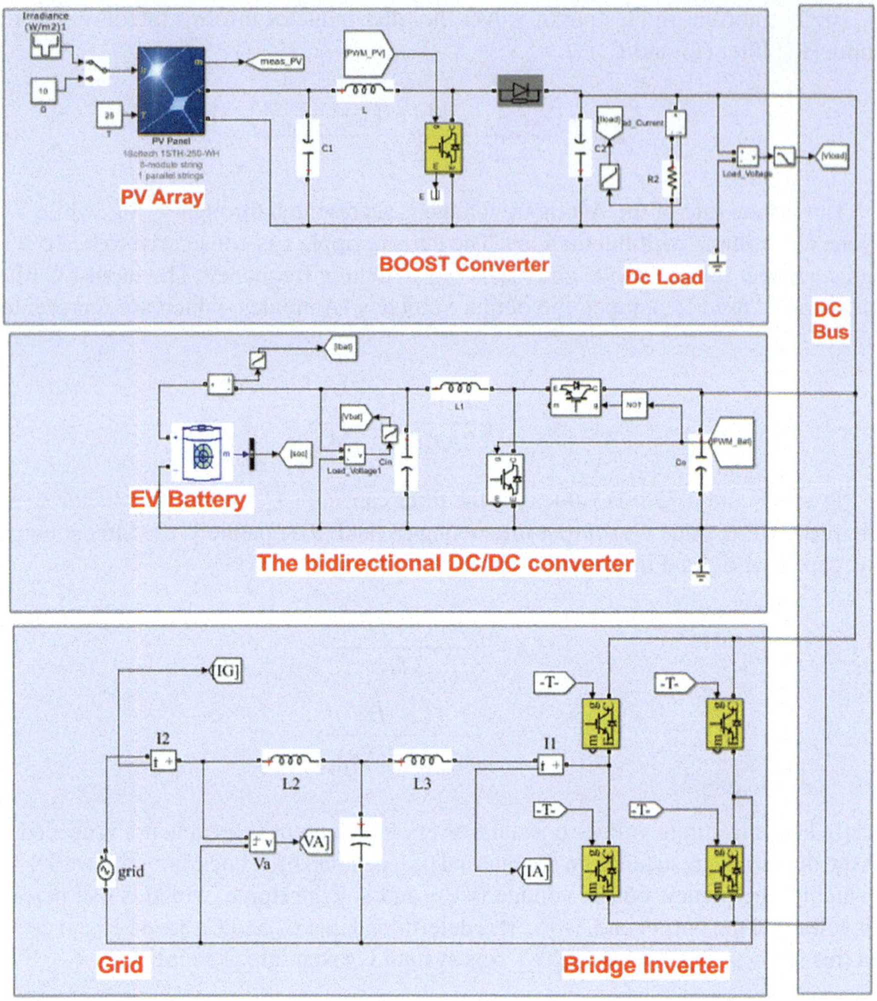


**Fig. 3** Solar photovoltaic (PV) charging circuit for electric vehicles, modelled in Simulink for G2V and V2G modes of operation

the DC bus is monitored and related with the DC bus standard voltage. The voltage error is handled through a PI regulator. The PI regulator produces the duty ratio and it is handled through the PWM generator. The PWM generator produces the pulse for the Bidirectional converter for controlling the voltage of the DC bus. The inverter of the grid side is controlled by means of the DQ current control concept for V2G and G2V operation.

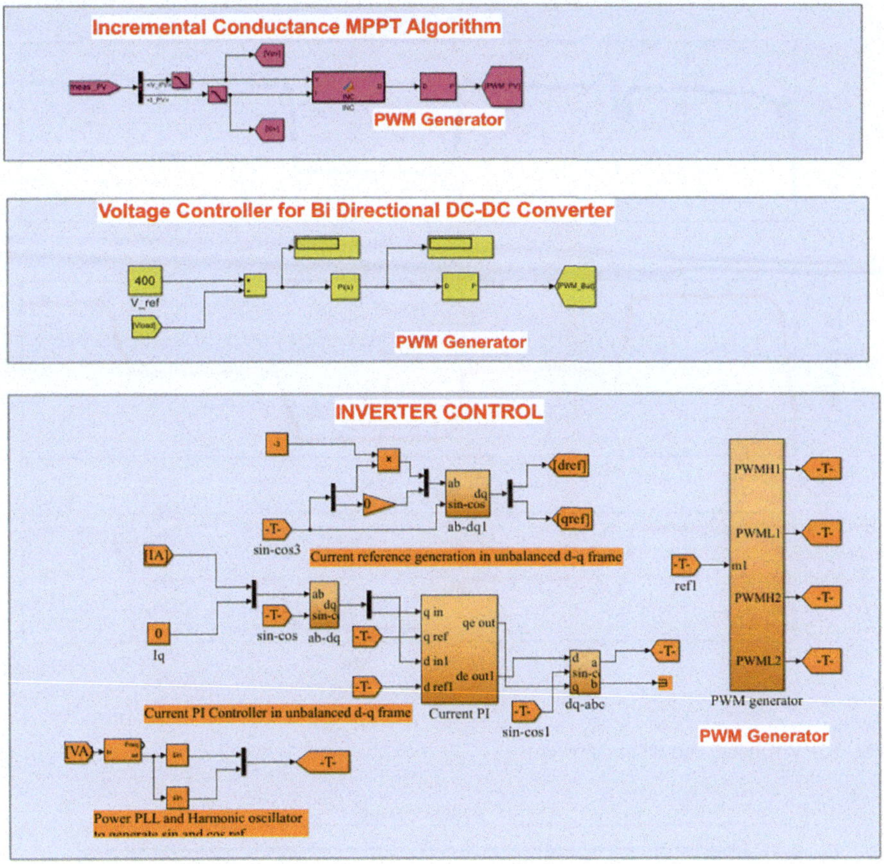


**Fig. 4** Control logic of Solar PV boost converter, EV battery bidirectional converter, grid inverter

## *4.1 Mode 1: Solar PV Used for Charge the EV Battery and Supply the DC Load*

The PV voltage is maintained at 240 V during 1000 and 500 W/m$^2$ and 80 V during 0 W/m$^2$ irradiance (Fig. 5). The PV current is 8.1 A during 1000 W/m$^2$, current is 4 A during 500 W/m$^2$ and current is zero at 0 W/m$^2$. The PV power is 2000 W during 1000 W/m$^2$, 1000 W for 500 W/m$^2$ and 0 W for 0 W/m$^2$.

The voltage of the Battery is upheld at 255 V, the battery is in charging mode during 1000 and 500 W/m$^2$ (Fig. 6). Battery in discharging mode during 0 W/m$^2$. The battery current is −6 and −2.5 A during 1000 and 500 W/m$^2$ and 1.8 A during 0 W/m$^2$.

The load of DC bus or voltage of the DC bus is upheld at 400 V, current of the DC load is maintained at 0.6 A and the Dc load power is maintained at 240 W (Fig. 7).

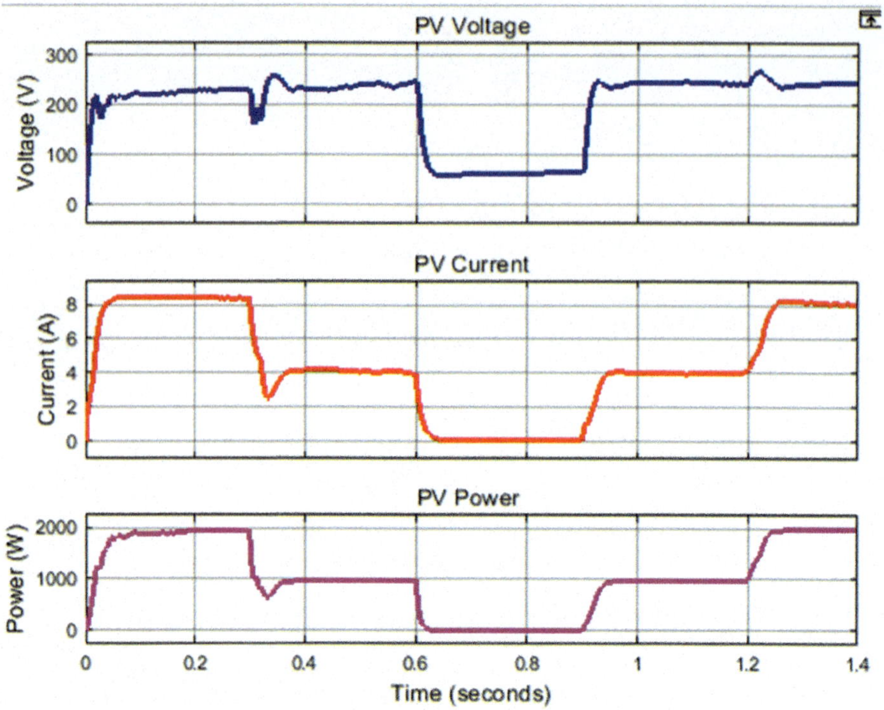


**Fig. 5** PV voltage, current and power

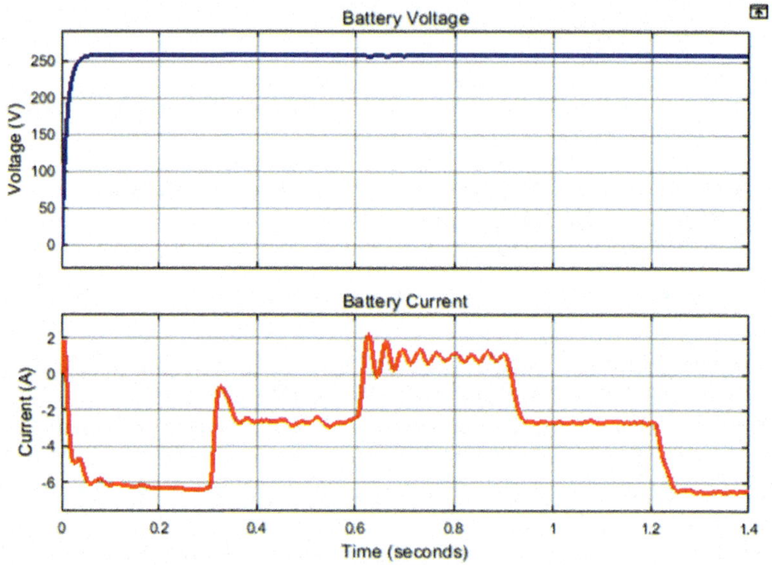


**Fig. 6** Battery voltage and current

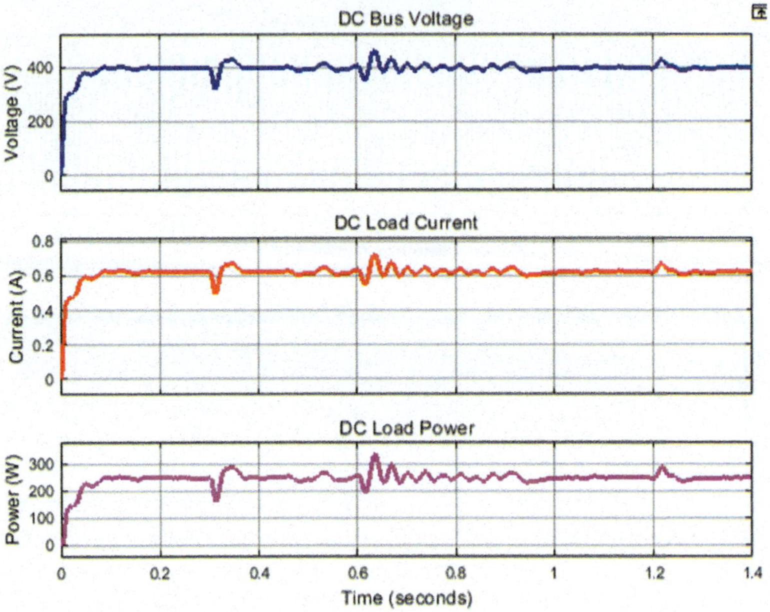


**Fig. 7** DC load voltage, current and power

The inverter or voltage of the grid is upheld at 230 V, inverter and current of the grid is maintained at 0 A (Fig. 8).

The solar PV system is linked to a charge controller in this mode of operation, which controls the voltage and current coming from the solar panels to make sure the electric vehicle battery is charged at the proper pace. Additionally, the charge controller prevents the battery from being overcharged or undercharged, both of which might harm it. In addition, a DC load—anything that uses DC electricity—is connected to the solar PV system. A DC load might be any appliance or gadget. In the event that the solar PV system is not producing enough energy, this load may be provided either directly by the solar panels or by the battery. This mode of operation may assist to maximise the use of solar energy and lessen dependency on the grid by utilising solar PV to charge the EV battery and power a DC load at the same time.

## 4.2 *Mode 2: Grid to Vehicle Mode with Solar PV*

The PV voltage is maintained at 240 V during 1000 and 500 W/m$^2$ and 80 V during 0 W/m$^2$ irradiance (Fig. 9). The PV current is 8.1 A during 1000 W/m$^2$, current is 4 A during 500 W/m$^2$ and current is zero at 0 W/m$^2$. The PV power is 2000 W during 1000 W/m$^2$, 1000 W for 500 W/m$^2$ and 0 W for 0 W/m$^2$.

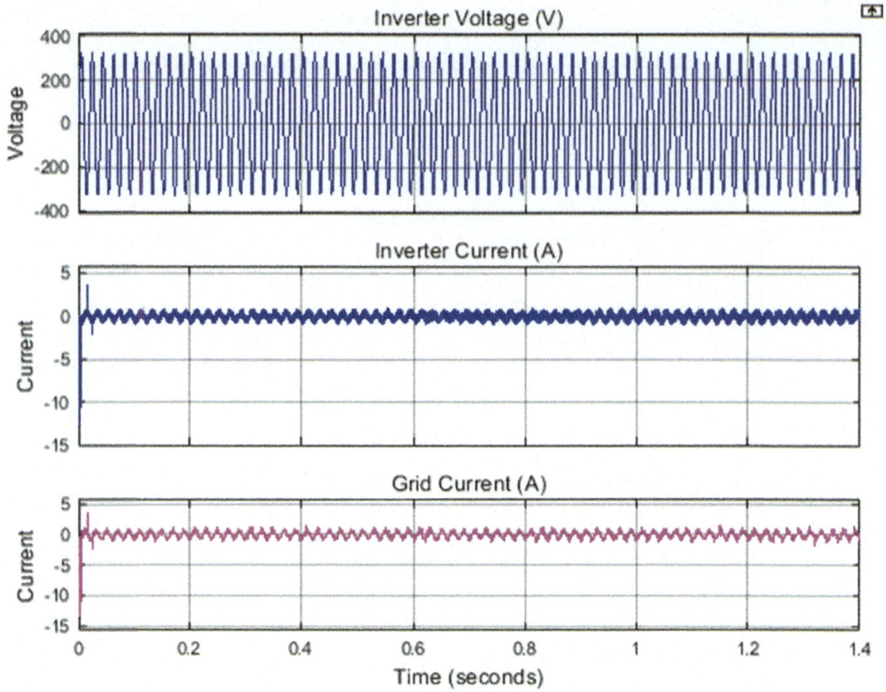


**Fig. 8** Inverter voltage, inverter current and grid current

The voltage of the battery is upheld at 255 V, the battery current is always in the negative region and the battery is in charging mode via the grid as well using solar PV (Fig. 10).

In this mode of operation, the solar PV system is connected to a charge controller that regulates the voltage and current from the solar panels to ensure that the EV battery is charged at the appropriate rate. The charge controller also monitors the state of charge of the battery and adjusts the charging rate accordingly. When the solar PV system is generating more energy than is needed to charge the EV, the excess energy can be fed back into the grid through a grid-tied inverter. This allows the solar PV system to offset grid energy usage and potentially generate income through net metering or feed-in tariffs.

### *4.3 Mode 3: Vehicle to Grid Mode with Solar PV*

The PV voltage is maintained at 240 V during 1000 and 500 W/m$^2$ and 80 V during 0 W/m$^2$ irradiance (Fig. 11). The PV current is 8.1 A during 1000 W/m$^2$, current is 4 A during 500 W/m$^2$ and current is zero at 0 W/m$^2$. The PV power is 2000 W during 1000 W/m$^2$, 1000 W for 500 W/m$^2$ and 0 W for 0 W/m$^2$.

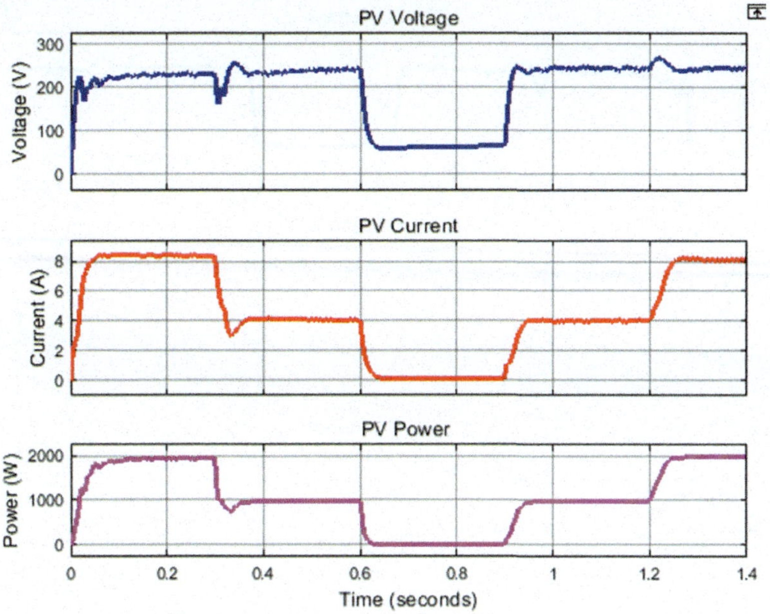


**Fig. 9** PV voltage, current and power

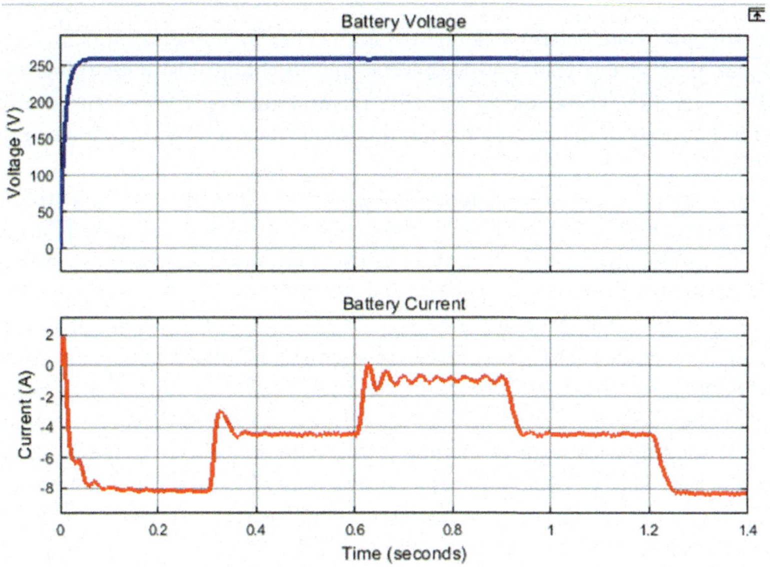


**Fig. 10** Battery voltage, current

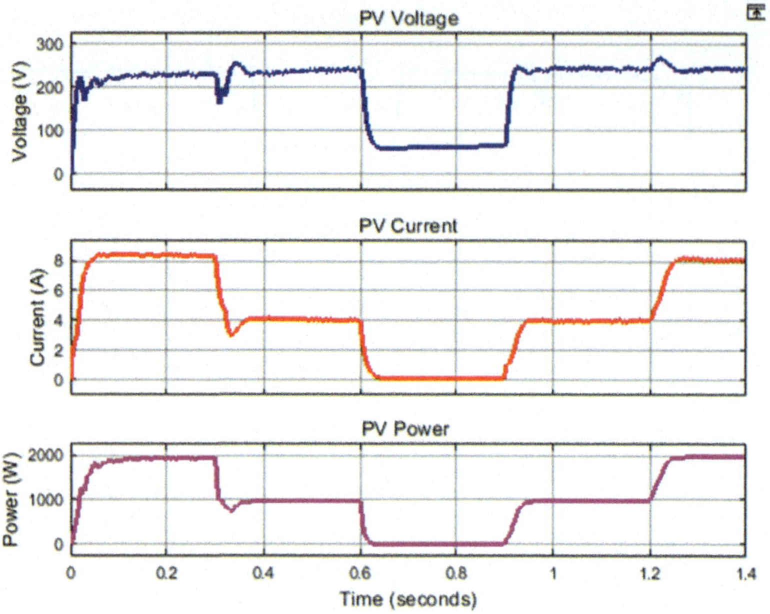


**Fig. 11** PV voltage, current and power

The voltage of the battery is upheld at 255 V, the battery current is always in the negative region during 1000 and 500 W/m$^2$ and the battery is in charging mode via solar PV (Fig. 12). The battery current is positive during 0 W/m$^2$ and discharging operation happens in the battery during this mode and supply the power to the grid and DC load.

The voltage of the inverter is upheld at 230 V, the inverter and the current of the grid are maintained at 3 A peak and voltage and current both in phase (Fig. 13).

In V2G mode, the EV battery may be utilised as a portable energy storage device, enabling the utilisation of the stored energy when the grid is experiencing high demand. During sunny seasons, the solar PV system may also be used to directly charge the EV battery. The EV is attached to a bidirectional charger that can charge the EV battery and discharge it back to the grid in order to allow V2G functioning. A system that governs the energy flow to and from the EV battery operates the charger depending on variables including grid demand, energy pricing, and battery state of charge. The V2G technology may use energy from the EV battery to assist balancing the grid during times of heavy demand. As a result, the EV may function as a distributed energy resource and enhance grid stability.

Table 1 presents data on photovoltaic (PV) power generation under different levels of irradiance. The irradiance is measured in W/m$^2$, while the PV power generated is measured in W. The table also provides information on the maximum power point ratio (%) and efficiency (%) of the PV panels under different irradiance levels. The data shows that as irradiance decreases, the PV power generated also decreases.

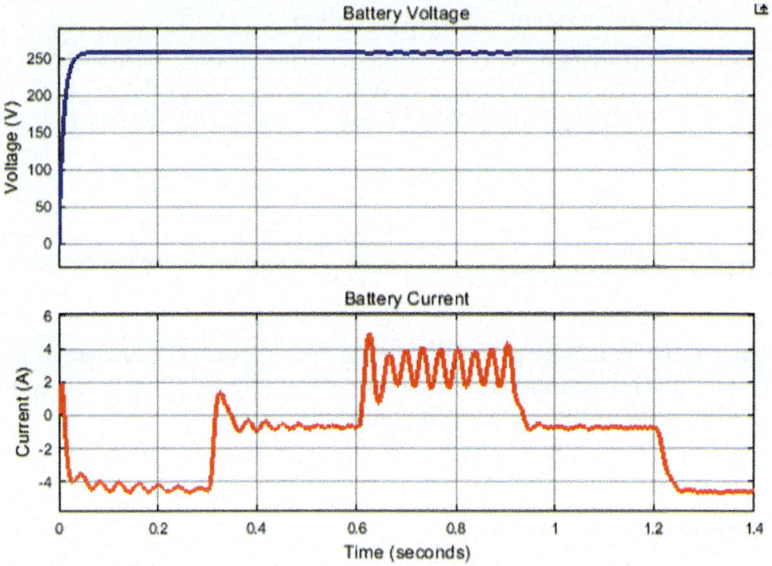


**Fig. 12** Battery voltage, current

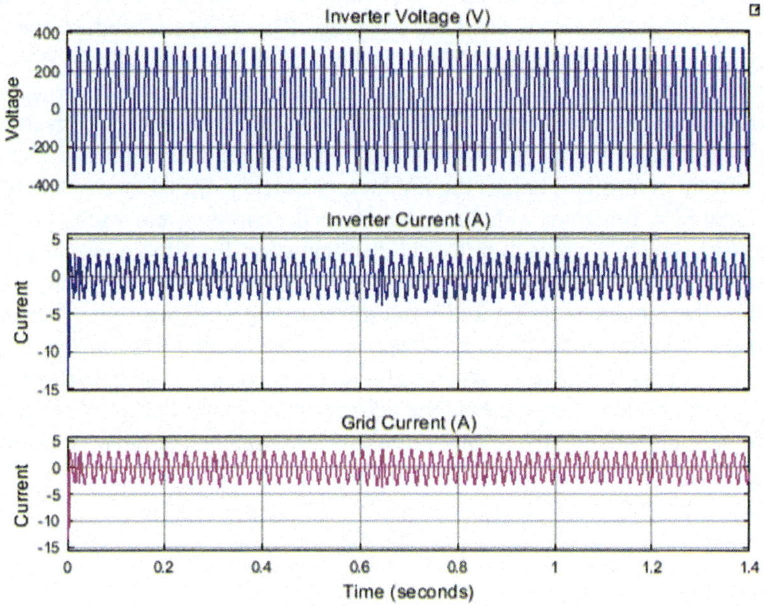


**Fig. 13** Inverter voltage, inverter and grid current

**Table 1** Comparison of PO MPPT and incremental MPPT

| Irradiance (W/m$^2$) | PV power (W) | | Maximum power point ratio (%) | | Efficiency (%) | |
|---|---|---|---|---|---|---|
| | PO MPPT | INC MPPT | PO MPPT | INC MPPT | PO MPPT | INC MPPT |
| 1000 | 1990 | 1998 | 99.5 | 99.9 | 96 | 98.5 |
| 800 | 1592 | 1599 | 99.5 | 99.9 | 96.5 | 98.6 |
| 500 | 993 | 997 | 99.3 | 99.7 | 97 | 98.4 |
| 400 | 794 | 799 | 99.25 | 99.87 | 97.2 | 98.9 |

Additionally, the maximum power point ratio and efficiency of the panels decrease slightly with decreasing irradiance. However, the efficiency remains relatively high, ranging from 96 to 98.9%, even at low irradiance levels.

## 5 Conclusion

In this research, a circuit for charging an electric car is outlined. using energy derived from the sun. Normal test conditions include measuring the Power–Voltage and Current–Voltage features of the solar PV array. The design procedure for filter of passive components is defined. With the help of a converter, the 245 V voltage at the output from the solar PV system will be increased to 400 V. To satisfy the requirements of charging an electric car battery, a bidirectional dc–dc converter filters the output voltage and steps down the filtered voltage. The employed DQ frame-based current controller charges the EV battery quickly. Based on measurements, it is clear that SOC functions well throughout both the charging and discharging stages. An electric vehicle's connection to and disconnection from the power grid may be represented by out-of-phase and in-phase voltage and current. The state-of-charge (SOC) of the battery is monitored while discharging and charging. The regulator performs a fine job of keeping the reference voltage stable while discharging and charging, with just a little amount of steady-state inaccuracy. Solar charging mitigates voltage problems and congestion in the delivery network produced by an increase in generating units and energy consumption as a consequence of a rise in the quantity of electric vehicles (EVs) charging from the grid.

## References

1. Liu N, Chen O, Lu X, Liu J, Zhang J (2015) A charging strategy for PV-based battery switch station considering service availability and self-consumption of PV energy. IEEE Trans Ind Electron 62(8):4878–4889
2. Edenhofer O (2015) Climate change 2014: mitigation of climate change, vol 3. Cambridge University Press, Cambridge, UK

3. Lucas A, Bonavitacola F, Kotsakis E, Fulli G (2015) Grid harmonic impact of multiple electric vehicle fast charging. Electr Power Syst Res 127:13–21
4. Tan L, Wu B, Yaramasu V, Rivera S, Guo X (2016) Effective voltage balance control for bipolar-DC-bus-fed EV charging station with three-level DC–DC fast charger. IEEE Trans Ind Electron 63:4031–4041
5. Yong JY, Ramachandaramurthy VK, Tan KM, Mithulananthan N (2015) A review on the state-of-the-art technologies of electric vehicle, its impacts and prospects. Renew Sustain Energy Rev 49:365–385
6. Hu X, Zou C, Zhang C, Li Y (2017) Technological developments in batteries. IEEE Power Energy Mag 20–31
7. Goli P, Shireen W (2014) PV integrated smart charging of PHEVs based on DC link voltage sensing. IEEE Trans Smart Grid 5:1421–1428
8. Dharmakeerthi CH, Mithulananthan N, Saha TK (2012) Modeling and planning of EV fast charging station in power grid. In: Power and energy society general meeting, 2012 IEEE
9. Haghbin S et al (2013) Grid-connected integrated battery chargers in vehicle applications: review and new solution. IEEE Trans Ind Electron 60(2):459–473
10. Nafeh AE-SA (2011) An effective and safe charging algorithm for lead-acid batteries in PV systems. Int Energy Res 35(8):733–740
11. Xiaopeng C et al (2012) An overview of lithium-ion batteries for electric vehicles. In: IPEC, 2012 conference on power & energy. IEEE
12. Karimi G, Li X (2013) Thermal management of lithium-ion batteries for electric vehicles. Int J Energy Res 37(1):13–24
13. Botsford C, Szczepanek A (2009) Fast charging vs. slow charging: pros and cons for the new age of electric vehicles. EVS24, Stavanger, Norway

# Ultrasonic Analysis of Lorentz Force for Gas Density Monitoring Using EMAT Sensor

**Soumya Debashis Das, Siddharth Mohanty, Sudhansu Panigrahi, Rajsekhar Sahu, Ahwan Pradhan, and Gyan Ranjan Biswal**

**Abstract** Gas-Insulated Switchgear (GIS) in substations encounters round the clock ppm/ppb level leakages hence increasing the necessity of a robust, reliable, and sensitive device for gas density monitoring. Electromagnetic-acoustic transduction is a novel alternative to piezoelectric transduction for resonator sensors, with unique features and benefits. The paper presents a modeling that defines the electromagnetic-acoustic interaction in the GIS (gas-insulated switchgear) vessel for the detection of $SF_6$ gas leakage. Detection of gas density using an acoustic wave was found to be suitable because of its good electrical properties in gases. It also presents the approach in which the analytical EM (electromagnetic) solvers for Lorentz force density were connected to the ultrasonic simulation. The arrangement is unique as well and helps in finding gas leakage at ppm level because in the substation the live conductors and other electrical components are surrounded by $SF_6$ gas. In this study, the performance and stability of the EMAT sensor for gas leakage were evaluated in the presence of parameters such as its geometry, pressure, temperature, humidity, and molar mass.

**Keywords** EMAT sensor · Substation automation · Gas density sensors · Acoustic wave · Gas-insulated switchgear (GIS) · Eddy current, ppm/ppb

## 1 Introduction

Gas-Insulated Switchgear (GIS) based on $SF_6$ is commonly utilized in High-Voltage (HV) applications. Because of the environmental impact of $SF_6$ gas, strict rules have been implemented to monitor their use and to restrict their emission from

S. D. Das · S. Mohanty · S. Panigrahi · R. Sahu · G. R. Biswal (✉)
Department of EEE, VSS University of Technology, Burla, India
e-mail: gyanbiswal@vssut.ac.in

A. Pradhan
Department of EE, VSS University of Technology, Burla, India



O. H. Gupta et al. (eds.), *Soft Computing Applications in Modern Power and Energy Systems*, Lecture Notes in Electrical Engineering 1107,
https://doi.org/10.1007/978-981-99-8007-9_7

switchgear. An Electromagnetic-Acoustic Transducer (EMAT) sensor was discovered to be a promising technique for detecting gas leaks. With its benefits of extended propagation distance, low attenuation, and high testing efficiency, ultrasonic guided wave testing has become widely employed as Nondestructive Testing (NDT) and structural health monitoring in recent years. A variety of NDT techniques are commonly used, including piezoelectric, laser, and electromagnetic-acoustic transducers. Among these techniques, electromagnetic-acoustic transducers can detect non-contact defects by electromagnetic coupling with the tested object. The complaint-free nature of EMATs makes them particularly useful in harsh conditions, such as hot, cold, clean, or dry.

One goal for sensor applications is to track a specific resonant peak frequency as a function of changes in the targeted gas medium. EMATs can primarily excite the longitudinal Bulk Acoustic Wave (BAW), also known as Longitudinal EMAT (L-EMAT), and the shear BAW, also known as shear EMAT (S-EMAT). One significant EMAT design feature is that it is used in both sensor and stiffness constant extraction applications where it can pick which BAW mode to stimulate. As a result of this property, a resonating plate's harmonic spectrum is clear, with no or few overlapping modes. This results in clear resonating peaks and the possibility of selecting higher harmonics for parameter extraction. To comply with stricter future regulations on $SS_6$ gas usage and to improve the reliability, cost-effective and easy-to-use in situ density sensors are required [1].

The eddy currents generated by the EMAT on the sample surface can be applied to conductive materials or materials that can be easily created by depositing a metallic coating on them. This eddy current is distributed at a very thin layer of the material, called skin depth, according to electromagnetic induction theory. Due to this eddy current in a static magnetic field, Lorentz's force is experienced. The distribution of this Lorentz force is primarily controlled by the magnet and electric coil designs, but it is also influenced by the qualities of the test material, the relative position of the transducer to the test part, and the transducer's excitation signal as shown in Fig. 1. The exact nature of the elastic disturbances is determined by the spatial distribution of the Lorentz force. The Lorentz force mechanism is used in the majority of effective EMAT applications.

## 2 Methodology

The major constituents of EMAT are a generating coil, a receiving coil, and a permanent magnet. The generating coil has a radio-frequency power source, and the receiving coil is directly connected with an RF amplifier while the magnet is responsible for producing the magnetic field [2].

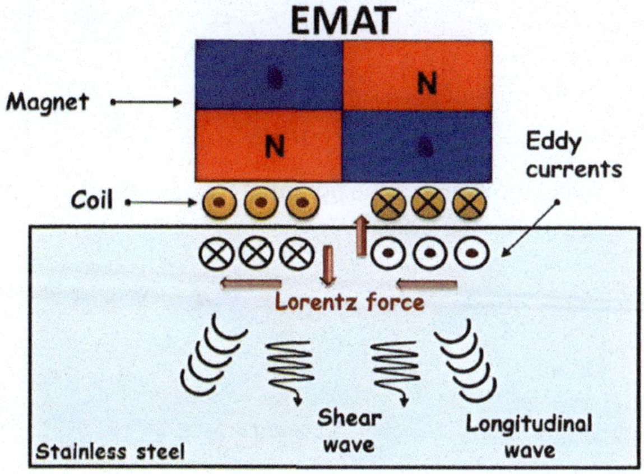


**Fig. 1** Schematics of an EMAT sensor

## *2.1 Mathematical Model*

The distributed point source method (DPSM) is used to model the acoustic sensor. In an EMAT system, Navier's equation (in its dynamic state) can be used to describe the complete motion of waves. In Eq. (1), the body force is depicted by F, the displacement vector is depicted as *s*, *K* and *L* are Lame's elastic constants, and the material density is depicted as *Q*.

$$(\lambda + \mu)\nabla(\nabla.s) + \mu\nabla^2 s + \rho F = \rho u \tag{1}$$

Considering the Cartesian coordinate system for a coil, the dependency of the excitation current on the magnetic potential vector is summarized as in Eq. (2).

$$\vec{A}(\vec{r}) = | \begin{matrix} \frac{\mu}{4\pi} \cdot \sum_i \vec{j}_{xi} \cdot \frac{e^{-jkR_i}}{R_i} \\ \frac{\mu}{4\pi} \sum_i \vec{j}_{yi} \cdot \frac{e^{-jkR_i}}{R_i} \\ \frac{\mu}{4\pi} \sum_i \vec{j}_{zi} \cdot \frac{e^{-jkR_i}}{R_i} \end{matrix} \tag{2}$$

where

$$R_i = \sqrt{(x - C_{xi})^2 + (y - C_{yi})^2 + (z - C_{zi})^2} \tag{3}$$

The active source coordinates in Eq. (3) are represented as $C_{xi}$, $C_{yi}$ and $C_{zi}$. Also "*x*, *y*, and *z*" are relational to the passive points. Considering the *x*-direction, the magnetic potential can be computed using Eq. (4).

$$A(x) = \frac{\mu}{4\pi} \begin{bmatrix} \frac{e^{-jkR_{11}}}{R_{11}} & \cdots & \frac{e^{-jkR_{1N_s}}}{R_{1N_s}} \\ \cdots & \cdots & \cdots \\ \frac{e^{-jkR_{N_{p1}}}}{R_{p1}} & \cdots & \frac{e^{-jkR_{N_pN_s}}}{R_{N_pN_s}} \end{bmatrix} \cdot j_x = [W_{xx}] \cdot j_x \tag{4}$$

When the effect of the potential is considered, the magnetic field can be understood from Eqs. (5), (6), and (7).

$$B = \nabla \times A \tag{5}$$

Or

$$B(\vec{r}) = \left| \begin{array}{l} \sum_i \frac{\alpha_i \mu}{4\pi} . [-(z - C_{zi}) \cdot J_{yi} + (y - C_{yi}) \cdot J_{zi}] \\ \sum_i \frac{\alpha_i \mu}{4\pi} . [-(x - C_{xi}) \cdot J_{zi} + (z - C_{zi}) \cdot J_{xi}] \\ \sum_i \frac{\alpha_i \mu}{4\pi} . [-(y - C_{yi}) \cdot J_{xi} + (x - C_{xi}) \cdot J_{yi}] \end{array} \right. \tag{6}$$

where

$$\alpha_i = -\frac{e^{-jkR_i}}{R_i^2 \left[ j.k + \frac{1}{R_i} \right]} \tag{7}$$

***Operation as Transmitter***

EMAT-based ultrasonic transmitter has been broken down into core components that are expected to operate in an uncoupled way [3, 4]. A finite-difference-based approach based on Maxwell's equations is used to determine the eddy currents generated inside the test specimen due to the EMAT coil. In the region of interest, it is assumed that the enormous static magnetic flux in the test specimen is uniform in both directions and intensity. The Lorentz forces generated within the test specimen are calculated using Eq. (8).

$$F = qv_p \times B \tag{8}$$

Finally, a finite difference approach is used to represent the ultrasonic wave propelled into test specimen by Lorentz body forces.

***Operation as Receiver***

An EMAT receiver for detecting acoustic waves in a specimen in a common arrangement is shown in Fig. 2. A meander line coil is suspended over the specimen with wire spacing "*b*" and lift-off "*h*" identical to that of an EMAT transmitter.

The received acoustic waves have an angle of $\theta$ concerning the *z*-axis, $B_o$ (*y*, *z*) is a static magnetic flux density, "*v* (*y*, *z*, *t*)" is the particle velocity, and "*u*" is the specimen's conductivity. Due to the mechanical vibrations of the medium and the

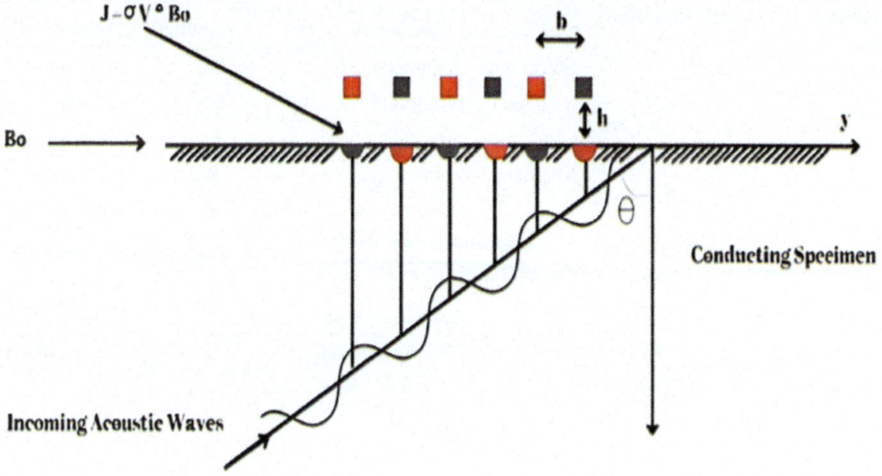


**Fig. 2** Signal received at EMAT receiver

existence of a static magnetic field, a free conduction current density distribution, $J = \sigma_V X B_o$, is established in the specimen. As a result, the current density generates a time-varying vector potential $A\ (y, z, t)$ inside and outside the specimen, which can be monitored as a voltage in an EMAT receiver. The induced current density distribution is also periodic, due to the anticipated spatial periodicity of the vibrating plane wave in the medium. To achieve maximal emf signal responsiveness, the induced currents must have the same frequency as pickup wire spacing ($b = X/2 \sin\theta$) as the sinusoidal plane wave assumption. The induced voltage $V_f$ for a finite cross-sectional one-turn coil is given by Eq. (9).

$$V_f(t) = \frac{\iint V_p(y, z, t)dydz}{\iint dydz} \tag{9}$$

where voltage $V_p(y, z, t)$ due to point pickup coil is given by Eq. (10).

$$V_p(y, z, t) = -\frac{\partial}{\partial t}\iint B.dS \tag{10}$$

## 2.2 Geometric Analysis for Designing the Sensor

The conventional magnet's volume is the same as that of the enhanced magnet. In the conventional configuration, the radii of the magnet are 15 mm. In the improved configuration, a ring magnet (Magnet 2) having an outer and inner radius of 15 and

**Table 1** Parameters used in FEM

| Objects | Parameters | Symbols | Values |
|---|---|---|---|
| Spiral coil | Diameter | $D_1$ | 0.51 mm |
| Magnet 1 | Radius | $R_{21}$ | 10 mm |
| | Height | $h_1$ | 20 mm |
| Magnet 2 | Inner radius | $R_{21}$ | 10 mm |
| | Outer radius | $R_{22}$ | 15 mm |
| | Height | $h_2$ | 20 mm |
| Aluminum | Mass density | $\rho$ | 2700 kg/m$^3$ |
| | Electrical conductivity | $\sigma$ | $3.77 \times 10^{-3}$ |
| | Young's modulus | $E$ | $70 \times 10^8$ Pa |
| | Poisson's ratio | $\mu$ | 0.33 |
| | Thickness | $H_2$ | 50 mm |

10 mm, respectively, and a solid magnet (Magnet 1) having a radius of 10 mm were being used by the 2-D Finite Element Method (FEM). Their total height is 20 mm. In both configurations, the size of the coil is the same having a diameter of 0.51 mm and 30 turns, and the material of the coil is copper wire. The parameters used were as shown in Table 1.

After numerical calculation, the magnetic flux density distribution on the specimen's surface was plotted for the two configurations. It should be noted that the graph is a 2-D axisymmetric graph with the $z$-axis rotated cross section as shown in the figure. The magnetic flux density vector on the aluminum plate's surface was calculated to precisely analyze the distribution of magnetic flux density magnitude concerning the traditional and modified EMAT.

It is imperative to note that the magnitudes vary. The radial component of magnetic flux density $B_{sy}$ is 0 at the center of the magnet, and near the edge of the magnet, there appears a spike peak. The magnetic flux density $B_s$ exhibits an arching peak in its axial component near the magnet's surface. It has been observed that near the magnet's edge, it drops sharply to 0 or even becomes negative.

## 2.3 *Generation of Eddy Current from Coil*

For the generation of eddy current, a quasi-static equation, i.e., Equation (11) for the magnetic potential $A$ is needed to solve the problem:

$$\sigma \frac{\partial A}{\partial x} + \nabla \times \left(\mu_0^{-1} \mu_r^{-1} \nabla \times A\right) = \sigma \frac{V_{\text{coil}}}{2\pi r} \tag{11}$$

Here $\mu_r$ denotes vacuum permeability, $\sigma$ denotes electrical conductivity, and $V_{\text{coil}}$ is the voltage across one turn of the coil [5]. The equation can then be reduced to time-harmonic form as shown in Eq. (12).

$$j\omega\sigma A + \nabla \times \left(\mu_0^{-1}\mu_r^{-1}\nabla \times A\right) = \sigma \frac{V_{\text{coil}}}{2\pi r} \tag{12}$$

By integrating Maxwell's stress tensor on the delimiting boundary $\partial\Omega$, one can obtain the total electromagnetic force acting on an area of space $\Omega$ depicted by Eq. (13)

$$F = \int_{\partial\Omega} T n dS \tag{13}$$

The force calculation tool conducts the integral along the boundaries of the selected region automatically, considering the problem in axisymmetric geometry, where electrical currents are induced in the metallic plate by the fluctuating magnetic field in the time-harmonic regime [6]. The currents, in turn, generate an opposite magnetic field that "shields" the plate from the magnetic field. The zone where electrical currents are formed is constrained relatively close to the surface. Consequently, the size shrinks with increasing frequency. At 10 and 300 Hz, respectively, Figs. 3 and 4 show the induced current density in the $SF_6$ gas medium. To evaluate the system's step response, a time-domain analysis is conducted in this model in a combined surface and arrow plot as shown in Fig. 5. Finally, the force calculation feature computes the total axial force between the coils and the plate as a function of time. The force acts in the opposite direction (negative).

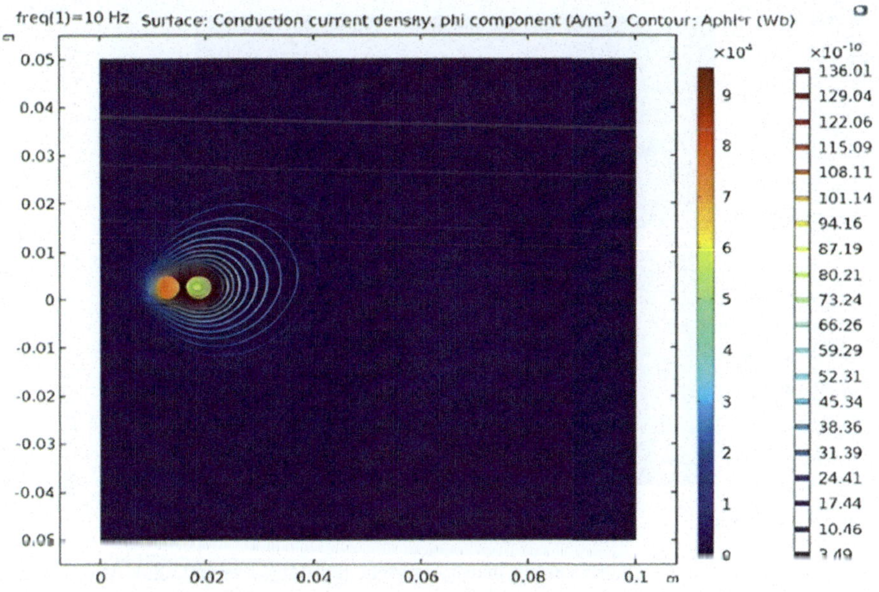

Fig. 3 The magnetic vector potential at a frequency of 10 Hz

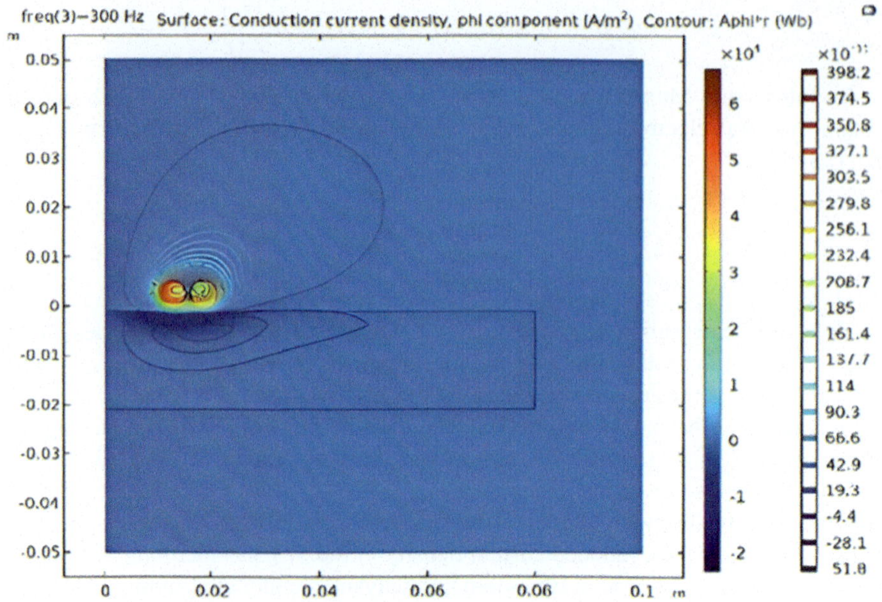


**Fig. 4** Plot of the quantities at a frequency of 300 Hz

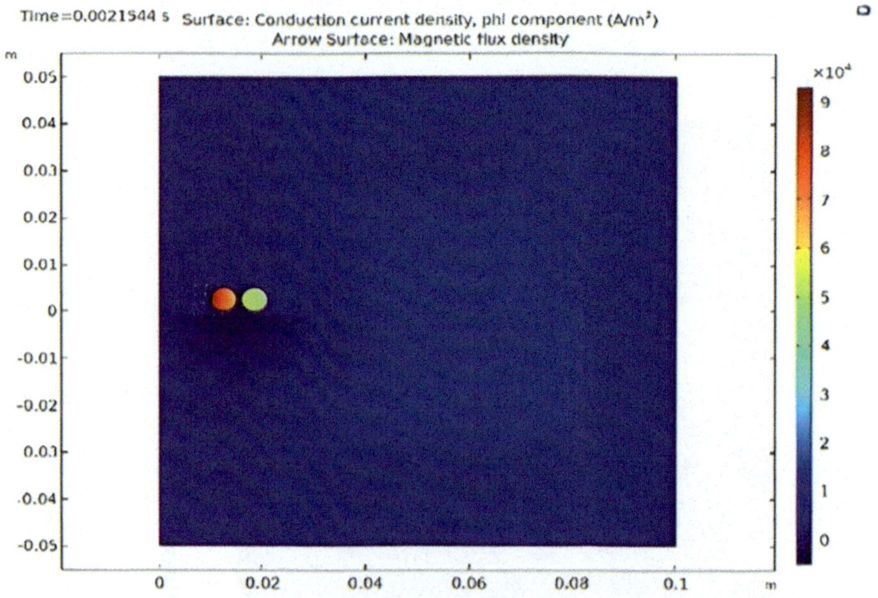


**Fig. 5** Snapshot of the induced current density (surface plot) and magnetic flux density (arrow plot) during the transient study

## 2.4 Simulation of Dynamic Characteristics of the EMAT

Three systems influence the ultrasonic proliferation of EMAT: Lorentz force, pertinent to non-ferromagnetic materials, dynamic attractive power, and magnetizing force, relevant to ferromagnetic materials. Subsequently, in aluminum, the Lorentz force rules the mathematical recreation of the transmission and gathering of mechanical waves in the presence of $SF_6$ gas [6, 7]. When an EMAT instrument is subjected to Lorentz force acting along the outspread bearing of the curve affected by the attractive polarization, it refers to the ultrasonic waves heading out opposite to the specimen surface as shown in Fig. 6. Using an aluminum plate as a test example, a 2-D pivotal symmetric FEM strategy has been developed and different systems have been ignored. This includes the curl test and the waveguide test of the ultrasound data [8, 9].

To receive this ultrasonic wave, the receiver of EMAT was assumed to be installed on a 200-mm-thick aluminum plate to simplify the numerical analysis. The dimensions and material properties of magnet and coil, respectively, will remain the same as described in the static simulation. The distance between the magnet and the upper surface of the aluminum plate is set to 0.5 mm. A three-cycle-long current signal having a peak value of 200 A, centered at 2 MHz, and modulated by a sine squared window drives the coil [10]. The excitation signal is loaded into the model at 6 μs. The step length is set to 0.1 μs to ensure a smooth and stable detection of any hidden crack as shown in Fig. 7.

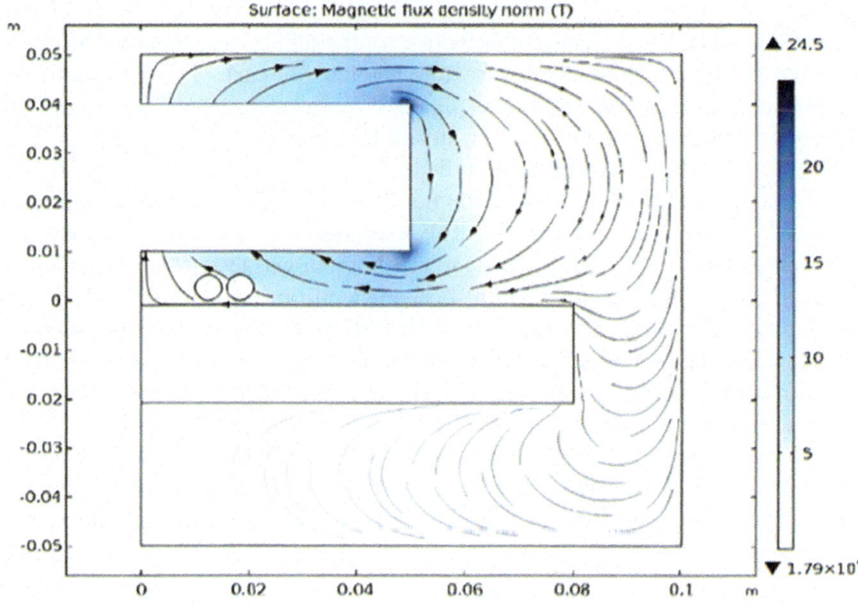


**Fig. 6** Distribution pattern of magnetic field in sustained mode

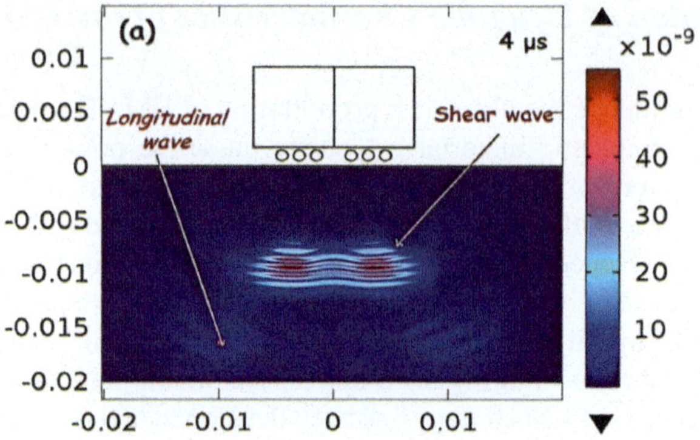


**Fig. 7** Detection of the crack using a generated ultrasonic wave

# 3 Results

## *3.1 Quantitative Analysis of the Simulated Sensor*

The crack height is computed by multiplying the shear wave velocity $C_s$ by the difference in ToF between the plain and defective shapes ($ToF_i$) using Eq. (14).

$$h_i = \frac{[C_s \times (\text{ToF} - \text{ToF}_i)]}{2} \tag{14}$$

An artificial crack was created at the bottom of the stainless-steel plate. The model was tested for three various sizes of fracture heights. For each of the recommended sizes, the amplitude of EMAT's output signal $V(t)$ was measured. It was found that the model can identify the presence of hidden fractures as well as changes in their height [11, 12]. This is owing to the reflection created by the presence of hidden cracks as shown in Fig. 8. A comparison between the targeted crack height and the hidden crack height assessed using the ToF technique was tabulated as in Table 2.

The sensitivity of EMAT to changes in crack width was tested. The crack height was set at 2 mm while the crack width was adjusted from 1 to 3 mm. The FEM and TBLO analysis illustrates how effectively the EMAT sensor can identify changes in fracture width, and this was tabulated as shown in Table 3. The results also demonstrate that a shift in the received signal occurs as the fracture width changes. By solving an inverse problem, the crack width was calculated. In the NDT sector, inverse issues are typically formulated to address optimization issues. A numerical model such as FEM is used to represent the forward problem. However, iterative methods are used to solve the inverse problem to deduce geometrical information about the defects.

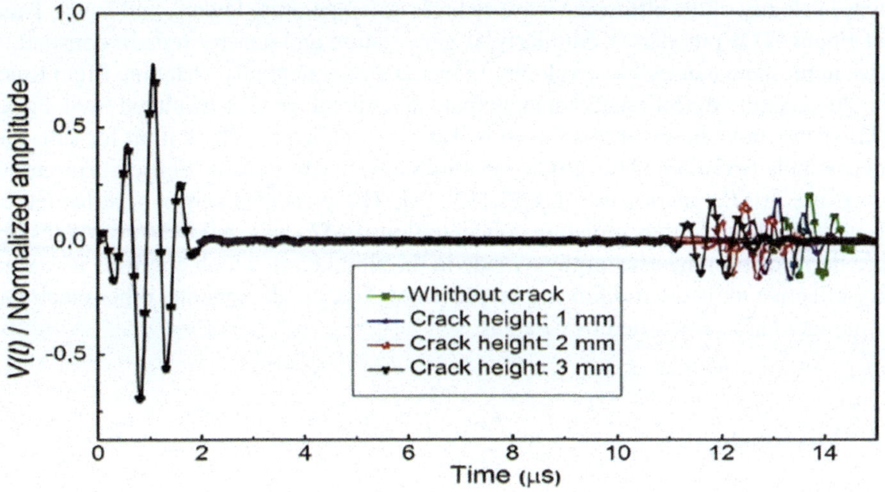


**Fig. 8** Deflection of the wave for various cracks

**Table 2** Crack height versus ToF

| Crack no. | $TOF_i$ (in μs) | Crack height (m) | | Mape (%) |
|---|---|---|---|---|
| | | Estimated | Desired | |
| 1 | 11.9 | $0.96 \times 10^{-3}$ | $1 \times 10^{-3}$ | 4 |
| 2 | 11.28 | $1.96 \times 10^{-3}$ | $2 \times 10^{-3}$ | 2.4 |
| 3 | 10.9 | $2.57 \times 10^{-3}$ | $2.5 \times 10^{-3}$ | 2.8 |
| 4 | 10.56 | $3.1 \times 10^{-3}$ | $3 \times 10^{-3}$ | 3.3 |
| 5 | 10.2 | $3.6 \times 10^{-3}$ | $3.5 \times 10^{-3}$ | 2.9 |

**Table 3** Width by TBLO

| Crack no. | Crack width (m) | | Mape (%) |
|---|---|---|---|
| | Estimated | Desired | |
| 1 | $0.95 \times 10^{-3}$ | $1 \times 10^{-3}$ | 5 |
| 2 | $1.55 \times 10^{-3}$ | $1.5 \times 10^{-3}$ | 3.3 |
| 3 | $2.08 \times 10^{-3}$ | $2 \times 10^{-3}$ | 4 |
| 4 | $2.45 \times 10^{-3}$ | $2.5 \times 10^{-3}$ | 2 |
| 5 | $2.92 \times 10^{-3}$ | $3 \times 10^{-3}$ | 2.7 |

## 3.2 *Parametric Analysis*

Pressure, temperature, and humidity are just a few of the parameters that have been investigated here. It was discovered that if pressure and density remain constant, wave velocity of the ultrasonic wave reduces as temperature lowers, and hence Time of Flight (TOF) increases. Similarly, if temperature and density remain constant, a rise in pressure causes wave velocity to increase exponentially, reducing flight time.

As moisture detection and management are critical for Gas-Insulated Switchgear (GIS), the moisture inside GIS must be kept within the prescribed limit for safe and dependable operation. This article describes an acoustic sensor for measuring trace moisture in $SF_6$ gas for use in GIS [13, 14]. The presented sensor's performance is simulated for various moisture concentrations in $SF_6$ gas. All the sensor's results are derived from the assumptions made in the simulation, which confirms its usage for effective moisture monitoring in GIS where live conductors and other electrical equipment are surrounded by $SF_6$ gas.

## 4 Discussion

Industrial structures are subject to microstructural alterations brought on by thermal aging, corrosion, and fatigue cracking. A hidden crack in a GIS is typically very hazardous as it is challenging to find using non-destructive evaluation. The method presented in this paper uses an ultrasonic signal to estimate the size of a hidden crack inside a GIS vessel. The Finite Element Method (FEM) is used to simulate the signal that EMAT has received. Then, using a combination of two techniques, it is possible to determine the hidden crack sizes. The initial technique is the Time-of-Flight (ToF) method, which calculates the height of the crack by comparing the ToF of the healthy and defective forms. The crack width is then calculated by employing the Teaching–learning-Based Optimization (TLBO) meta-heuristic algorithm for resolving inverse problems from the signal hence received. The obtained results demonstrate the EMAT sensor's sensitivity to variations in crack sizes. Additionally, the quantitative examination of the crack dimension amply demonstrates the effectiveness and dependability of the chosen strategy.

By using the EMAT sensor and the NDT technique, the solution to the problem of monitoring the density of $SF_6$ in GIS vessel can be achieved with the analysis and assumptions that have been developed for its development [15]. It was found that the EMAT sensor is both sustainable and efficient in detecting gas density as well as detecting any leaks from the GIS vessel [16–18]. Although there are many well-developed sensors on the market, the EMAT sensors were developed because $SF_6$ GIS was preferred over conventional transmission substations. The design and cost of land made it unsuitable for urban areas. The maintenance burden led to the use of $SF_6$ switchgear. Here, we find piezoelectric sensors that produce the same wave as EMAT sensors but are less robust due to their contact nature with the GIS vessel. Both the sensors have their benefits as acoustic signals are encountered as a result of their excellent electrical properties.

A piezoelectric sensor is the first of its kind to detect gas leakage in a dense environment, but some of its crystals are water soluble and dissolve in a high-humidity environment. It is also used for dynamic measurement, so it does not suit the static environment, which is why EMAT sensors had to be developed. This paper studies the frequency response of EMAT sensors under different gases to determine the effect of a different medium on the response. Fabrication and construction of EMAT for the detection of gas density are likely to be studied in the future.

## 5 Conclusion

The equipment for the generation of shear and longitudinal acoustic waves in static and pulsed magnetic fields has been optimized by analyzing the ultrasonic Lorentz force for gas density measurement using an EMAT sensor, mathematical and finite element modeling of magnetic systems, and the study of EMAT specificities. The results in a $SF_6$ gas medium, which is chosen as an ideal gas in this study, have also been explored without angular deflection and with angular deflection. Furthermore, data for the longitudinal ultrasonic field deflection for various molar masses at various pressures and temperatures have been reported. These findings are also reported for the $SF_6$ gas density measurement.

Based on the results of the simulation, it is evident that the system still needs considerable improvement. To boost their endurance, sensor wear pads will be upgraded to ceramic-based ones. Wear pad durability will improve if the EMAT sensor carrier is fine-tuned to optimize the transition between gas chambers.

**Acknowledgements** We thank the "*Automation and Sensors Lab*", Department of Electrical and Electronics Engineering, VSSUT, Burla, Odisha for providing such an atmosphere for carrying out this research work. We are immensely grateful to Dr. Gyan Ranjan Biswal, H.O.D, Department of Electrical and Electronics Engineering, for his support and guidance throughout the research.

## References

1. Lucklum F, Jakoby B (2009) Non-contact liquid level measurement with electromagnetic–acoustic resonator sensors. Meas Sci Technol 20(12):124002. https://doi.org/10.1088/0957-0233/20/12/124002]
2. Pucci L, Raillon R, Taupin L, Baqué F (2019) Design of a phased array EMAT for inspection applications in liquid sodium. IEEE Sens J 19(20):4460. https://doi.org/10.3390/s19204460
3. Mirkhani K, Chaggares C, Masterson C, Jastrzebski M, Dusatko T, Sinclair A, Shapoorabadi RJ, Konrad A, Papini M (2004) Optimal design of EMAT transmitters. NDT & E Int 37(3):181–193. https://doi.org/10.1016/j.ndteint.2003.09.005
4. He J, Dixon S, Hill S, Xu K (2017) A new electromagnetic acoustic transducer design for generating and receiving S0 lamb waves in ferromagnetic steel plate. IEEE Sens J 17(5):1023. https://doi.org/10.3390/s17051023
5. Pei C, Zhao S, Xiao P, Chen Z (2016) A modified meander-line-coil EMAT design for signal amplitude enhancement. Sens Actuators A 247:539–546. https://doi.org/10.1016/j.sna.2016.07.006

6. Samokrutov A, Alekhin S, Bobrov V, Bobrov S, Shevaldykin V (2014) Modelling of EMAT magnetic systems for ultrasonic waves generation in elastic layer, in static and pulsed magnetic fields. In: 11th European conference on non-destructive testing (ECNDT 2014)
7. Biswal GR, Mohanty P, Akram KJ, Padhy NP, Islam T (2020) Design and fabrication of an inexpensive capacitive humidity sensor for smart substation automation. IEEE Sens J 20(12):6215–6223. https://doi.org/10.1109/JSEN.2020.2974522
8. Das SD, Akram KJ, Biswal GR, Islam T (2021) A high precision cost-effective ultrasonic sensor for detection of gas leakage in gas insulated switchgear. In: 2021 International conference on sustainable energy and future electric transportation (SEFET). IEEE, pp 1–6. https://doi.org/10.1109/SeFet48154.2021.9375743
9. Ma Y (2021) Analysis of common faults in gas insulated substation. J Phys Conf Ser 1865(2):22–55. https://doi.org/10.1088/1742-6596/1865/2/022055
10. Kamei M, Takai O (2010) Influence of sensor information accuracy on condition-based maintenance strategy for GIS/GCB maintenance. IEEE Trans Power Deliv 26(2):625–631
11. Zargar ZH, Akram KJ, Biswal GR, Islam T (2020) A linear capacitive sensor for ppm moisture measurement in $SF_6$ gas-insulated switchgear. IEEE Trans Instrum Meas 70:1–8. https://doi.org/10.1109/TIM.2020.3040880
12. Min SD, Kim JK, Shin HS, Yun YH, Lee CK, Lee M (2010) Noncontact respiration rate measurement system using an ultrasonic proximity sensor. IEEE Sens J 10(11):1732–1739. https://doi.org/10.1109/JSEN.2010.2044239
13. Lu Q, Li Q, Hu L, Huang L (2021) An effective low-contrast SF6 gas leakage detection method for infrared imaging. IEEE Trans Instrum Meas 70:1–9. https://doi.org/10.1109/TIM.2021.3073443
14. Yang D, Tang J, Zeng F, Yang X, Yao Q, Miao Y, Chen L (2018) Correlation characteristics between SF6 decomposition process and partial discharge quantity under negative DC condition initiated by free metal particle defect. IEEE Trans Dielectr Electr Insul 25(2):574–583. https://doi.org/10.1109/TDEI.2018.006684
15. Tan H (2020) Live detection technology of SF6 equipment based on computer aided laser imaging technology. J Phys Conf Ser 1578(1):012061. IOP Publishing. https://doi.org/10.1088/1742-6596/1578/1/012061
16. Biswal GR, Das SD (2021) IoT driven smart plug for speed control of high current-rated household appliances. Indian Patent Application No 202111011525 A, Filing Date: 18 Mar 2021, Publication Date: 09 Apr 2021
17. Biswal GR, Islam T, Das SD, Akram KJ (2021) Ultrasonic sensor based gas density monitoring of SF6 gas insulated switchgear. India Patent 202111011524A, Apr 2021
18. Das SD, Biswal GR (2022) Development of an energy neutral free space optical link using off-the-shelf solar panel. In: AICTE-SPICON 2022, SPIT. Mumbai, India, 22–23 Dec 2022

# WSN Based Energy-Efficient Protocols for Smart Grid: A State-of-Art Review

**Sandeep Kanaujia and Ajay Sharma**

**Abstract** The topic of research is wireless sensor network in smart grid systems for monitoring of power system parameters' safety and stability. The Process of selecting a path for network traffic to travel from its source to its destination across network is difficult, as the unit of data are routed to the base station through a number of nodes. It is important that efficient packet routing with minimal energy consumption must be followed so that the life of the network and battery's remaining power can be extended. We begin by reviewing literature on location based, data centric and then details in hierarchical base protocols such as LEACH, power efficient PEGASIS, threshold sensitive TEEN, adaptive periodic threshold sensitive APTEEN and cluster based routing HEED. Despite the fact that it can help to increase energy efficiency, the review's findings indicate that energy consumption is the main issue with WSN.

**Keywords** Smart grid · WSN · Energy efficiency · Routing literature review

## 1 Introduction

The maintenance personal can obtain Power Grid information in real time, the intelligent extra High Voltage Power Grid safety monitoring system combines wireless sensor network with Information Communication Technology. In this system intelligent sensors are placed on the power transmission line [1]. The sensor includes multiple induction coil equipped with a backup function and a chargeable battery which can work for a long time and the sensing circuit allows the sensor to monitor power transmission line in real time. The system is equipped with an infrared temperature sensor, 3 Axis accelerator temperature sensors, humidity sensor and an illumination sensor. Through these sensors the system can monitor the conductor temperature

S. Kanaujia (✉)
Department of Electronics, United University, Prayagraj, India
e-mail: sandeepkanaujiamnnit@gmail.com

A. Sharma
Department of Electronics Engineering, UCER, Prayagraj, India



O. H. Gupta et al. (eds.), *Soft Computing Applications in Modern Power and Energy Systems*, Lecture Notes in Electrical Engineering 1107,
https://doi.org/10.1007/978-981-99-8007-9_8

of power lines. After the intelligence sensor conducts a sensing task, it transmits the sensing data to a Gateway installed on the tower. The Gateway uses a low power consumption embedded system as the computing platform [2].

Wireless sensor networks (WSN) are groups of gadgets connected wirelessly that are capable of gathering and exchanging data over wireless links. The gathered data will be delivered to the base station or sink for additional processing. Newly, small, cheap, and smart sensors are fitted with wireless interfaces with which they can transmit and receive together to form a network. The design of a WSN is greatly influenced by the application and must take into account factors like the environment, the design objectives of the application, cost, hardware, and system constraints [3]. The sensor nodes can be fitted with various types of sensors such as whether, acoustic, thermal, chemical, pressure, optical sensor and can be used for building up various types of powerful application. Apart from remote, under ocean monitoring and target tracking, each sensor has its own characteristics, requirements, and applications.

With the rapid development of MEMS (Micro Electromechanical Systems) and wireless communication, large-scale wireless sensor networks, millions of tiny sensor nodes dispersed across an area of interest, are now feasible to install. Energy efficiency for extending network lifetime is one of the most crucial topics in sensor networking because of the stringent energy limits. As a result, every component of the node, including the hardware and protocols, must be exceedingly energy-efficient [4].

## 2 Energy Efficiency

Energy efficiency, said simply, is the process of eliminating energy waste by using less energy to accomplish the same task. The word "energy efficiency" refers to using less energy to provide the same amount of useful output. In other words, lowering energy costs allow consumers to potentially save money. Energy consumption is a serious issue because nodes only have a little amount of non-replaceable, non-rechargeable power.

## 3 Routing Protocol

The foundations of WSN routing protocols are Location-based protocols, Data-centric protocols and Hierarchical Protocols, which are different dependent on network structure and which improve the energy efficiency in sensor networks. Hierarchical protocols will be covered in full in this article (Fig. 1).

**Location based routing protocols**

In Wireless Sensor Networks (WSN), where node locations are used for communication, location-based routing methods are used. Geographic routing protocol and

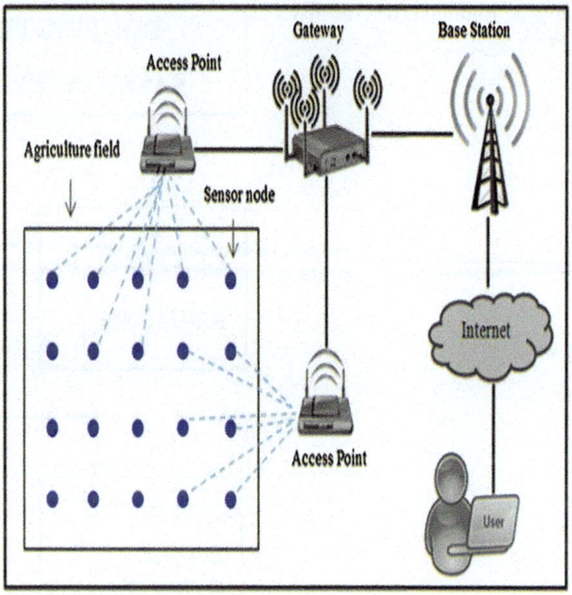


**Fig. 1** WSN overview

position-based routing protocols are other names for it. These protocols lengthen the lifespan of the network and minimize energy consumption but they have the drawback of using the same old path if the target node is in the same location, which uses more energy for certain nodes, while leaving others unused [5].

**Data-centric routing**

When information queries are sent across the network, they produce sampled data. This data is then named or labeled based on certain properties or attributes, which are used to route the data to its intended destination (Fig. 2).

This type of data processing and routing is commonly used in distributed systems and databases, where data is spread out across multiple nodes or servers. By sampling the data and routing it based on its properties, the system can efficiently process and retrieve large amounts of data from distributed sources.

Examples of attributes used to label data in this type of system could include data type, location, size, or timestamp. The routing algorithm would then use these attributes to determine the optimal path for the data to take, based on factors such as network congestion, available bandwidth, and latency. This is known as data-centric routing, which is diffusion-based. The network is interested in performing sensing tasks when routing is data-centric [6]. Here, as the number of nodes rises, there is a correlation between packet transmission and energy dissipation in a network, and that both values tend to decline over time. In future, sensor fields may employ signal amplifiers to properly relay weak signals from one end of the network to the other [7].

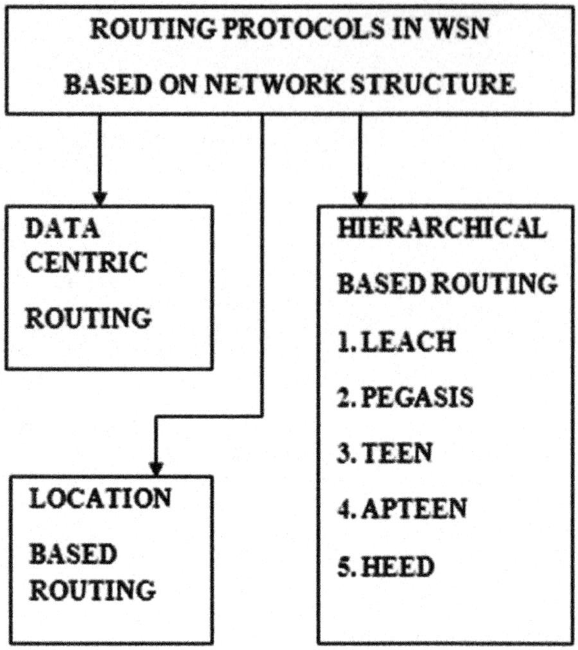


**Fig. 2** Routing protocols in WSN

**Hierarchical based protocols**

It is been carefully investigated the current routing protocols and the process of identifying and understanding the underlying processes or components that are responsible for a particular phenomenon or behaviour that effectively use energy [8]. There are numerous ways to approach the topic of energy use. A data transfer process in a wireless sensor network where, a sink node acts as a data collection point and typically receives data from other nodes in the network. The base station (BS) is often the endpoint or destination for this data, where it can be processed and analyzed. Clustering routing algorithms are among the most dependable protocols used [9].

WSN are extremely complex and applicable because scientists and engineers must deal with a wide range of problems caused by their complex nature. Numerous energy-efficient routing techniques have been established in recent years. Numerous studies have been conducted to address energy efficiency in WSN. To reduce energy usage, a variety of routing techniques are used in WSNs, including hierarchical and cluster-based networks [10]. CBRP, which uses energy and Centroid-based routing to transfer data packets to the nearest base station. Using energy-efficient composite event detection (ED), the volume of data can be handled while increasing energy economy. The majority of the methods in use today for energy consumption use composite event detection [9, 10].

The foundation of the hierarchical protocol is made up of the LEACH, PEGASIS, TEEN, APTEEN, and HEED protocols. The most energy-efficient routing protocols are hierarchical [11]. Based on this supposition, PEDAP, GSTEB, and TBC perform significantly better than LEACH and HEED [12].

## 4 Literature Review

Numerous energy-efficient algorithms have been created for the significance of energy efficiency in WSN as an outcome. None of these, however, suffices. To address a range of problems, including energy consumption [13], load balancing [14], and transmission costs [15], numerous routing algorithms have been devised. There are numerous protocols that have already been put into use with the data fusion transmission technique that is a tree-based hierarchical topology. This family's fundamental protocols are LEACH and HEED [11]. Because the length of the message transmitted from leaf to BS remains constant, every node transmits the same amount of data regardless of the amount of data it receives from its leaf node and sends forward in the same manner. Energy is used by either "productive" or "wasteful" work in a sensor node [12].

The review is done on hierarchical based routing, and it explores various cluster based routing protocol approaches, including the following:

### *4.1 LEACH: Low Energy Adaptive Cluster Head Protocol*

WSN sensors rely on batteries for power, which makes it unchangeable in many circumstances where there is insufficient energy. One of the routing protocols used in this investigation is the LEACH protocol. Utilizing the LEACH protocol can cut down on energy use and increase network longevity. The cluster-based WSNs in the WSN are designed in a variety of ways. To communicate the data to the BS, only identical data and identical events are used. Cluster heads are utilized to cut down on energy use. The cluster head LEACH protocol, reduces energy usage while extending network lifetime [16]. When the cluster head sent the request messages, they were fixed in the routing table, bringing all of the nodes in the cluster up to date in accordance with the TDMA list. Clustering routing methods are employed at the different layers such as network, transport, and data to solve the "hot spot" issue and boost output performance.

In Low Energy Adaptive Clustering Hierarchy (LEACH), exchange of data is taken into account between a fixed square's worth of randomly distributed sensor nodes and an external receiver. It consists of algorithms for adjusting clusters, a distributed cluster creation technique that allows large numbers of nodes to self-organize, and circular movement of cluster head positions to uniformly spread the energy used across all nodes [17].

Leach-C is an enhanced version of leach. Nodes informed the BS about the amount of energy left at the start of each round. Leach-c performance does not to exceed or go beyond a certain limit or standard when the BS is located close to the sensor nodes, which is a drawback [18].

## 4.2 PEGASIS: Power Efficient Gathering in Sensor Information System

Packets from all nodes are transferred progressively to the BS through the routing technique PEGASIS known as the Power Efficient Gathering in Sensor Information System [19]. The LEACH convention transports data from the aggregate to the sink via resident data calculation, increasing the network's life. One problem is that it works well in homogenous networks but not in those with heterogeneous networks. Two phases of heterogeneity nodes are provided by (Stable Election Protocol) SEP. The first is an advance node, while the second is a standard node. Advance nodes have more energy than regular nodes.

PEGASIS out performs LEACH by saving energy during numerous processes. As compared to sending to a cluster-head in LEACH, most nodes travel far shorter distances during local gathering. During each exchanging data round, only one junction communicates to the Base Station [19].

## 4.3 TEEN: Threshold Sensitive Energy Efficient Sensor Network Protocol

TEEN is WSN's energy-efficient dispel convention, the TEEN protocol is utilized to save energy usage while enhancing network consistency and longevity.

The TEEN protocol's disadvantage is that none of the dead nodes make it to the last point, therefore we cannot access any network data and nodes are unable to connect with the cluster head as a result of this cause. TSEP is employed in this to introduce another routing protocol. Threshold Sensitive Stable Election procedure (TSEP), a reactive procedure that makes use of three type levels in heterogeneity. In contrast to proactive networks, reactive networks react right away to changes in pertinent criteria of interest [20]. As the three stages of heterogeneity nodes in the TSEP technique, use advance junctions, intermediate junctions, and normal junctions. The TEEN procedure is used to increase energy regulations. TEEN is appropriate in time-sensitive severe requests which is the future problem.

## 4.4 APTEEN: Adaptive Periodic Threshold Energy Efficient Network Protocol

The Adaptive Periodic Threshold Efficient Energy network (APTEEN) is a proposed routing protocol.

By choosing of the CHs, the cluster head in APTEEN broadcasts some information first during each cluster period like Qualities in which the admin is attentive in fact on a group of physical properties.

This parameter has two thresholds: Schedule, a time division multiple access periods that allots a time slot to each node, and monitor Time, the longest interval a node can send between two consecutive reports. In this study, the author observed that APTEEN delivers a lower energy dissipation value and a greater count of alive junction at any one time after simulating multiple different protocols. For network longevity and energy, the TEEN, LEACH, and APTEEN protocols are contrasted [19].

## 4.5 Energy Efficient Distributing Clustering

One of the key goals of WSN is to increase energy efficiency.

The author provides two recommendations for lowering the use of power of a system by making use of supply and putting an end to delays. The distribution clustering methodology comes first. In this method, the clusters determine the ideal cluster head based on end-to-end delay and energy consumption. Then, in order to reduce end-to-end latency, the Inter clustered routing algorithm makes use of a novel energy cost function. Numerous heuristic tactics are put out to address energy consumption and end-to-end delays; however, these solutions are not appropriate for long-term coverage. Because they are effective at reducing energy usage and put an end to latency, clustering techniques are used. The distributed scheduling technique for joining clusters that underlies the Tree utilizes less energy [21].

The author integrates all of the sensor data when the mingling of packets reaches the destination node to minimize the delay. There are many routing algorithms used for clusters, and one of them works well in three-hop clusters because it makes an energy-efficient connection between the clusters and the sink. Clustering is both hybrid and energy-efficient (HEED) [22]. The intersection, residual energy usage, and other parameters are randomly used to choose the cluster head limits. The researcher tackles the two issues surrounding energy-efficient routing in power hinder escalation in an Asynchronous network sensor with numerous Gateways [23]. The primary one seeks to develop the trees that describe effective routing, and the secondary one seeks to demonstrate how to enlarge the cluster head member utilizing cluster fuses.

### 4.6 Additional Cluster Based Protocols Network

In history, numerous clustering-based routing protocol concepts have been made. Data from sensor nodes that have been detected in close vicinity to the same collection are sent to the head. In order to ignore the last packet and merge the data from different quantification to the information sink, the collection head discards the two sets of information. The network's resilience and energy effectiveness are enhanced as a result.

Energy is a limited resource that quickly depletes in sensor nodes, especially in hostile environments where it is challenging to maintain or replenish the sensor's batteries once they have been deployed. As a result, the purpose of applications like air force surveillance and structural health monitoring (SHM) is to design an energy optimization routing approach for wireless networks [24].

The study of the clustering method is used to optimize the power. Energy generating sensor nodes that are near the destination and have a high likelihood of receiving energy are selected as the cluster head junction during clustering. The final non-elected node stores a portion of its harvested energy for use when it becomes the cluster leader. The cluster head nodes can thus endure longer by using this stored gathered energy. The shortest path is taken to route the information from cluster head nodes to the destination.

Additionally, in SHM applications, optimizing sensor node location is essential because doing so results in cost savings if the required information quality can be obtained with fewer sensor nodes. Therefore, the Energy boost with greater data Quality for SHM Application in Wireless Networks [25] implements the exhaustive search approach to carry out the sensor node position optimization.

## 5 Conclusion

Given how quickly technology is developing, WSN is becoming more and more common, and sensors are being used in a number of contexts such as smart grids. Despite being widely used, sensors are considered to be very sophisticated because to their low energy and memory requirements. The issue of energy efficiency is serious.

There have been several attempts to make the algorithm more energy-efficient, but the methods have not been successful since the algorithm is either very complex or overly basic for WSN. The data used in this investigation are recent.

The systematic literature review that was conducted to develop the research subject and review the literature is discussed in this survey. The purpose of this review is to evaluate WSN's energy effectiveness in various fields such as smart grid technology. The WSN environment is complex, and only a few study approaches are believed to be successful. Numerous research philosophies provide very energy-efficient techniques that increase network longevity and evenly distribute load across nodes. Only a small number of the load balancing and energy-efficient methods that have been considered are thought to function well for WSN energy. In future more number of routing methods can be developed for the conservation of energy in the smart grid.

## References


1. Jin X, Yibin Z, Wang X (2012) Strategy and coordinated development of strong and smart grid. https://doi.org/10.1109/ISGT-Asia.2012.6303208
2. Kant K, Gupta OH (2023) DC microgrid: a comprehensive review on protection challenges and schemes. https://doi.org/10.1080/02564602.2022.2136271
3. Yick J, Mukherjee B, Ghosal D (2008) Wireless sensor network survey. Comput Netw 52(12):2292–2330. https://doi.org/10.1016/j.comnet.2008.04.002
4. Shang F (2009) An energy-efficient adaptive clustering algorithm for wireless sensor networks. 6(2):117–126
5. Grover J, Shikha, Sharma M (2014) Location based protocols in wireless sensor network—a review. In: 2014 5th International conference computing communication and networking technologies (ICCCNT), pp 3–7. https://doi.org/10.1109/ICCCNT.2014.6962990
6. Krishnamachari B, Estrin D, Wicker S (2002) IEEE INFOCOM 2002 1 modelling data-centric routing in wireless sensor networks. In: IEEE infocom, vol 2, pp 39–44
7. Zabin F, Misra S, Woungang I, Rashvand HF, Ma NW, Ahsan Ali M (2008) REEP: data-centric, energy-efficient and reliable routing protocol for wireless sensor networks. IET Commun 2(8):995–1008. https://doi.org/10.1049/iet-com:20070424
8. Padigala JB, Umar S (2013) A review of QoS in wireless sensor networks. Int J Comput Sci Eng 3(8):273–278
9. Roy S, Das AK (2014) Energy efficient cluster based routing protocol (EECBRP) for wireless sensor network. In: Proceedings of the 1st international conference on networks & soft computing (ICNSC) 2014, pp 25–29. https://doi.org/10.1109/CNSC.2014.6906660
10. Khan MK, Shiraz M, Ghafoor KZ, Khan S, Sadiq AS, Ahmed G (2018) EE-MRP: energy-efficient multistage routing protocol for wireless sensor networks. Wirel Commun Mob Comput 2018. https://doi.org/10.1155/2018/6839671
11. Aslam M, Javaid N, Rahim A, Nazir U, Bibi A, Khan ZA (2012) Survey of extended LEACH-based clustering routing protocols for wireless sensor networks. In: Proceedings of the 14th IEEE international conference on high performance computing and communication HPCC-2012—9th IEEE international conference on embedded software and systems ICESS-2012, no July, pp 1232–1238. https://doi.org/10.1109/HPCC.2012.181
12. Gotefode K, Kolhe K (2015) Energy efficiency in wireless sensor network using Fuzzy rule and tree based routing protocol. In: International conference on energy systems and applications ICESA 2015, no Icesa, pp 712–717. https://doi.org/10.1109/ICESA.2015.7503442
13. Zhou H-Y, Luo D-Y, Gao Y, Zuo D-C (2011) Modeling of node energy consumption for wireless sensor networks. Wirel Sens Netw 3:18–23. Published Online January 2011. https://doi.org/10.4236/wsn.2011.31003
14. Kumar N, Singh Y (2017) Trust and packet load balancing based secure opportunistic routing protocol for WSN. In: 2017 4th IEEE international conference on signal processing, computing and control (ISPCC), pp 463–467. https://doi.org/10.1109/ISPCC.2017.8269723
15. Bahale SV, Pund MA, Sci C, Amravati B (2017) Performance evaluation of SCERP: a cluster based network lifetime improvement protocol for WSN. Int Res J Eng Technol 4(4):2460–2466. https://www.irjet.net/archives/V4/i4/IRJET-V4I4615.pdf
16. Shen J, Wang A, Wang C, Ren Y, Wang J (2015) Performance comparison of typical and improved LEACH protocols in wireless sensor network. In: Proceedings of the 2015 1st international conference on computational intelligence theory, systems and applications (CCITSA), pp 187–192. https://doi.org/10.1109/CCITSA.2015.44

17. Pang BM, Shi HS, Li YX (2012) An energy-efficient MAC protocol for wireless sensor network. In: Lecture notes on electrical engineering, LNEE, vol 143, pp 163–170. https://doi.org/10.1007/978-3-642-27323-0_21
18. Handy MJ, Haase M, Timmermann D (2002) Low energy adaptive clustering hierarchy with deterministic cluster-head selection. In: 2002 4th International workshop on mobile and wireless communications network (MWCN), pp 368–372. https://doi.org/10.1109/MWCN.2002.1045790
19. Lindsey S, Raghavendra CS (2002) PEGASIS: power-efficient gathering in sensor information systems. In: IEEE aerospace conference proceedings, vol 3, pp 1125–1130. https://doi.org/10.1109/AERO.2002.1035242
20. Kashaf A, Javaid N, Khan ZA, Khan IA (2012) TSEP: threshold-sensitive stable election protocol for WSNs. In: Proceedings of the 2012 10th international conference on frontiers of information technology FIT 2012, pp 164–168. https://doi.org/10.1109/FIT.2012.37
21. Li Y, Guo L, Prasad SK (2010) An energy-efficient distributed algorithm for minimum-latency aggregation scheduling in wireless sensor networks. In: Proceedings of the international conference on distributed computing systems, pp 827–836. https://doi.org/10.1109/ICDCS.2010.11
22. Lin CH, Tsai MJ (2006) A comment on 'HEED: a hybrid, energy-efficient, distributed clustering approach for ad hoc sensor networks.' IEEE Trans Mob Comput 5(10):1471–1472. https://doi.org/10.1109/TMC.2006.141
23. Cohen R, Kapchits B (2011) Energy-delay optimization in an asynchronous sensor network with multiple gateways. In: 2011 8th Annual IEEE communications society conference on sensor, mesh and ad hoc communications and networks, SECON 2011, pp 98–106. https://doi.org/10.1109/SAHCN.2011.5984953
24. Elsersy M, Elfouly TM, Ahmed MH (2016) Joint optimal placement, routing, and flow assignment in wireless sensor networks for structural health monitoring. IEEE Sens J 16(12):5095–5106. https://doi.org/10.1109/JSEN.2016.2554462
25. Geoffrine Judith MC, Geetha V (2019) Energy optimization with higher information quality for SHM application in wireless sensor networks. IEEE Sens J 19(9):3513–3520. https://doi.org/10.1109/JSEN.2019.2892870

# Wireless Sensor Network for Condition Monitoring of Axel Counter Device in Railways

**Rajev Kumar Sharma, Ajay Sharma, and Sandeep Kanaujia**

**Abstract** Axel counter device plays a pivotal role in safe operation of train. ACD is fixed on track-side to count the axels of train boggies. It works on the principle of cutting of magnetic flux between two coils. It counts the axels with passing of wheel-flange through the passage of two coils of ACD fitted on track. It is employed in station area or yard area which simultaneously monitors the occupancy of different track sections as approach track, main line, loop line, berthing line, common loop line, point zone, washing line, Siding line etc. ACD with high availability is also installed outside station area to push the train in the block section to give line clearance from rear station for the next train in sequence to avoid any unusual incident in the block section. This helps in designing a standard interlocking system to give signal on best paths for train movement in busy routes for premium trains (like Rajdhani Express train on Indian Railway Network). Presently ACDs are monitored with wired network such as Copper quad cables, OFC, Two core Power cables, give poor insulation to OHE induced voltage in Railway Electrified area. Also wires and cables are prone to be damaged during excavation activities along the track, and use of these cables results in loss of power and redundancy in data. This paper proposes a technology to replace physical wired network with Wireless Sensor Network technology. Adhoc wireless sensor network improves the availability of train data and reduces redundancy, failures and reduces power requirement. With the help of various routing methods, an energy efficient WSN is developed which can be employed in accessible areas, reducing human interaction.

**Keywords** DP (Detection Point) · ACD (Axel Control Devices) · MSSDAC (Multi Section Digital Axel Counter) · OHE (Over Head Equipment) · OFC (Optical Fibre Cable) · IoT (Internet of Thing)

R. K. Sharma (✉) · A. Sharma · S. Kanaujia
Department of Electronics and Communication Engineering, United University, Prayagraj, India
e-mail: rajeevsharmaiete@gmail.com

A. Sharma
e-mail: ajaysharma@united.ac.in


O. H. Gupta et al. (eds.), *Soft Computing Applications in Modern Power and Energy Systems*, Lecture Notes in Electrical Engineering 1107,
https://doi.org/10.1007/978-981-99-8007-9_9

## 1 Introduction

Railway ACD is used to identify the track occupancy or track clearance of the section. Axel counters are mounted on the rail. ACD is used in pair at a single location, operated at different frequencies to check the reference count direction (RCD). Combination of two detection points defines the limit of the section. When the number of axel-count-pulses entering-DP of the section is equal to the number of axel-count-pulses at exit DP of the section then the section is declared free, otherwise the section is considered as occupied is installed on the railway track so that it can monitor station line Main line, Loop Line, Berthing Line, point-zone, stabling line Siding , Washing line etc. to check the occupancy of each section.

The main parts of Axel Counter Devices are (i) Axel Detector (EM-sensor) (ii) Track side unit (iii) Central Evaluator (iv) Line Verification Box, (v) Relay Unit (vi) Event Logger (vii) Diagnostic Terminal.

## 2 Requirements for Remote Supervision

To achieve a preventive maintenance of this vital component, real time supervision of information is required [1, 2]. Various parameters like battery voltage, frequency, insulation, poor dip-count, are to be monitored on top priority. Poor insulation of copper cable connecting the trackside equipment to the central evaluator gives poor dip or corrupted bit count of the wheel, which is very well improved with wireless data transmission. The remote supervision through ad hoc-WSN will help the Railway system to be more robust, error free and reliable and fail-safe operation of traffic. The remote supervision proposed here will not have any influence on main functionality of an axel counter maintaining the Integrity of the Specifications.

## 3 Literature Review on Condition Monitoring

A drastic development in condition monitoring of various railway subsystems has taken place in the recent past. The innovating medium on the railway signalling system [3] has been introduced using a typical control system which takes data from switch, track circuit, Signal aspect and finally transfer data to route control and traffic control. Design optimization of magnetic Sensors [4] is proposed for moving wheel detection. Magnetic phase reversal principle has been utilized by the trac device which detects the physical crossing of the wheel through detection point and counts axel of the train. A railway axel counter prototype using a microcontroller [1] has been introduced that fulfils various train safety standards, like IEC61508, SIL (safety Integrity Level) etc; here a modular structure with several control boards for input, output, display, optocoupler, monitoring, modem is proposed for the train control system. Railway Axel counter remote supervision system [2] is presented with wifi, GPRS, FreeRTOs. A train tracking and collision avoidance system [3] has been introduced using android and WSN [5]. A model control room in android platform is introduced so the data of the moving train is exchanged using internet and WSN; here an algorithm using Zigbee module stops the train in case of head on collision if trains are moving on the same track in the opposite direction. A typical control mechanism [3] is proposed in this paper, which synchronizes signal control, route control and traffic control. A different simulation model [6] and performance study of WSN is elaborated using MATLAB code. A signal quality analysis is given to study interference versus BER/FER. A handbook of CAMTECH on the installation procedure of GGtronics & CEL make the Axel counter device [7, 8] to give a clear overview of internal components where sensors can be introduced.

## 4 Basic Working Principle of an Axel Counter

It works on deviation of magnetic field [4] between two coils (one is transmitting coil and other is receiving coil).

When the flange of the wheel crossed the passage between two coils mounted on the track itself, it changes the direction of the flux. The phase reversal of the induced voltage on the receiver coil, gives a dip in voltage proving the existence of wheel at DP (Fig. 1).

The Fig. 2 shows the principle of detection of wheel through induced field voltage waveform received at the receiver coil.

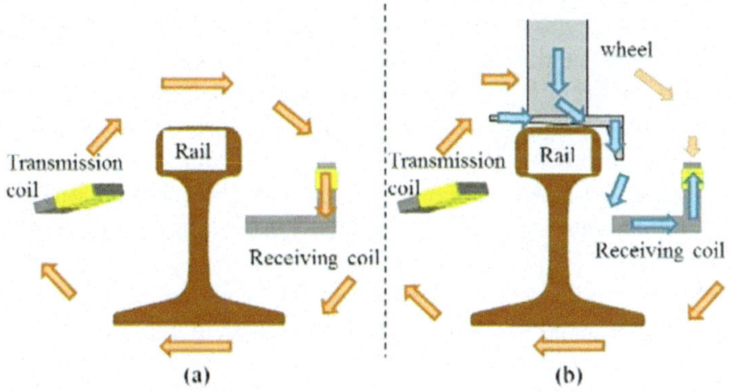


**Fig. 1** Basic principle of wheel detection

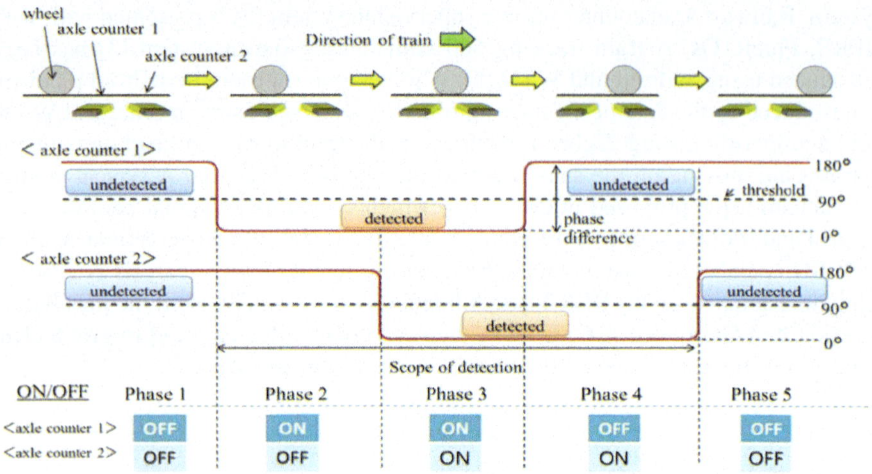


**Fig. 2** Phase reversal and dip detection waveform

## 5 Proposed Architecture Wireless Sensor Network

The wireless Sensor Node is placed at each detection point in the section. Figure 3 shows a section of two DPs constituting a small section of the railway track using single detection.

The counts are checked at both ends—entry and exit points. However, for complex railway yards having more than 50 routes, a higher level of topology can be designed in bigger railway yards as seen in Fig. 1. The sensor nodes are located exactly in close vicinity of DP keeping schedule of the dimension from the centre of the track to avoid infringement with traffic on that part of the track. As DPs are configured to the control track section, WSN nodes are also kept in tandem with them. Now this group of WSN -node makes an adhoc wireless sensor network to operate under a suitable network protocols shown in figure. The layout shows two Axel Counter Track Sections (ATS-1 & ATS-2) controlled by different combinations of associated DPs. Figure 4 illustrates a section of two DPs [7] constituting a small section of the railway track [8].

Here a Wireless node is attached to each DP and fed with individual energy backup, constituting an ad hoc wireless network given in Table 1.

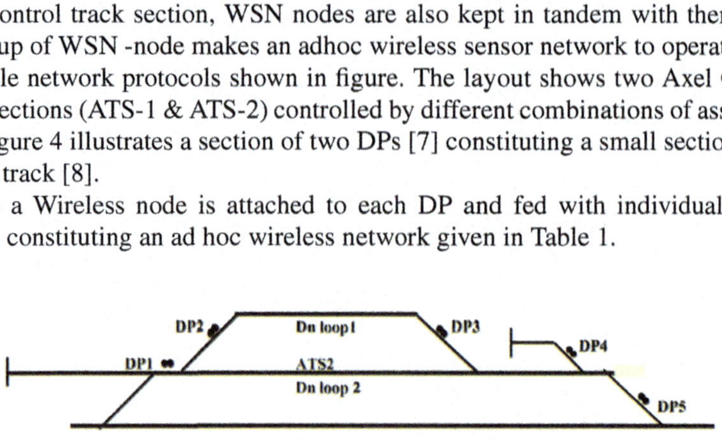


**Fig. 3** Axel counter tack sections comprising various routes

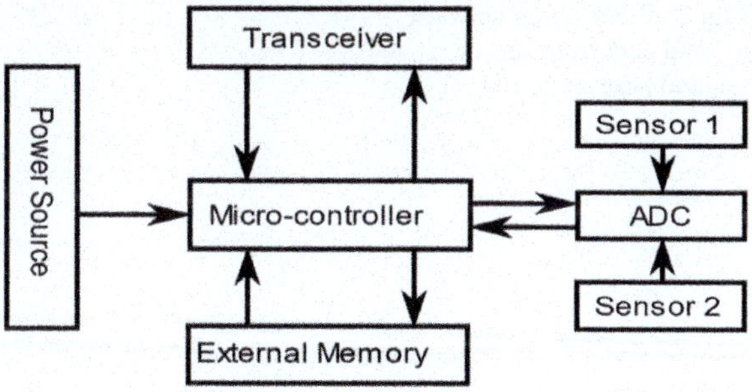


**Fig. 4** Wireless node basic building block

**Table 1** Detection point versus wireless sensor node configuration table

| S. No. | Axel counter track section | Controlling DP | Controlling nodes |
|---|---|---|---|
| 1 | ACTS-1 | DP2-DP3 | N2-N1 |
| 2 | ACTS-2 | DP1-DP5 | N1-N5 |
| | | DP1-DP2 | N1-N2 |
| | | DP4-DP5 | N4-N5 |
| | | DP3-DP5 | N3-N5 |

The block diagram of the WSN Node consists of the Power Source, Transceivers, Micro-controller, External Memory, Sensor, and ADC, as shown in Fig. 4. Each Point Zone has a group of DPs attached with a sensor node constitutes a cluster, each cluster is finally connected to the base station at station. The evaluator will compare the counts of each DP and declare if ATS is occupied or not. The wireless node comprises of a power source, Microcontroller, Sensor, ADC, external memory and transceiver as shown in Fig. 4.

The central evaluator is connected to the Field unit through a copper cable as shown in Fig. 5. The main disadvantage is the Physical media of carrying the low power bits which is prone to error in received data resulting in wrong count of Axels. The power requirement is also 12 W per detection point, 24 W for central evaluator (16 DP model) station master panel. Cable requirement is Half Quad (Cooper cable) from the Central Evaluator t of each DP and 2 core 6 sq mm or 10 sq mm copper cable from the battery bank to the Central Evaluator, and 2 core shielded cables from that the CT rack to central evaluator, which consume heavy construction cost.

The proposed WSN adhoc network which replaces the copper wire (to connect the Field DP to central Evaluator) using WSN-Cluster heads as shown in Fig. 5.

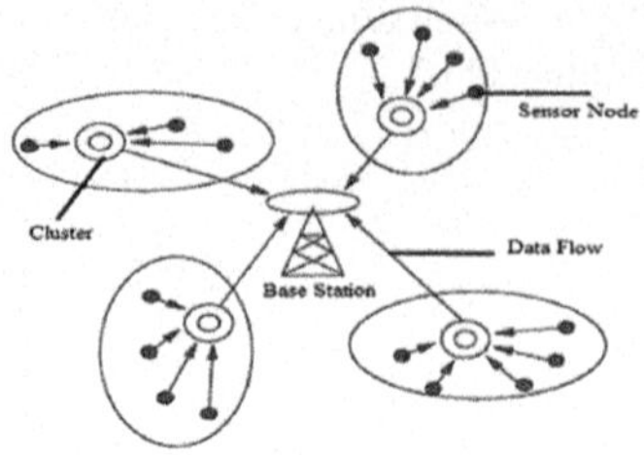


**Fig. 5** WSN-based network to send data from sensor node to base station

# 6 Advantages of Proposed System

Railway signalling system is a complex architecture which involves signal aspect control, point machine control, and track circuit control to give exact selection of a particular track to the train loco pilot. The quick decision of the exact path for the train through bigger yards and busy routes (like Rajdhani express train on Indian Railway network) is a cumbersome exercise. In the present study the interlocking system is governed with Copper cables and relays which are prone to heavy loss of signalling data and frequent failure of track, signal or point operation, resulting in delay of the train or unusual incidents like head on collision, derailment, accidents due to wrong selection of track [3] etc. The proposed system enhanced with the Ad hoc Wireless sensor Network has the following advantages [4].

## *6.1 Low Construction Cost*

Cable laying is a tedious exercise in the yard area and in block section involving a huge capital expenditure. Thousands of cables are laid underground in the yard area, which makes the civil structure like track formation weak due to frequent digging under the track. However WSN node deployment required onetime investment without disturbing track formation, resulting in low construction cost.

## *6.2 Error Free Signalling Data*

Copper cables are very much prone to cut due to earth work executed by civil agencies throughout the year. This results in the following: the poor merging of the cable results in insulation loss that results in heavy voltage loss, which is the main reason of error in the received information at the interlocking system. Human error during cable connection are minimized.

## 6.3 Low Down Time of Train

The restoration of failure due to cable cut is a tedious task, sometimes it becomes impossible to detect the location of the fault, especially at night. The engineer has to walk along the track with the cable route tracer device to detect the cable-cut, and again lot of time is involved to reconnect the cable which results in train detention. Long detention of the premium passenger train is very difficult to manage the railway traffic to be pushed in sequence for higher transport management. The restoration of the Adhoc WSN node is much easier and takes lesser time without hampering train operation.

## 6.4 Low Demand of Block Period

Block is the time slot required by the restoration agency from traffic control. During this period no train movement is permitted on the faulty track section, resulting in huge revenue loss to railways due to under-utilization of the line capacity, which is overcome upto a great extent in case of the proposed WSN network.

## 6.5 Energy Efficient system

A heavy power source is required for continuous operation. Integrated power supply with redundant voltage sources with dedicated power backup is an essential for continuous train operation via copper cables. However WSN node utilises an optimum power source through the energy efficient routing protocol.

## 6.6 Ease of Installation and Maintenance

In comparison to the existing signalling system on physical media, WSN based adhoc network is easy to deploy the sensor node reducing the initial construction cost. Repair and maintenance is much easier without hampering train operation because no excavation work along the track is required

## 7 Conclusion

This paper introduces an energy efficient and secure Wireless Sensor Network in the Signalling subsystem, which will enhance the safety of the train movement and help the engineer to attend failures more efficiently. However WSNs are an emerging technology which will continue to grow exponentially in the coming years, Network optimization in terms of bandwidth, latency, economics, power efficiency through energy-efficient routing, load balancing to distribute power consumption, network aggregation to reduce traffic load, and minimization of up-time of sensors, requires new network technologies, different routing algorithms, new MAC protocols, which play very crucial role in the design of the Ad hoc WSN. Therefore this new communication paradigm introduces a new set of design constraints, where the design must be extremely low-cost both for installation and operation.

**Future scope**: Modern communication technology may be used to integrate the WSN with the outer world through Internet. Field data can be stored for fault analysis over handheld device through IoT.

## References

1. Nikolic MV, Kosic BD, Mianovic MD, Antonic NM, Stojkovic ZM, Kokic IZ (2014) Railway axel counter prototype. In: 22nd telecommunication forum TELFOR 2014
2. Kokic IZ, Nikolic MV, Kosic BD, Mianovic MD, Antonic NM, Stojkovic ZM (2014) Railway axel counter remote supervision system. In: 22nd telecommunication forum TELFOR 2014
3. Grover J, Anajli (2015) Wireless sensor network in railway signalling system. In: Fifth international conference on communication system and network technology 2015
4. Yusukawa S, Takagi N, Dong G, Wakao S, Takahashi M, Yagi M, Okutani T (2015) Design optimization of magnetic sensor for train detection. IEEE Trans Magn 51(3)
5. Immanuel Rajkumar R, Sundari G (2017) Android & WSN based train tracking and collision avoidance for railway sector. In: IEEE international conference on computational intelligence and computational research (ICCIC)
6. Ali QI, Abdulmaowjod A, Mohammed HM (2011) Simulation & performance study of wireless sensor network (WSN) using MATLAB. Iraq J Electr Electron Eng 7(2)
7. http://www.ggtronics.com
8. http://www.celindia.co.in/multisection-digital-axel-counter

# PID Based Optimal Neural Control of Single Wheel Robot (SWR)

**Ashwani Kharola, Vishwjeet Choudhary, Rahul, Sankula Madhava, Abhijat, and Aditya Nagalia**

**Abstract** This study investigates the application of Proportional-integral-derivative (PID) based Artificial neural network (ANN) controller for the stabilisation of highly nonlinear and multivariable Single wheel robot (SWR). The nonlinear governing equations of motion for the proposed system were derived using Newton's second law. The gains of PID controller were optimised using auto-tuning function. The results of PID were used for training of ANN controller designed using optimal number of neurons in the hidden layer. The performance of the proposed controller was measured in terms of settling time, overshoot ranges and steady state error responses. The results indicate superior performance of ANN controller designed using 30 neurons in the hidden layer.

**Keywords** SWR · Nonlinear control · PID · ANN · Modelling · Simulation

## 1 Introduction

Wheeled robotic systems are popular nonlinear systems which find immense applications in various fields of Science and Engineering [1]. The unique capability of these systems to take sharp turns and navigate through congested areas makes them more agile compared to conventional legged robotic systems [2]. Plenty of work has been done by researchers for adequate control of these highly nonlinear systems [3]. For an instance, a hybrid controller based on a combination of Particle swarm optimisation technique and PID for stabilisation of mobile humanoid robot has been proposed

A. Kharola (✉) · S. Madhava · Abhijat · A. Nagalia
Department of Mechanical Engineering, Graphic Era (Deemed to be University), Dehradun, India
e-mail: ashwanidaa@gmail.com

V. Choudhary
Department of Mechanical Engineering, BITS-Pilani, Pilani, India

Rahul
Department of Electronics, Sri Venkateswara College, University of Delhi, New Delhi, India
e-mail: rahul_electronics@svc.ac.in



O. H. Gupta et al. (eds.), *Soft Computing Applications in Modern Power and Energy Systems*, Lecture Notes in Electrical Engineering 1107,
https://doi.org/10.1007/978-981-99-8007-9_10

in [4]. Incorporation of particle swarm optimisation technique with PID aided in minimising the computational complexity and reduced the number of iterations. The results indicate that combining the above two techniques resulted in reducing the stabilisation time and overshoot responses by 25%.

Further, a novel five degree of freedom two-wheeled robot suitable for both industrial and service applications has been proposed in [5]. The authors adopted Lagrangian modelling approach to develop a mathematical model of the proposed system and further simulated it in Matlab/Simulink platform. The proposed system has been finally stabilised using PID and fuzzy logic control strategies. Moreover, different control strategies namely PID, Genetic algorithm tuned PID and Model predictive control have been considered for realistic control of Segway system in [6]. The study further adopted an online optimisation technique which aided in solving the dynamic optimisation problem in model predictive control effectively.

A fuzzy fractional-order PID controller for stabilisation of two-wheel robot on an inclined surface has been proposed in [7]. The primary objective of the study was to realise the wheel position control and stabilise the non-vertical direction of the body. The controller has been further compared with conventional fractional-order PID controller and the results indicate superior performance of fuzzy fractional-order PID controller. In this study, a PID controller has been designed for control of SWR moving on a horizontal surface. A mathematical model of the proposed system has been initially developed using Newton's second law of motion. The gains of PID controller have been optimised using auto-tuning function in Matlab. The outcomes of PID controller were further used for training of ANN controller. In order to optimise the performance of ANN controller, the number of neurons ($n$) in the hidden layer has been varied i.e. $n = 10$, 20 and 30 respectively. Finally, the performance of the proposed controller has been monitored through responses obtained for settling time, overshoot and steady state error.

## 2 Mathematical Model and Simulink of SWR

A SWR comprises of robot chassis mounted on a single wheel. The robot chassis is inclined at an angle ($\theta$) from vertical as shown in Fig. 1.

The objective is to balance the robot wheel and chassis at desired position ($x$) and angle ($\theta$) respectively under the influence of external control force ($F$) generated through controller. The final derived equations for linear acceleration of the wheel ($\ddot{x}$) and angular acceleration of the chassis ($\ddot{\theta}$) are shown below [8].

$$\ddot{x} = \frac{-M^2L^2g\theta}{M_s(ML^2 + I) - M^2L^2} + \frac{(M^2L^2 + I)F}{M_s(ML^2 + I) - M^2L^2} \tag{1}$$

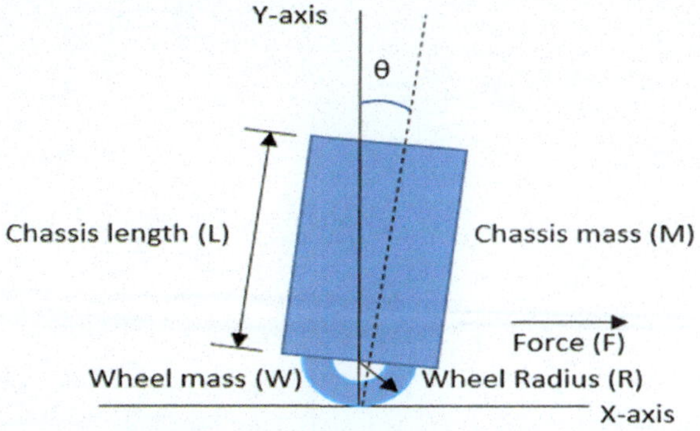


**Fig.1** Schematic view of SWR

**Table 1** Attributes considered for designing of SWR

| Symbol | Value |
|---|---|
| Mass of chassis ($M$) | 1 kg |
| Mass of wheels ($W$) | 0.25 kg |
| Length of chassis ($L$) | 0.1 m |
| Radius of wheel ($R$) | 0.05 m |
| Wheel moment of inertia ($I$) | 0.0123 kg $m^2$ |
| Pendulum moment of inertia ($J$) | 0.1984 kg $m^2$ |
| Acceleration due to gravity ($g$) | 9.81 $m/s^2$ |

$$\ddot{\theta} = \frac{MgLM_s\theta}{M_s(ML^2 + I) - M^2L^2)} - \frac{MLF}{M_s(ML^2 + I) - M^2L^2} \quad (2)$$

where, $M_s = M + W + \frac{J}{R^2}$.

The attributes along with symbols used in the above equations are highlighted with the help of Table 1.

The above equations were further used for developing a Simulink model of SWR as shown in Fig. 2.

## 3 Designing of PID Based ANN Controller for SWR

PID is a closed-loop control feedback mechanism commonly used in various industrial and control applications [9]. It finds vast application in problems involving continuous modulated control [10]. A PID controller evaluates an error value $e(t)$ as the difference between desired output and measured output and further applies a correction based on proportional ($K_p$), integral ($K_i$) and derivative ($K_d$) terms [11]. Mathematically, a PID controller can be represented with the help of Eq. (3).

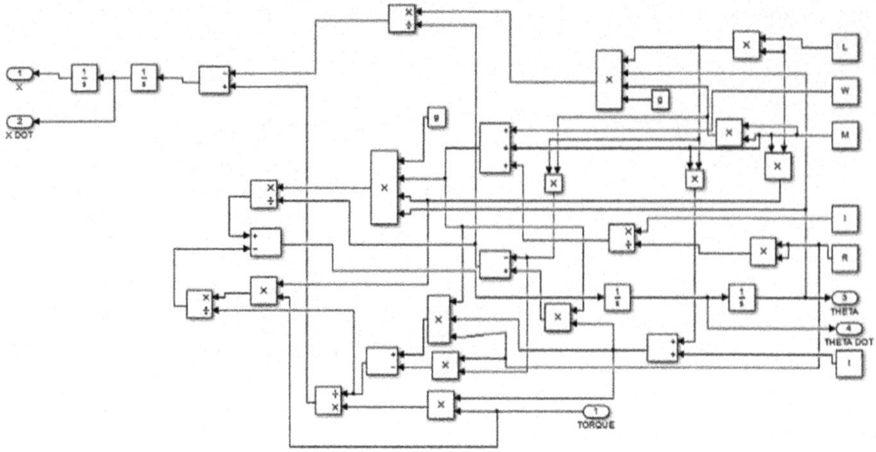


**Fig. 2** Simulink model of SWR

**Table 2** PID gains obtained after auto-tuning

| Controller | $K_p$ | $K_i$ | $K_d$ |
|---|---|---|---|
| Chassis | −851.832 | −1142.019 | −60.813 |
| Wheel | −101.019 | −0.786 | −22.244 |

$$u(t) = K_p e(t) + K_i \int_0^t e(\tau)d\tau + K_d \frac{de(t)}{dt} \tag{3}$$

In the above equation $K_p$, $K_i$ and $K_d$ are the coefficients of proportional, integral and derivative gains respectively and $u(t)$ denotes the control variable. In this study, two separate PID controllers were designed for control of robot chassis and wheel respectively. The PID gains obtained after tuning are shown with the help of Table 2.

The ANN controller has been trained using Levenberg–Marquardt algorithm using 790 data samples which were randomly divided into training (552 nos.), validation (119 nos.) and testing (119 nos.) sets [12]. ANN controller has been trained using different number of neurons in the hidden layer and it has been observed that the best validation performance of 0.0052208 has been obtained using 30 neurons in the hidden layer at 14 epochs as shown in Fig. 3 [13]. Similarly, the values obtained for gradient, *Mu* and validation checks are shown with the help of Fig. 4. The gradient represents slope of the curve and helps in determining the direction of the curve a few points ahead, *Mu* is a control parameter which directly affects the convergence rate of the network [14]. Whereas, validation checks help in stopping iteration when the error value consecutively increases for 6 epochs.

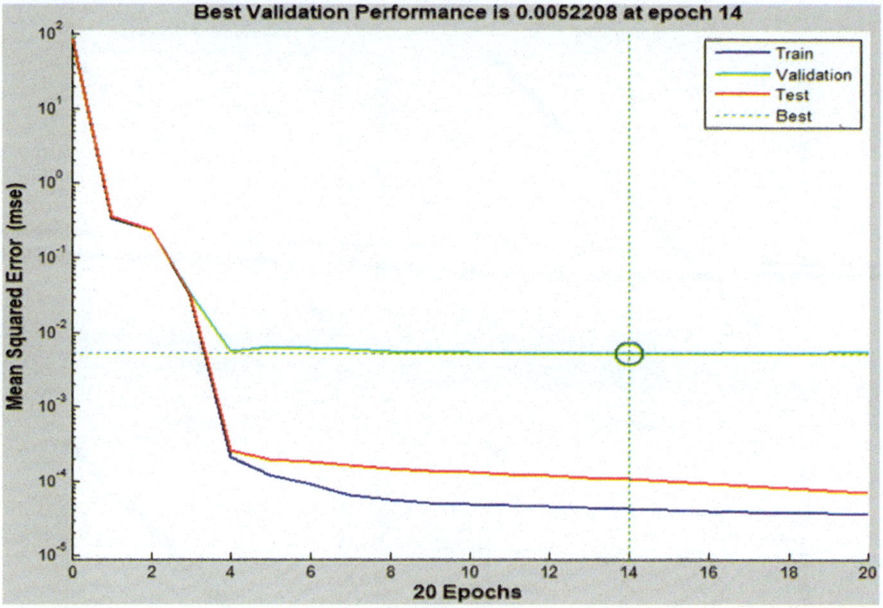


**Fig. 3** Best validation performance obtained after 14 epochs

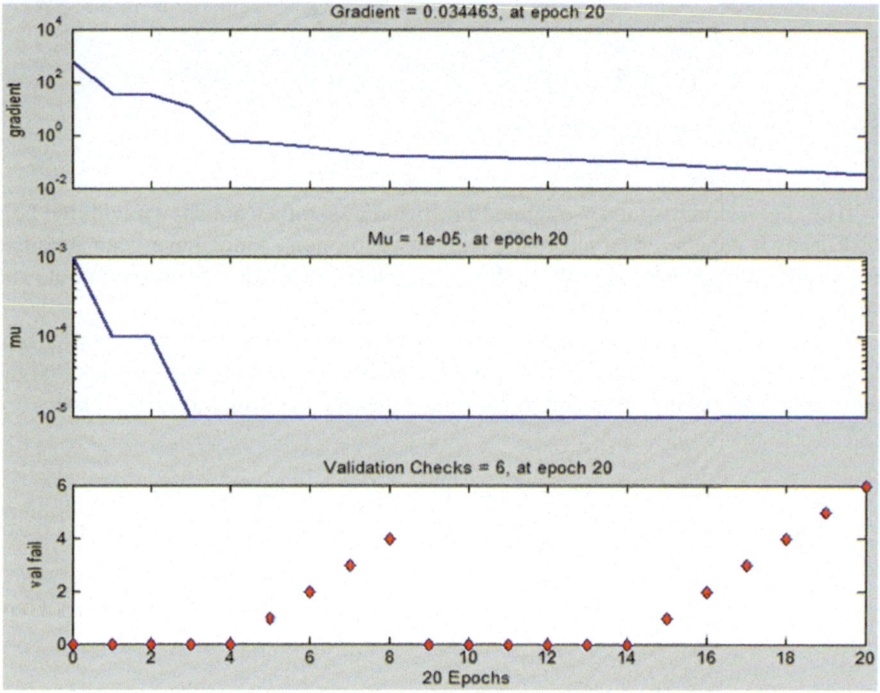


**Fig. 4** Values of gradient, *Mu* and validation checks obtained after training

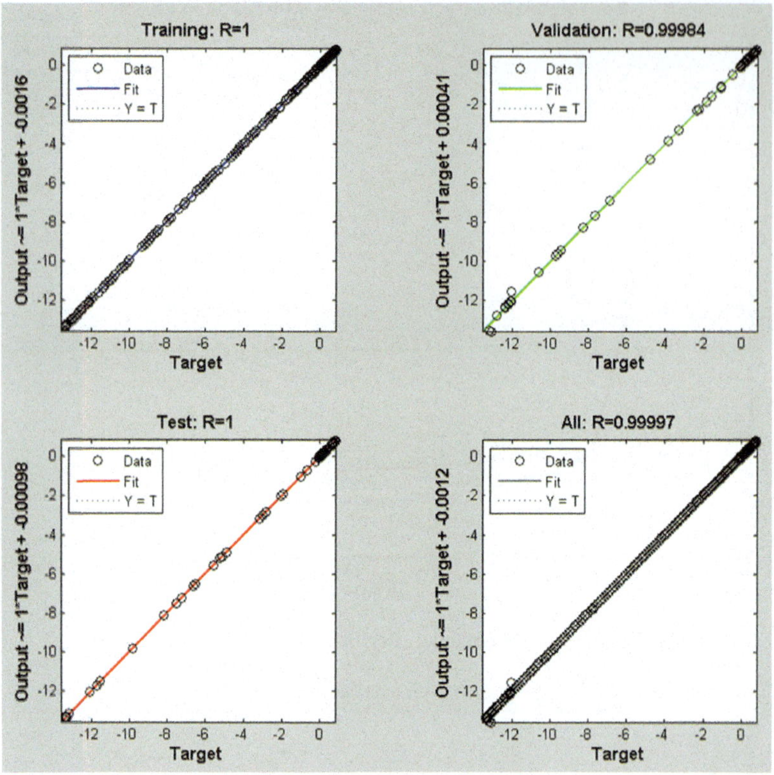


**Fig. 5** Regression results obtained after training

The regression responses obtained for different samples are shown with the help of Fig. 5. It can be observed that excellent regression values have been obtained for all the categories of samples indicating a close correlation between outputs and targets [15].

The error histogram which represents histogram of errors along 20 smaller bins is shown with the help of Fig. 6. It can be clearly observed from the results that the zero error line i.e. line with zero error falls under the bin with value 0.007141.

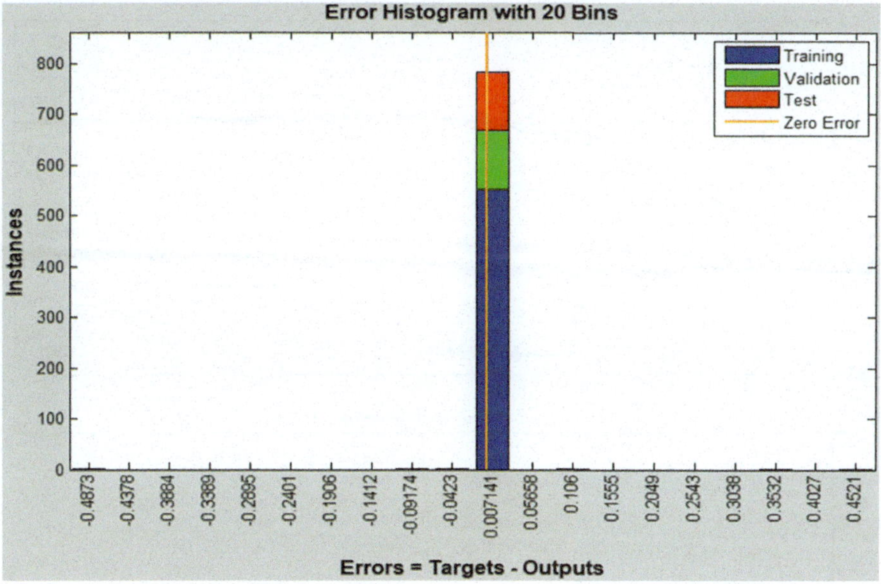


**Fig. 6** Error histogram obtained after training

## 4 Results and Comparison

The responses obtained for SWR ($t = 5$ s) using ANN controllers with different number of neurons in hidden layer are shown below with the help of Fig. 7.

A comparison of simulation results for ANN controller designed with different number of neurons in the hidden layer is shown with the help of Table 3.

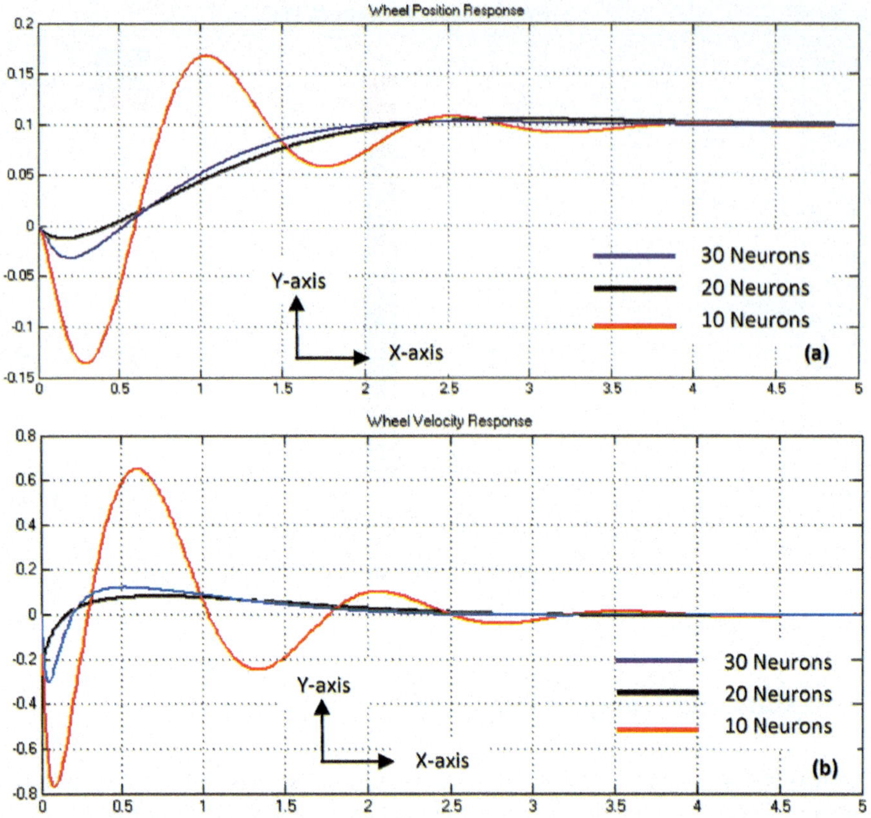


**Fig. 7** Simulation responses for **a** Wheel position, **b** Wheel velocity, **c** Chassis angle, **d** Chassis angular velocity

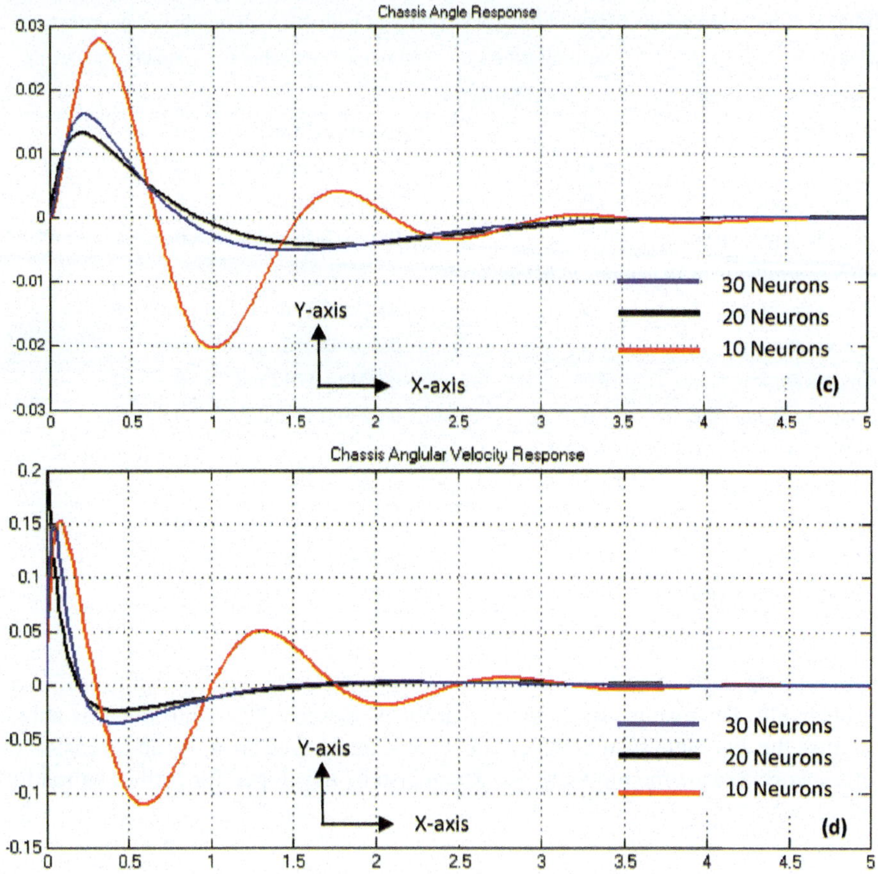


**Fig. 7** (continued)

## 5 Conclusion

The study successfully considers a PID based optimal ANN controller for stabilisation of highly nonlinear SWR. A mathematical model of the proposed system has been developed using Newton's second law. The gains of PID controller have been successfully optimised using auto-tuned function in Matlab. The results clearly indicate superior performance of ANN controller designed using 30 neurons in the hidden layer both in terms of settling time and maximum overshoot responses. The settling time taken by ANN controller designed using 30 neurons in hidden layer for stabilisation of wheel position, wheel velocity, chassis angle and chassis angular velocity was 2.0, 1.0, 3.0 and 1.5 s respectively. Thus, the proposed ANN controller was able to stabilise the complete system within 3.0 s. Further, the maximum overshoot range responses obtained using proposed controller for output wheel position were −0.025 to 0.1 m, for wheel velocity were −0.3 to 0.1 m/s, for chassis angle

**Table 3** Simulation results obtained for ANN controller with different number of neurons

| Output | Settling time (s) | Maximum overshoot | Steady state error |
|---|---|---|---|
| *ANN controller with 30 neurons in hidden layer* | | | |
| Wheel position | 2.0 | −0.025 to 0.1 m | 0 |
| Wheel velocity | 1.0 | −0.3 to 0.1 m/s | 0 |
| Chassis angle | 3.0 | −0.005 to 0.015 rad | 0 |
| Chassis angular velocity | 1.5 | −0.03 to 0.15 rad/s | 0 |
| *ANN controller with 20 neurons in hidden layer* | | | |
| Wheel position | 2.5 | −0.01 to 0.1 m | 0 |
| Wheel velocity | 1.5 | −0.2 to 0.8 m/s | 0 |
| Chassis angle | 3.5 | −0.004 to 0.012 rad | 0 |
| Chassis angular velocity | 2.0 | −0.02 to 0.17 rad/s | 0 |
| *ANN controller with 10 neurons in hidden layer* | | | |
| Wheel position | 3.6 | −0.14 to 0.17 m | 0 |
| Wheel velocity | 3.5 | −0.75 to 0.65 m/s | 0 |
| Chassis angle | 4.0 | −0.02 to 0.028 rad | 0 |
| Chassis angular velocity | 3.0 | −0.12 to 0.15 rad/s | 0 |

were −0.005 to 0.015 rad and for chassis angular velocity were −0.03 to 0.15 rad/s respectively. The results also indicate that the proposed ANN controller was able to stabilise the complete system with almost zero steady state error. As an extension for future work a real-time model of SWR can also be developed for further validation.

## References

1. Zhang C, Liu T, Song S, Wang J, Meng MQH (2022) Dynamic wheeled motion control of wheel-biped transformable robots. Biomim Intell Robot 2(2):1–8
2. Liu D, Wang J, Lei T, Wang S (2022) Active suspension control with consensus strategy for dynamic posture tracking of wheel-legged robotic systems on uneven surfaces. ISA Trans 131:628–638
3. Pajaziti A, Gara L (2019) Navigation of self-balancing mobile robot through sensors. IFAC-PapersOnLine 52(25):429–434
4. Kashyap AK, Parhi DR (2021) Particle swarm optimization aided PID gait controller design for a humanoid robot. ISA Trans 114:306–330
5. Goher KM, Fadlallah SO (2019) Control of a two-wheeled machine with two-directions handling mechanism using PID and PD-FLC algorithms. Int J Autom Comput 16:511–533
6. Prabhakar G, Selvaperumal S, Pugazhenthi PN, Umamaheswari K, Elamurugan P (2020) Online optimization based model predictive control on two wheel Segway system. Mater Today Proc 33(7):3846–3853
7. Zhang J, Zhao T, Guo B, Dian S (2021) Fuzzy fractional-order PID control for two-wheeled self-balancing robots on inclined road surface. Syst Sci Control Eng 10(1):289–299
8. Huang CN (2010) The development of self-balancing controller for one-wheeled vehicles. Engineering 2:212–219

9. Narasimha PT, Devakirubakaran S, Muthubalaji S, Srinivasan S, Karthikeyan B, Palanisamy R, Bajaj M, Hossam Z, Salah K (2022) Power management in hybrid ANFIS PID based AC–DC microgrids with EHO based cost optimized droop control strategy. Energy Rep 8:15081–15094
10. Carlucho I, Paula MD, Acosta GG (2019) Double Q-PID algorithm for mobile robot control. Expert Syst Appl 137:292–307
11. Juneja PK, Sunori SK, Sharma A, Sharma A, Kumar P, Joshi V, Raizada R (2021) Comparison of closed loop performance of consistency process dynamics for various PID controller algorithms. J Phys Conf Ser 2070(1):1–12
12. Al-Mayyahi A, Wang W, Birch P (2015) Levenberg-Marquardt optimised neural networks for trajectory tracking of autonomous ground vehicles. Int J Mechatron Autom 5(2):140–153
13. Ogunbo JN, Alagbe OA, Oladapo MI, Shin C (2020) N-hidden layer artificial neural network architecture computer code: geophysical application example. Heliyon 6(6):1–9
14. Lv L, Chen J, Zhang L, Zhang F (2022) Gradient-based neural networks for solving periodic Sylvester matrix equations. J Franklin Inst 359(18):10849–10866
15. Zhao D, Wang Y, Liang D, Ivanov M (2020) Performances of regression model and artificial neural network in monitoring welding quality based on power signal. J Market Res 9(2):1231–1240

# Impact of Instrument Transformer Secondary Connections on Performance of Protection System—Analysis of Field Events from Indian Power Sector

**Pankaj Kumar Jha, M. S. Hada, and Sandeep Yadav**

**Abstract** Protective relays are commonly connected to the secondary windings of instrument transformers i.e., current transformers (CTs), and/or capacitive voltage transformers (CVTs). The purpose of CT and CVT is to provide galvanic isolation from high voltages and reduce primary currents and voltages to a nominal quantity recognized by the protective relays. Selecting the correct instrument transformers for an application is imperative: failing to do so may compromise the relay's performance, as the output of the instrument transformer may no longer be an accurately scaled representation of the primary quantity. Having accurately rated instrument transformer is of no use if these devices are not properly connected. The performance of the protective relay is reliant on its programmed settings and on the current and voltage inputs from the instrument transformers secondary. This paper will help in understanding the fundamental concepts of the connections of instrument transformers to the protection relays and the effect of incorrect connection on the performance of protective relays. Multiple real-world case studies of protection system mis-operations due to incorrect connections of instrument transformers are discussed in detail in this paper. All the case studies presented in this paper are the actual field events of high-voltage AC transmission networks of Indian power transmission sectors. Apart from the connection issue of instrument transformers to protective relays, this paper also discusses about the effect of multiple grounding of CTs and CVTs secondary on the performance of the protection system. We hope that the case studies presented in this paper will help the readers to analyze the problem through real-world challenges in the complex power system networks.

P. K. Jha (✉)
Power Grid Corporation of India Ltd, New Delhi, India
e-mail: pankaj.jha@powergrid.in

M. S. Hada
Power Grid Corporation of India Ltd, Gurugram, India

S. Yadav
Power Grid Corporation of India Ltd, Faridabad, India

 
O. H. Gupta et al. (eds.), *Soft Computing Applications in Modern Power and Energy Systems*, Lecture Notes in Electrical Engineering 1107,
https://doi.org/10.1007/978-981-99-8007-9_11

**Keywords** Auto transformer · CT · CVT · Instrument Transformer · Grounding · Distance Relay · Directional Relay · Differential Relay · REF · VT Fuse Failure · Disturbance Report · DR · Voltage Selection relay · Shunt Reactor

# 1 Introduction

Protective relays are commonly connected to the secondary windings of instrument transformers i.e., current transformers (CTs), and/or capacitive voltage transformers (CVTs). The purpose of CT and CVT is to provide galvanic isolation from high voltages and reduce primary currents and voltages to a nominal quantity recognized by the protective relays. With today's focus on digital technology, the importance of instrument transformer secondary connections is often overlooked. Selecting the correct instrument transformers for an application is imperative: failing to do so may compromise the relay's performance, as the output of the instrument transformer may no longer be an accurately scaled representation of the primary quantity. Having an accurately rated instrument transformer is of no use if these devices are not properly connected. The performance of the protective relay is reliant on its programmed settings and on the current and voltage inputs from the instrument transformers secondary.

Section-2 of this paper describes industry practice on instrument transformer secondary connection. Section-3 will focus on methodical approach of commissioning and retrofitting philosophies as adopted by POWERGRID. Section-4 discusses real-life example events from Indian power sector. Section-5 of this paper will focus briefly on CVT transients.

*In this paper, the case studies are prepared using the actual field event data of EHVAC transmission of Indian power transmission networks. All the case studies describe, mis-operation of protection system due to incorrect connection of instrument transformer secondaries. The disturbance records and other field event data are the actual extracted data from the numerical IEDs at the EHVAC substations. At the end of each case study, remarks have been provided which describes that how each of these mis-operations could have been avoided by following the correct instrument transformer secondary connections to the protective relays and by following the correct commissioning procedures.*

# 2 Fundamental Concepts and Review of Relevant IEEE Standards

This section specifically details application guidance on grounding and conventional wiring practices for Instrument Transformers and their secondary circuits. Guidelines provided in this section will help the utilities and practicing engineers in designing a dependable and secure protection system.

## *2.1 Neutral Connection and Grounding of Instrument Transformer Secondary Circuits*

IEEE Std C57.13.3™-2014 stresses the importance of and provides guidance on grounding instrument transformer secondary circuits [1]. Readers may refer to this resource for more detailed information.

### 2.1.1 Connect Only One Grounding Point Per Secondary Circuit

The instrument transformer secondary circuits are generally grounded at one point. The circuits shall preferably be connected to ground at the first point of application of the circuit. Figure 1 shows the single point grounding in one and half circuit breaker scheme. If the secondary circuits are grounded at more than one location, the potential differences between the grounding locations during faults can result in current flows through the relays and meters connected to the circuit. These additional currents could cause the relays to perform incorrectly and could result in providing incorrect measurements by meters. If the neutral conductor of the secondary circuit is connected to ground at multiple locations, large amounts of current could flow in the neutral conductor that could overheat this conductor and damage it.

### 2.1.2 Connect Used Instrument Transformer to Ground Near the Relays at the Switchboard Terminal and Connect Unused Instrument Transformer to Ground at Instrument Transformer Location

The following two practices are used for grounding instrument transformer secondary circuits:

- The circuit is connected to ground where the instrument transformers are located.
- The circuit is connected to ground where relays and other devices connected to the circuit are located.

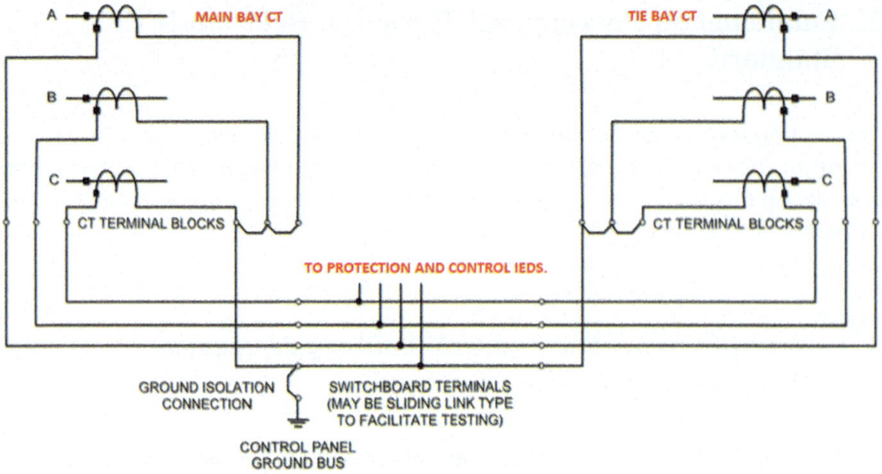


**Fig. 1** CT secondary circuit for One and Half CB scheme

Connecting the instrument transformer secondary circuit to ground at or near the instrument transformer location limits the voltage stresses on the secondary windings of the instrument transformers. Another reason to ground instrument transformer at the first point of application is because it then becomes convenient to locate and isolate the ground during testing. The grounding location should be accessible to facilitate the temporary disconnection and reconnection of the ground during testing. The use of fuses, contacts, auxiliary relays, or any switching devices that may unexpectedly open or leave open the ground connection is not recommended [1].

In many situations, some of the secondary windings of instrument transformers are not used and no circuit is connected to the windings. In such cases, the secondary winding of the instrument transformer is connected to ground where the instrument transformer is located.

#### 2.1.3 Ground Conductor Size

The grounding conductor size shall be as large or larger than the phase conductors [1].

### *2.2 Cable Selection for Instrument Transformer Connection*

IEEE Std 525™-2007 provides detailed guidelines for design & installation of cable system in the substation [2]. Readers may refer to this resource for more detailed information.

## 3 Commissioning and Retrofitting Philosophies of Powergrid

Pre-commissioning checks/tests are the activities carried out to ascertain the correctness, completeness of installation and healthiness of the equipment before its charging. POWERGRID follows a well-documented and structured method for commissioning and retrofitting of protection system. During the commissioning of the protection system, following are ensured by the commissioning team:

a. CVT and CT secondary neutral or common lead are earthed at one place only at the terminal blocks where they enter the panel. Such grounding is made through links so that grounding may be removed from one group without disturbing the continuity of grounding system for other groups.
b. Continuity check, cross ferruling and measurement of insulation resistance of each wire before connecting to the protection system.
c. Verification of the factory supplied connection.
d. Continuity test of each CT core with earth.
e. Ensuring that all unused CT cores are shorted and earthed.
f. Correctness of CT star point and verification of CT ratio in relay and meters with respect to CT name plate data.
g. Measuring and confirming that the voltages and currents measured at various instrument transformers and transducers are being supplied to each and every protective relay, meter, recorder, etc.
h. Trigger fault record in all the numerical relays and check magnitude and phase of the current and voltages.

POWERGRID has formulated well-documented procedure for retrofitting of protection system. Generally, all the steps taken during commissioning of new system shall also be applicable to the retrofitting cases.

POWERGRID has created a detailed risk-assessment matrix for retrofitting of protection system. The risk-assessment matrix is created for all the possible retrofitting cases.

POWERGRID has created a detailed risk-assessment matrix for retrofitting of protection system for all the possible retrofitting cases.

## 4 Analysis of Field Events

The events in this section are a result of problems identified with the wiring on the instrument transformer circuits, system related settings, such as the instrument transformer ratio or the tripping due to problem in voltage selection relays.

## 4.1 Problem in VT Selection Relay Leads to Tripping of Bus Reactor

At one of the 765 kV substations of POWERGRID, line reactor charged as bus reactor tripped after closing the Main CB of the adjacent diameter. Backup impedance relay of this reactor tripped due to problem in the measured voltage by the relay.

Voltage selection relay was used for supplying the selected voltage to the backup impedance relay. During analysis, it was observed that, after closing of the main CB of adjacent diameter, 75A relay picked up and 75B relay drops off (as per voltage selection logic of the substation).

On further analysis it was observed that, the voltage selection scheme was correct and the selection relay operated correctly, however some of the contacts of 75B relay were defective and even though 75B relay was in reset condition, some of its contacts were in make state and this leads to erroneous voltage to backup impedance relay (refer Figs. 2, 3).

*This type of tripping was due to erroneous voltage to the Backup Impedance Relay during normal service of the shunt reactor. To avoid occurrence of similar events in future, POWERGRID created additional logic in Backup Impedance Relays, wherein the Timers of the distance Zones will become active only, if backup impedance relay detects faults in the system by Over-Current or earth Fault Start Logic. No such inadvertent tripping was reported by any of the POWERGRID sites after the implementation of this logic.*

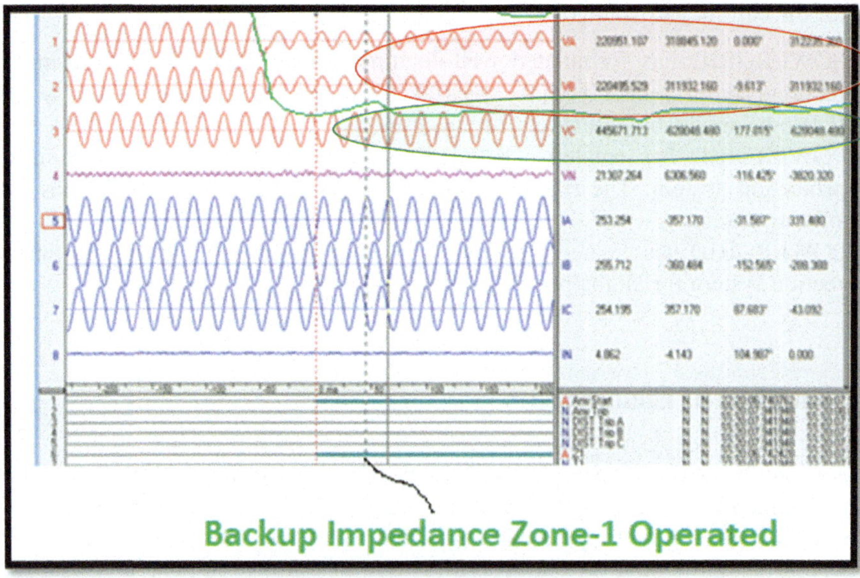


**Fig. 2** Operation of backup impedance relay due to distorted voltage waveform

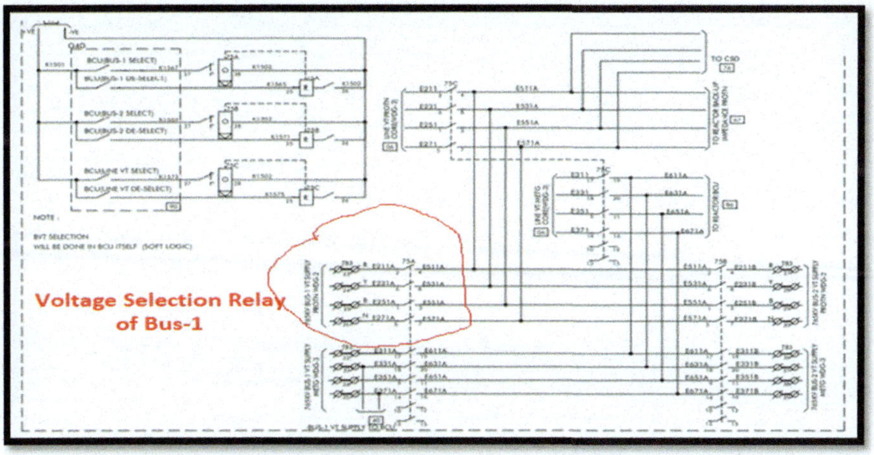


**Fig. 3** Voltage selection scheme of the bus reactor

### *4.2 400/220 kV 500MVA Auto Transformer Tripped on Differential Protection After Closing of Tie Circuit Breaker of HV Side*

400/220 kV 500MVA auto transformer at one of the POWERGRID substation was in service through its main CB on 400 kV side (one & half CB scheme) and main CB of 220 kV side (DMT Scheme).

During analysis of the disturbance report, it was observed that the differential current was present in all three phases of the auto transformer. The differential current keeps on increasing with an increase in loading condition. When, the differential current crosses the set differential current threshold in the relay, the differential protection operates and leads to tripping of the auto transformer (refer Fig. 4).

Since, the auto transformer was already in service through its main CB of HV side and tie CB was charged recently, it was suspected that the polarity of tie CT was not correct. On further, analysis of DR, it was observed that the sum of main & tie CT current of HV side was not the correct representation of 220kV side of auto transformer current. Further, analysis of DR confirms that the tie CT star point was not correct and leads to operation of differential protection of auto transformer.

*This case could have been avoided, by simply checking and analyzing the manually triggered DR of differential relay after closing of tie CB which is also a standard pre-commissioning procedure of POWERGRID.*

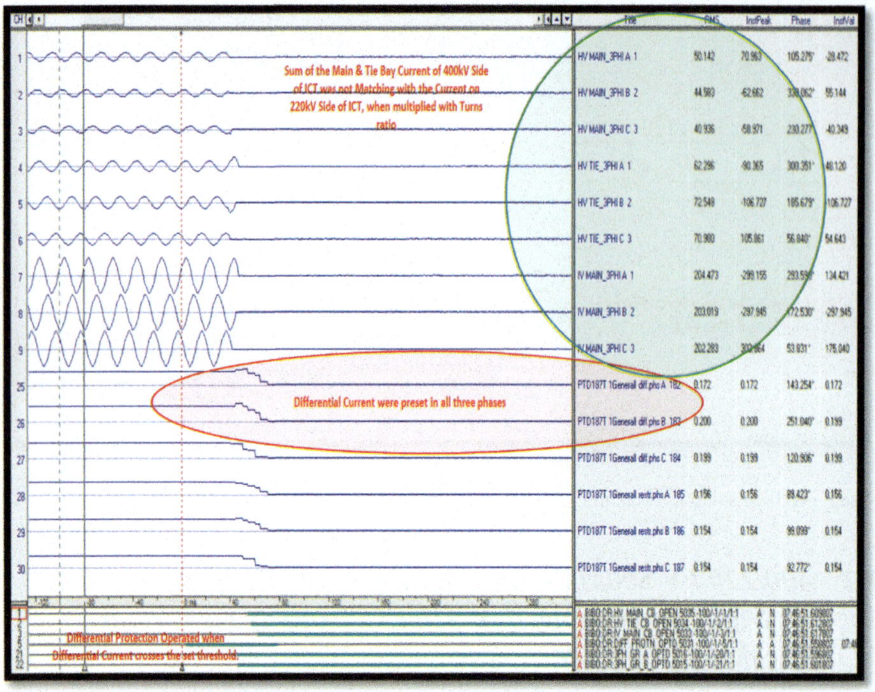


**Fig. 4** Differential protection operated, incorrect star point of CT

## *4.3 Line Tripped on Single Line to Ground Fault but the Phase to Neutral Voltage Were not the True Representation of Single Line to Ground Fault*

One of the 400 kV line of POWERGRID tripped on single line to ground Fault (CN fault) after successful auto reclose from both ends. During, this tripping of transmission line, it was observed that apart from $V_{CN}$, the $V_{AN}$ and $V_{BN}$ also became distorted during both instances of CN fault (refer Fig. 5).

After a detailed investigation at site, multiple grounding in CVT secondary circuit were found. It was observed that the secondary circuit of CVT was grounded at relay location in the relay panel as well as at CVT Junction Box in the switchyard. This second grounding of CVT was removed and wiring was made proper as per IEEE Std C57.13.3–2014. After modification, no such distortion in voltage waveforms was observed during subsequent line faults on this transmission line.

*This problem could have been rectified, if the proper wiring guidelines as per IEEE Std C57.13.3–2014 were followed at the commissioning stage. However, issue was rectified by proper DR analysis after the first occurrence of the problem and leads to avoidance of any mis-operation of the distance protection of this transmission line.*

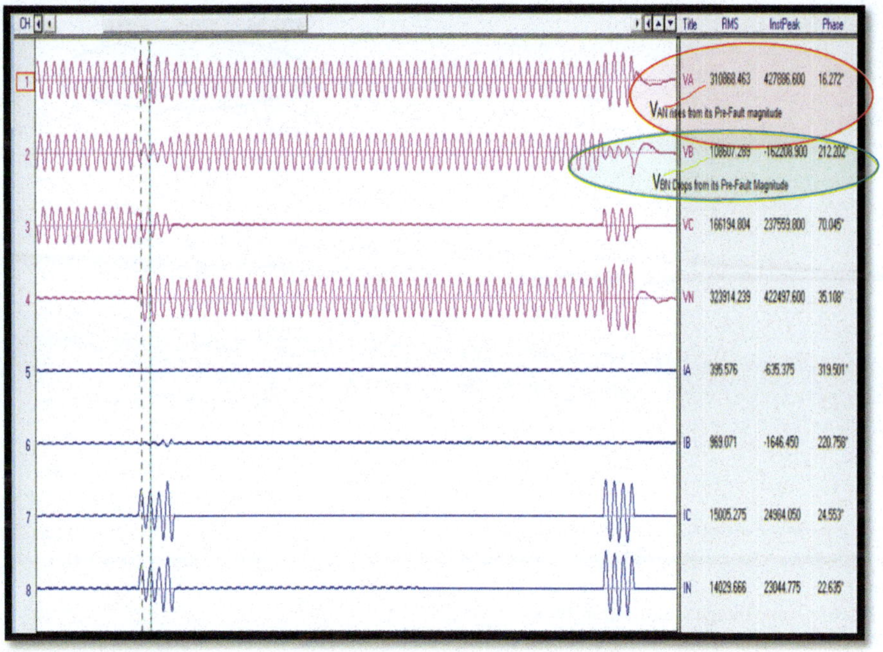


**Fig. 5** Distorted $V_{AN}$ and $V_{BN}$ for CN fault on transmission line

*Proper disturbance report analysis after every tripping is very helpful in avoiding the mis-operation of relays during faults on the feeder.*

## 4.4 Line Tripped on Operation of Over-Voltage During Planned Bus Shutdown at 765 kV Substation Due to Wiring Error.

One of the 765 kV line of POWERGRID tripped due to operation of over-voltage protection during normal system condition. During analysis of the disturbance report (DR), it was observed that, over-voltage protection operated in one of the relays of this 765 kV transmission line after opening of the last circuit breaker of the bus (planned shutdown of the bus was being availed).

Since, the over-voltage protection operated in only main-2 relay, it was clear that the problem was with the wiring. During analysis, it was observed that R phase voltage dropped from its pre-fault magnitude and a slight increase in Y and B-phase voltages was observed (refer Fig. 6). The R phase voltage reduction occurred from the time of BUS shutdown. Such voltage response is observed when R phase of line CVT is connected with R phase of bus CVT through voltage selection relays. In this

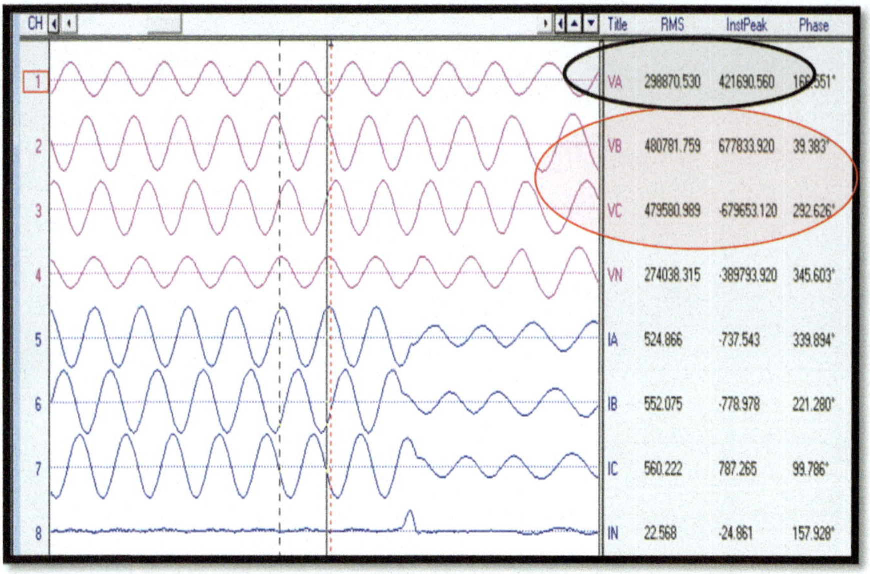


**Fig. 6** Over-voltage condition in B and C- phase

case, bus CVT secondary voltage drops to zero (last CB of bus opened for SD), R phase of line CVT drops due to the shorting and a slight increase in other phase voltages is observed due to double grounding (line CVT secondary grounding and bus CVT secondary grounding). The scheme was checked and it was found that R phase line CVT and bus-1 CVT are selected via tie BCU output contacts and extended to tie CSD relay for auto transformer switching (refer Fig. 7).

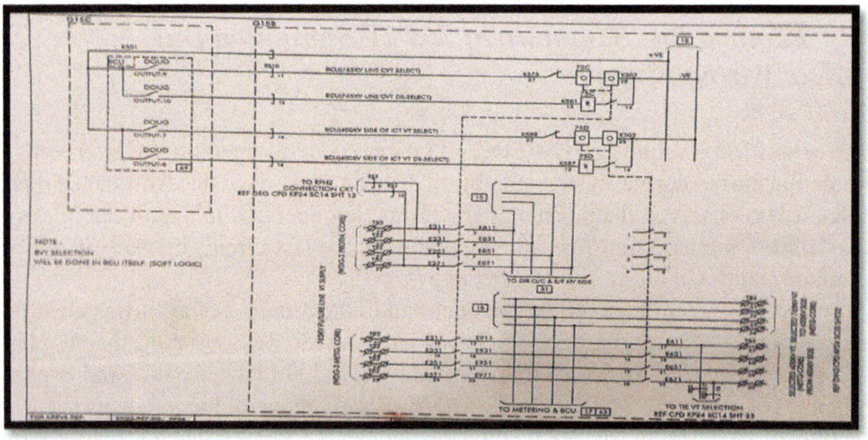

**Fig. 7** Voltage selection scheme

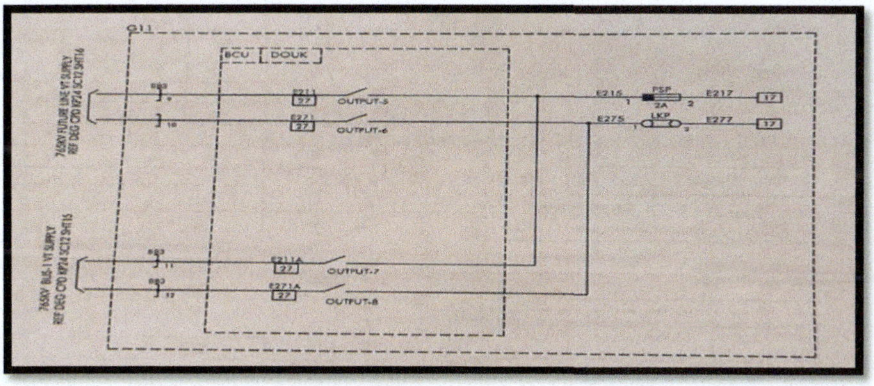


**Fig. 8** Line and bus CVT voltage connection

In the wiring of the selected voltage, bus-1 output wires were connected directly to input and the BCU contacts were found to be bypassed. The output of the above scheme was permanently shorted to bus 1 R phase voltage (refer Fig. 8). As soon as the main bay was taken into shutdown, the line voltage was selected which caused mixing of R phase and neutral of line and bus-1. As soon as the bus went under shutdown this mixing resulted in lowering of R phase voltage to relay and over-voltage in other phase which led to tripping.

*This type of problem could be avoided by proper scheme checking at the pre-commissioning stage.*

## 4.5 Line Tripped Due to Faulty CT Cable on Detection of Phase-To-Phase Fault in the Distance Relay During Phase to Ground Fault on the Transmission Line

One of the 400 kV line of POWERGRID tripped due to pickup of phase-to-phase loop during phase to ground close in fault on the transmission line. This line was in service for more than ten years and no such tripping was noticed during several phase to ground fault on this transmission line during the past more than ten years. Hence, this was a very peculiar problem because the tripping could not be attributed to either wiring error or to the relay settings. During, analysis it was observed that only one out of the two distance protection relays of this line has detected the phase-to-phase fault.

The DR of both Main-1 and Main-2 Relay was analyzed and it was observed that in main-1 relay the current in neutral was the same as that of A-phase current. Further, the voltage dip in only A-phase was observed and hence this relay declares phase to ground fault (AN fault) correctly (refer Fig. 10).

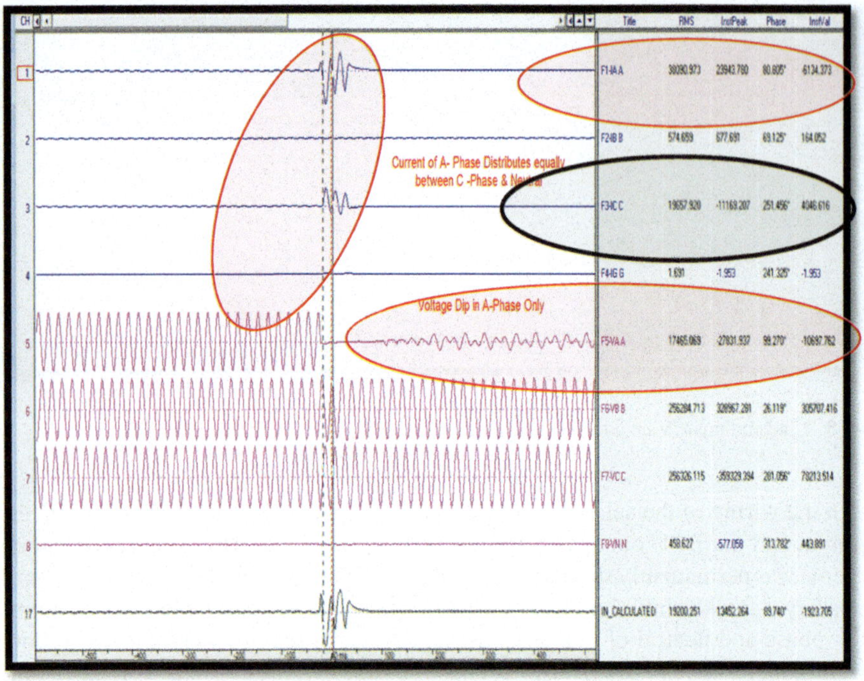


**Fig. 9** A-phase current distributes equally between C-phase and neutral in main-II relay

While in the main-2 relay, the current in neutral was approximately half as that of A-phase current and remaining current was flowing through C-phase. However, voltage dip in only A-phase was observed but because the current was detected in A, C and N in this relay hence this relay doesn't declare phase to ground fault (AN fault) correctly (refer Fig. 9).

Both relays have sensed a dip in VAN and an increase in A-Phase current but in one of the relays, A-phase current returns through neutral while in second relay A-phase current distributes equally between C-phase and neutral. This distribution of current was the reason for incorrect operation of the distance relay.

*To identify the problem, the scheme was checked thoroughly with the available scheme drawing at the substation and nothing was found abnormal. Finally, it was decided to carry out insulation resistance measurement of CT cable from CT junction box to relay panel. During, IR measurement it was observed that the insulation resistance of C-phase was very low as compared with A, B and neutral wire. Further, C-phase cable was showing direct continuity with ground. This defective CT cable was replaced with healthy cable and the problem was rectified.*

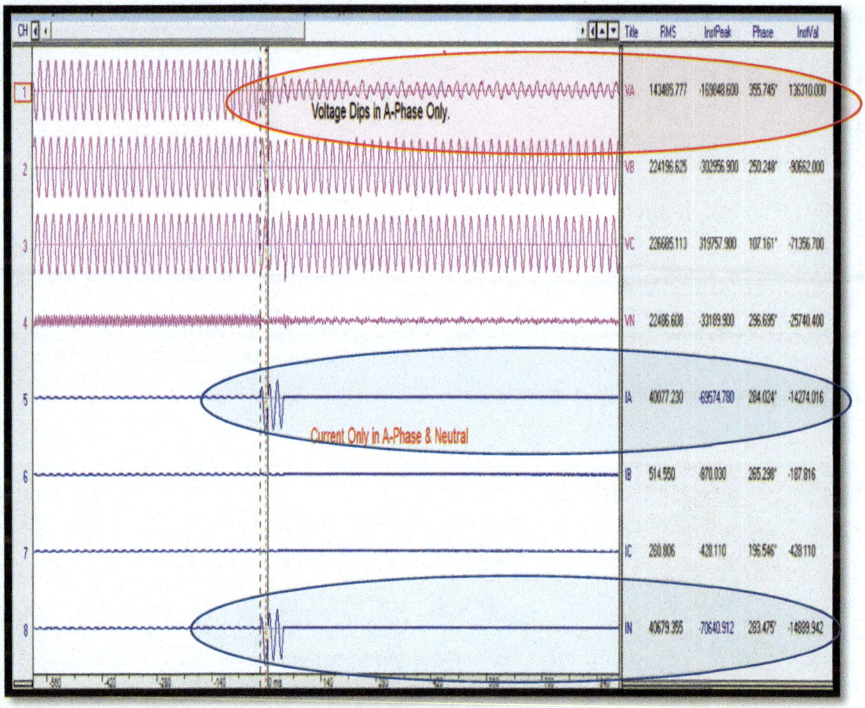


**Fig. 10** Current only in A-Phase and neutral in Main-I Relay

## 4.6 Auto Transformer Tripped on Operation of REF Protection Due to Damaged CT Lug

At one of the POWERGRID substation, 765/400 kV auto transformer tripped on operation of REF (restricted earth fault) protection. As per the scheme high impedance type REF protection was used in this auto transformer which was a bank of three single phase units. From the SCADA events of the substation, it was observed that REF protection of B-phase unit has operated. The Disturbance report of B-phase REF relay was extracted and from the DR, it was observed that the current in the REF coil of the relay has increased suddenly and crossed the threshold limit (refer Fig. 11). Since, the auto transformer was already in service since long hence it was clear that the sudden increase of current was due to some external factor.

Since, the REF relay was high impedance type and the relay has detected the current suddenly during normal operation, it was clear that some of the current has stopped reporting to the relay correctly. In high impedance relay, all CTs are paralleled in switchyard and only two wires are connected at relay terminal hence it was decided to check the REF circuit at switchyard level. During checking, neutral CT lug was found broken and CT wire was found removed from TB (refer Fig. 12).

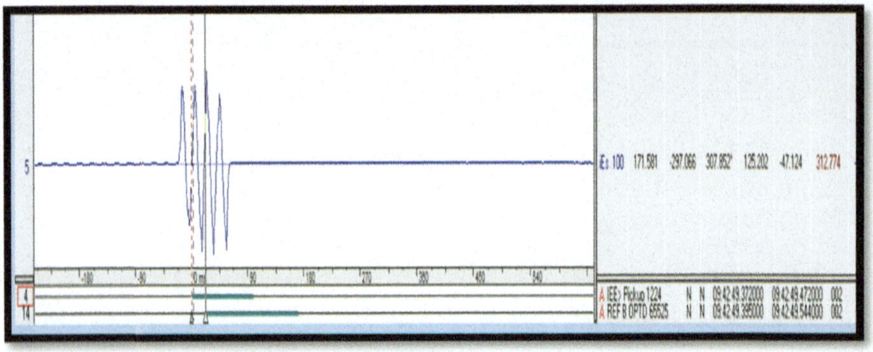


**Fig. 11** B-phase REF protection operated

**Fig. 12** Broken lug of neutral CT of REF

*This problem could have been avoided by proper crimping of ct wires and proper wire termination at TB.*

## 4.7 Auto Transformer Tripped on Operation of Differential Protection Due to Incorrect CT Ratio

At one of the POWERGRID substation, 400/220 kV 500MVA auto transformer tripped due to operation of differential protection (refer Fig. 13). The tertiary of this auto transformer was charged to feed the substation auxiliary supply at the substation. At the time of incident, it was observed that, differential protection of auto transformer operated due to internal problem in a motor connected at the LV side.

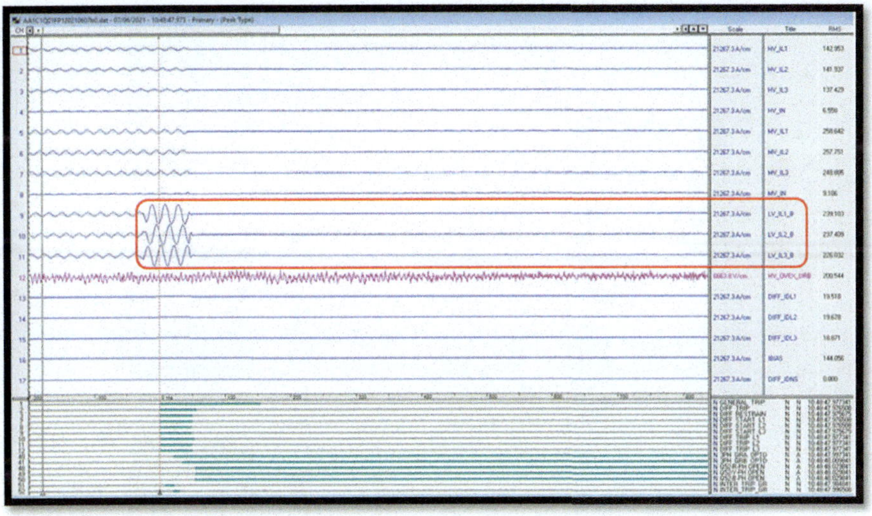

**Fig. 13** Differential protection operated due to problem in the LV side

As per DR, Current in HV (400kV) and MV (220kV) side was normal but LV side (33kV) current was showing high value (refer Fig. 14).

During further analysis, it was observed that the LV current was connected to the transformer differential protection module of the differential relay (refer Fig. 15).

On further analysis, it was observed that the actual CT ratio of 33kV (Channel-7, 8 & 9) is 50/1 but it was set to LV side full load current i.e., 1837/1 (refer Fig. 16) which leads to approx. 37 times of actual LV current measured by differential relay.

*This tripping was due to incorrect configuration of the differential protection relay and setting of incorrect CT ratio. The problem could have been avoided by following*

| No. | Name | RMS | Angle |
|---|---|---|---|
| 1 | HV_IL1 | 143.057(A) | 284.9° |
| 2 | HV_IL2 | 141.571(A) | 162.4° |
| 3 | HV_IL3 | 137.616(A) | 41.5° |
| 4 | HV_IN | 6.68(A) | 320.0° |
| 5 | MV_IL1 | 255.395(A) | 104.8° |
| 6 | MV_IL2 | 253.495(A) | 342.6° |
| 7 | MV_IL3 | 245.691(A) | 221.9° |
| 8 | MV_IN | 9.599(A) | 130.7° |
| 9 | LV_IL1_B | 1195.957(A) | 101.4° |
| 10 | LV_IL2_B | 1178.811(A) | 342.3° |
| 11 | LV_IL3_B | 1199.174(A) | 222.6° |
| 12 | DIFF_IDL1 | 96.706(A) | 44.6° |
| 13 | DIFF_IDL2 | 94.441(A) | 43.6° |
| 14 | DIFF_IDL3 | 94.884(A) | 52.2° |
| 16 | IBIAS | 143.636(A) | 168.5° |
| 16 | DIFF_IDNS | 0.0(A) | 270.0° |

**Fig. 14** Analog current data of differential relay

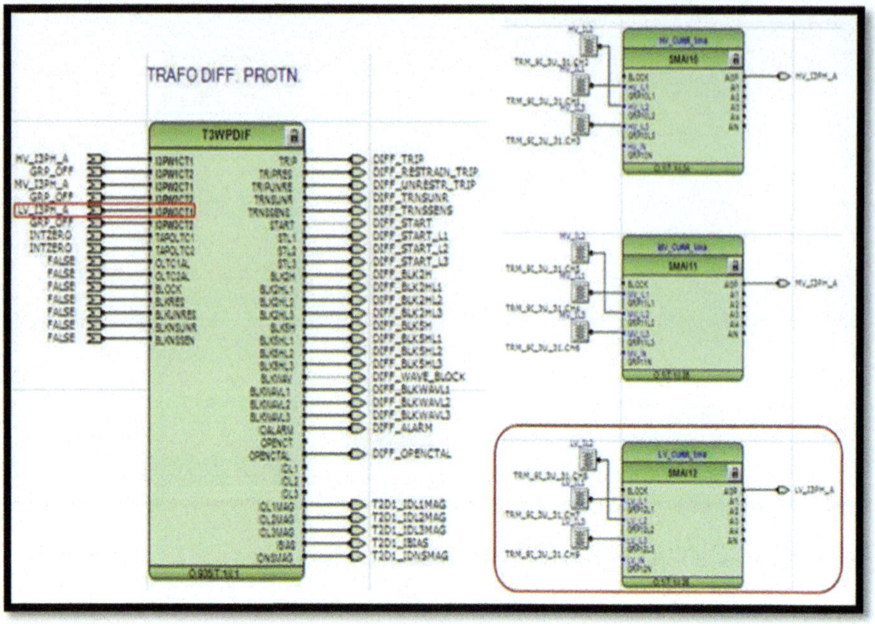


**Fig. 15** Incorrect configuration of differential relay

| | |
|---|---|
| NAMECH7 | LV_IL1 |
| ChannelType7 | CurrentProtection |
| RatedTrans7 | 1.0 A |
| CTStarPoint7 | ToObject |
| CTsec7 | 1 A |
| CTprim7 | 1837 A |
| NAMECH8 | LV_IL2 |
| ChannelType8 | CurrentProtection |
| RatedTrans8 | 1.0 A |
| CTStarPoint8 | ToObject |
| CTsec8 | 1 A |
| CTprim8 | 1837 A |
| NAMECH9 | LV_IL3 |
| ChannelType9 | CurrentProtection |
| RatedTrans9 | 1.0 A |
| CTStarPoint9 | ToObject |
| CTsec9 | 1 A |
| CTprim9 | 1837 A |

**Fig. 16** Incorrect CT ratio in the differential relay setting

*the standard pre-commissioning procedure of POWERGRID and by following the standard setting and configuration template of POWERGRID. Problem was rectified by changing the relay configuration and settings.*

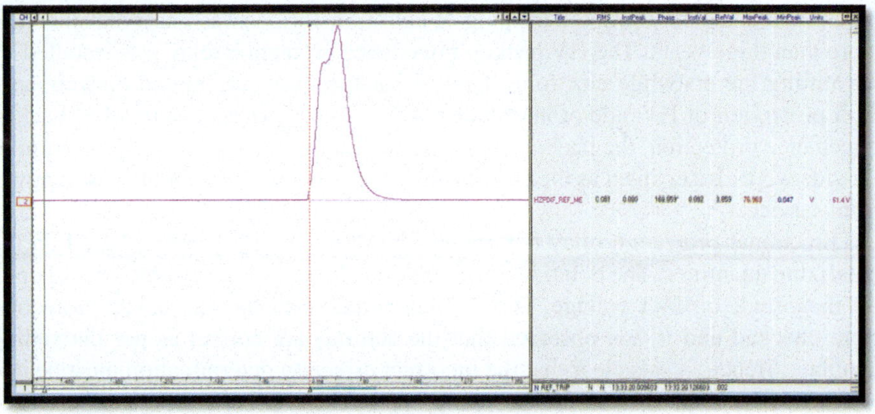

Fig. 17 REF protection operated after charging of the bus reactor

## 4.8 Shunt Reactor Tripped on Operation of REF Protection Due to Problem in CT Wiring

At one of the substations of POWERGRID, 125MVAR bus reactor tripped on the operation of REF protection immediately after charging (refer Fig. 17).

High impedance type REF protection was used for bus reactor at this substation. The HV and neutral side CT were connected in parallel and two wires were finally connected to the relay. During the checking of the scheme it was observed that the REF protection operated because of the reverse polarity of neutral CT at the switchyard.

*This problem could have been avoided, if proper wiring methodologies as per commissioning procedure of powergrid were followed*

## 4.9 Incorrect Direction Determination by Directional Overcurrent Protection Because of Missing CVT Neutral and Leads to Tripping of Auto Transformer

At one of the POWERGRID substation, 400/220 kV auto transformer tripped on operation of LV side over current protection during a fault on LV network. The operation of LV overcurrent for fault on LV network was a cause of concern and needs analysis. For analyzing the incorrect operation of LV backup EF protection of auto transformer, the relay internal events and other DR of previous incidences were extracted. From the previous events, it was observed that the LV backup EF protection was getting picked up during any fault on the 220 kV side of the auto transformer. However, most of the time, since the fault was cleared before the operation of LVEF protection hence no unwanted tripping of auto transformer was observed since last more than three years. The HV backup impedance protection relay was installed by retrofitting the erstwhile electromechanical backup over current and backup earth fault protection of HV side of auto transformer three years ago. In this HV backup impedance protection, the backup over current and backup earth fault protection of LV side were also enabled as the existing electromechanical protection was removed from service.

Directional protection relay determines the direction of operation based on the polarizing quantities. The polarization quantity to the relay is determined based upon the magnitude of CVT voltage. The settings of auto transformer backup protection were checked and it was observed that the settings are correct as per the setting template. Hence, it was clear that the incorrect direction determination must be due to the incorrect connection of CVT wiring to the relays. On analyzing the retrofitted scheme of auto transformer backup impedance relay the problem was due to the missing CVT neutral to the relay.

To ascertain the CVT connection to the relay, the CVT connection from the switchyard up to the relay was checked and the following were observed:

1. Before retrofitting, three over current and one earth fault relay in each side were being used as backup protection of auto transformer.
2. Core-1 of the bus CVT was being used for the backup directional overcurrent protection. Three wires, one each from A, B and C-phases were terminated to the three TBs at relay panel and the neutral of core-1 was earthed at the CVT MB itself. Three overcurrent relays were connected between these three wires as AB, BC and CA directional overcurrent relays.
3. Core-2 of the bus CVT was being used for backup directional earth fault protection. This core of the CVT was connected in open delta manner and the two ends of this open delta were terminated to the two TBs at relay panel. One directional earth fault relay was connected between these two wires in the relay panel.
4. During, retro-fitment work, three wires of core-1 (point-2 above) were connected as A, B and C-phase and one wire of core-2 (point-3 above) was connected as neutral to the relay.

From Fig. 18, it is clear that, core-1 of bus CVT was being used for the directional overcurrent protection. Further the neutral of CVT Core-1 was earthed in the CVT MB itself and six wires, two from each phase were connected to six different TBs in the relay panel. Out of these six, wires, the wire nos. E213, E233 and E253 were being used for connecting the CVT input of A, B and C-phase to the erstwhile directional overcurrent protection.

Core-2 of bus CVT was connected in open delta form and two wires i.e., wire no. E373 and E374 were connected to the two different TBs in the relay panel. These two wires become E377 and E378 after fuse and the erstwhile directional earth fault protection relay was connected between these two wires in the relay panel.

At the time of retrofitting, during the connection of CVT wires to the backup impedance relay, wire nos. E213, E233 and E253 were connected to the relay as the

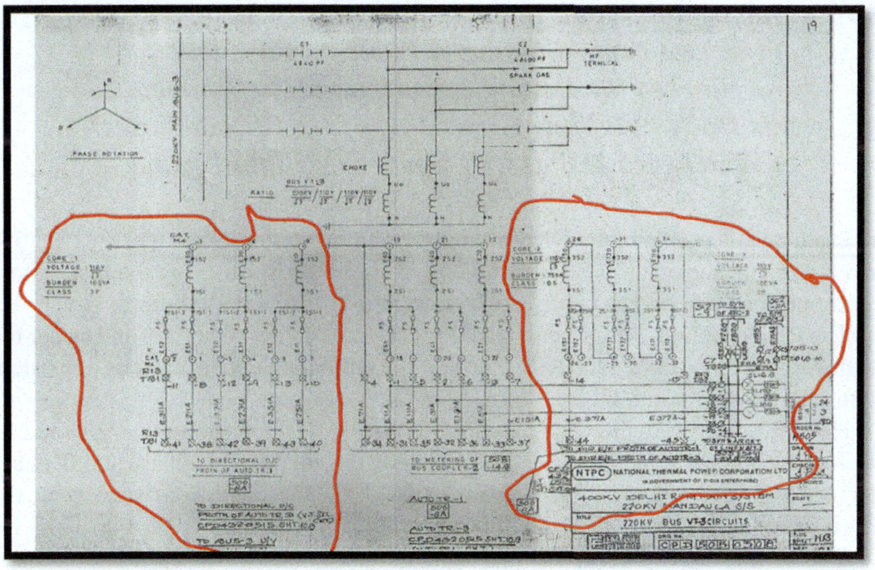

**Fig. 18** Connection of CVT core-1 and core-2 in CVT MB in switchyard

A, B and C-Phase CVT input while wire no. E377 was connected as neutral of new relay.

Although E213, E233 and E253 are of A, B and C-Phase CVT input respectively but E377 was not the correct neutral to the new relay. This incorrect connection of neutral wire to the new relay was the reason for incorrect direction determination by the relay.

The connection of CVT secondary wires was modified and four wires from core-1 of bus CVT were connected to the relay with only single point neutral grounding. After, the modified connection relay was kept under observation to ascertain the correct operation of the relay. During recent faults on downstream feeders, it was observed that auto transformer backup protection relay is now detecting the correct direction of the fault and LVEF relay was not picking up during LV side fault.

*This problem could have been avoided if proper wiring and pre-commissioning procedures were followed at the retrofitting stage. Further, the internal events of the numerical relays help in the identification of this type of problem and these internal events must be checked after any out of zone faults after any retrofitting or commission work.*

## *4.10 Phase-phase Fault in One of the 400 kV Line, One Out of the Two Protection Relay Declares Fuse Fail and Blocked the Distance Protection. The Problem was Due to Multiple Grounding of CVT Secondaries at Switchyard MB as well as at relay/control Panel*

To evaluate this problem, the DR was analyzed and it was observed that the negative and zero sequence magnitude of voltages and current was satisfying the fuse failure conditions of the Relay as per the relay algorithm for VT Fuse failure.

The Relay settings were found to be correct based on best practices and as per the setting template of POWERGRID. Further, the faulted phase current was consistent with AB, phase-phase fault.

However, analyzing the voltages captured during the fault shows that instead of a decreasing voltage signature, in A- and B-phase, the voltage of B-phase and C-phase decreases but the voltage of A-phase increases and this was inconsistent w.r.t. the Ph–Ph fault and leads to declaration of fuse failure by the relay (refer Fig. 19).

The wirings were checked as per scheme and it was observed that there was grounding in the CVT secondary at switchyard MB as well as at the relay location

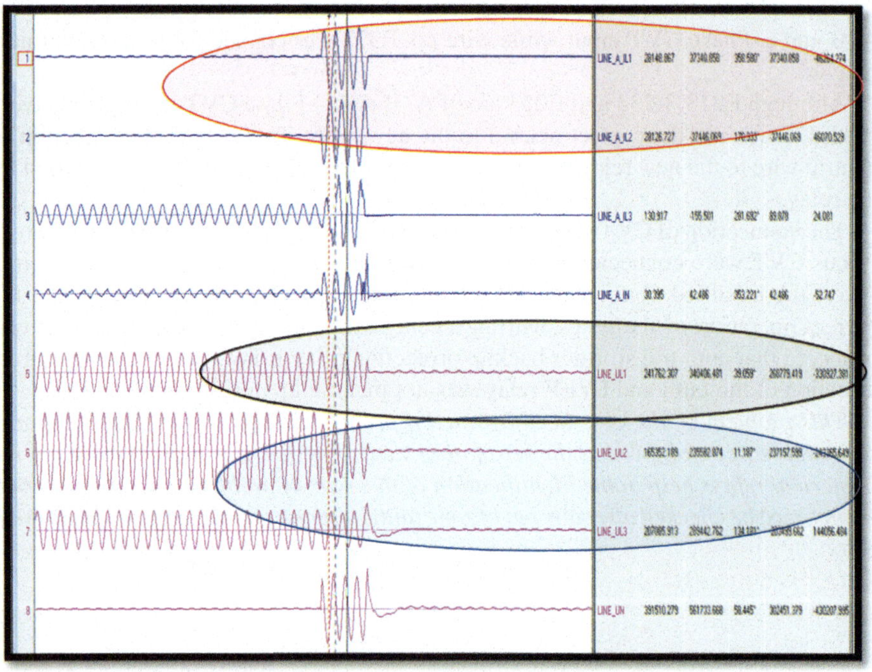

**Fig. 19** DR captured by the relay during phase-phase fault

and this double grounding was the main cause for mis-operation of distance protection relay.

## 5 Instrument Transformer Secondary Connection and Cvt Transients

The transient response of a capacitive voltage transformer is the ability to reproduce rapid changes in the primary voltage. It's defined as the remaining secondary voltage after a specific time due to a short circuit on the primary voltage. Several factors influence this, they are; the equivalent capacitance of the stack, the tap voltage, the connected burden, and the type of ferro-resonant suppression circuit. Out of these factors, only the connected burden is under the control of the utility engineers.

Distance relays calculate the impedance to the fault and operate if the calculated impedance is within its reach setting. Correct analog inputs are necessary for a relay to make the impedance calculation. CVT transients cause incorrect information to be presented to the relay for a short period of time. Since zone 2 and zone 3 timers are much longer than the CVT transient period, zone 2 and zone 3 elements are not affected by CVT transients. Zone 1 elements operate with no intentional delay. Therefore, their operation is affected by the CVT transient. Zone 1 relay reach settings need to be reduced based on the SIR of the line. The latest version of microprocessor relays uses filtering techniques to lessen the effects of CVT transients. Case study 4.10 describes the effect of multiple **grounding** of CVT secondary on the measured voltage by the protective relays. This case study describes, how due to incorrect connection of CVT secondary, to the distance protection relay, inadvertently gets blocked during Phase-Phase fault on the transmission line.

## 6 Conclusion

CT and CVT connection errors can lead to undesired operations of protection systems. However, many of these operations can be avoided by adhering to industry standards and implementing tried-and-true field testing and commissioning practices.

Fundamental concepts of the connections of instrument transformers to the protection relays and the effect of incorrect connection on the performance of protective relays are discussed in detail in this paper. Multiple real-world case studies of protection system mis-operations due to incorrect connections of instrument transformer are presented in this paper.

We learned through the discussion of these case studies that, how a missing neutral wire, led to incorrect tripping of auto transformer due to incorrect direction determination by the backup earth fault relay. We have also learnt that how due to multiple grounding in CVT secondary the distance protection relay gets blocked during actual

Ph–Ph fault on the transmission line. In addition to these, we also learnt that how incorrect CT star point, incorrect CT ratio, incorrect relay configuration and the avoidable errors at retrofitting stages lead to tripping at a later stage during normal service of the feeders.

We hope that the case studies presented in this paper will help the readers to analyze the problem through real-world challenges in the complex power system networks.

## References

1. IEEE C57.13.3–2014, IEEE guide for grounding of instrument transformer secondary circuits and cases
2. IEEE Std 525™-2007, IEEE guide for the design and installation of cable systems in substations
3. Zimmerman K, Costello D (2009) Lessons learned from commissioning protective relaying system. In: Proceedings of the 62nd Annual Conference for Protective Relay Engineers, College Station, TX
4. Peterson O, Scharlach R, Baltazar C (2009) Proper testing of protection system ensures against false tripping and unnecessary outages. In: Proceedings of the Pulp and Paper Industry Technical Conference, Birmingham, AL
5. Boecker M, Corpuz G, Candelaria J, Hargrave A, (2020) Finding common ground: a case study of phantom currents and ground loops. In: Presented at the 47th Annual Western Protective Relay Conference, Virtual
6. Schweitzer engineering laboratories, Inc., "SEL-311C Relay protection and automation system instruction manual. (2016). Available: selinc.com
7. Disturbance records from differential protection relays for 400/220kV auto transformer (ABB RET670, Relay manuals available online)
8. Disturbance records from distance protection relays for 765kV and 400kV transmission lines (Siemens 7SA522, ABB REL670, ALSTOM P444, GE D60, all these relay manuals are available online)
9. Disturbance records from high impedance differential protection relays shunt reactors (Siemens 7SJ80, Relay manuals available online)
10. Costello D (2008) Lessons Learned Through Commissioning and Analysing Data from Transformer Differential Installations. IEEE Trans Ind Appl 44(4):979–1003
11. Blackburn JL (2014) Protective relaying: Principles and applications, 4th ed. CRC Press
12. Protection and automation application guide (Available: gegridsolutions.com)
13. WAVEWIN Data management and analysis system software (latest release available: http://www.abb.com/substationautomation)
14. Fault record evaluation software of siemens SIGRA (latest release available: http://siemens.com)
15. Anderson PM (1999) Power system protection. IEEE Press/McGraw-Hill, New York
16. Determining distance relay Zone-1 reach settings to prevent CCVT transient overreach, Gregory A. Franklin, Randy Horton, IEEE 978–61284–738–2/11
17. ABB instrument transformer application guide, Edition 3, 1 HSM 9543 400—00en

# Reducing the Burden on the Utility Grid by Implementing the Demand Response Strategy with Home Loads and Solar PV Using TLBO Technique

**Raseswar Sahoo, Sagarika Rout, and Gyan Ranjan Biswal**

**Abstract** The demand response (DR) strategy gives electricity users the chance to contribute to the reliability of the power system by controlling their electricity usage. A scheme for optimal management of electricity consumption in the home environment is introduced in this work. The work is primarily focused on electricity price minimization and reduced burden on the power grid. The proposed methodology introduced the scheduling of home appliances optimally according to the real-time pricing (RTP) structure as well as the consumers' preferences. The solar Photovoltaic (PV) assists the demand response strategy in the home environment. Furthermore, the scheduling problem is mathematically formulated with Teaching Learning based optimization (TLBO) method. The results show the proficiency of the proposed method in comparison to the conventional uncontrolled energy consumption.

**Keywords** Demand response · Home load · Scheduling · TLBO

## 1 Introduction

The demand response (DR) enables consumers to participate actively in demand side management (DSM) by lowering or shifting the energy consumption in response to time-based rates during peak periods. The strategy implements a cost-efficient energy environment in coordination between utility and consumers. The various dynamic tariff structures such as time-of-use pricing, critical peak pricing, real-time pricing, and critical peak rebates are the most suitable ways for customers to reshape the energy pattern in demand response. The scheduling of household appliances by considering customers' satisfaction level and energy bill is reduced [1].

R. Sahoo · S. Rout (✉)
Department of EE, Veer Surendra Sai University of Technology, Burla, Odisha 768018, India
e-mail: srout_ee@vssut.ac.in

G. R. Biswal
Department of EEE, Veer Surendra Sai University of Technology, Sambalpur 768018, India

 
O. H. Gupta et al. (eds.), *Soft Computing Applications in Modern Power and Energy Systems*, Lecture Notes in Electrical Engineering 1107,
https://doi.org/10.1007/978-981-99-8007-9_12

In [2] authors considered the deferrable loads of a home for reshaping the residential load curve. The population-based meta-heuristic technique is used to schedule the home load while taking consumer preference into consideration with two different pricing mechanisms for demand response. In [3], the scheduling of home loads is considered to minimize peak demand and consumption prices. For scheduling appliances, three techniques are used. The effectiveness of developed techniques in terms of bill reduction and peak reduction is demonstrated by experimental analysis. In [4], Particle swarm optimization (PSO) is used in the DR strategy for home apparatuses. It considers the switches of the smart home appliances operation from peak hours to doing so during off-peak hours, which ultimately lowers the energy bill while maintaining low computational costs.

In [5], the appliance scheduling issue for residential consumers in the smart grid is examined. Two types of loads are taken into consideration time-shiftable and power-adjustable for the decision variable with starting operating time. The best solution was developed by using a specific weighted sum method in order to change the defined multi-objective mixed integer optimization problem to a single weighted sum objective with comfort enhancement and energy cost deprecation as the scheduling objectives. For all types of loads in the house, the discomfort index was introduced based on their ideal start time and actual start time.

The authors explained that home load arrangement using the branch and cut technique of mixed integer linear program (MILP) is solved. The ideal starting time of both kinds of home loads, which are deferrable and adjustable, is a decision constraint. The goals of the optimization technique include reducing the peak demand and the energy use cost by using the Time of Use (TOU) price tariff [6].

The authors in [7]; suggested a scheduling method together with the Time of Use tariff to address the industrial area's scheduling issue for plants. The suggested scheme lowers the electricity cost and can only be used in scaled-down problems due to excessive computation costs. The problem is constructed as a continuous-time model. In order to manage discrete events in a more natural and straightforward manner, this led to discrete-time models.

To schedule and keep track of the electrical appliances, a control algorithm is suggested for the home energy management system (HEMS) [8]. The strategy is based on the energy cost. When switching household smart loads, if the energy is costly, the demand is satisfied with the battery bank. This study takes into account solar panels as well, and the suggested architecture is put into practice on a real-time testbed. The findings show that there is a reduction in electricity costs with minimum comfort violation.

An effective energy management technique was created by Luna Adriana et al. for the purpose of integrating RESs and lowering electricity costs [9]. Decreasing the amount of imported energy is the main target of RES integration. In their job, RESs like solar panels and WTs with battery storage systems are taken into consideration. However, the installation cost of RESs is not considered in this study, even though simulation findings support their suggested system model. In [10], a home load energy management controller is proposed with swarm intelligence to schedule the home loads. The objective is to minimize the cost of energy use by harnessing the power from electric vehicles with the least burner to the conventional grid. In [11], the DR problem while considering numerous homes with the same living arrangements is examined. An efficient strategy has been proposed for reshaping the pattern of appliance energy consumption. The scheduling arrangement, therefore, represents a similar functioning as the utility demand curve. Customers can reduce their energy costs and only bring in a small amount of energy from the external grid in this manner. Additionally, effective scheduling might guarantee a steady and dependable microgrid.

The paper is structured with the home load energy management system described in Sect. 2 followed by problem formulation in Sect. 3. The optimization technique has been explained in Sect. 4. The results and discussion is introduced in Sect. 5 whereas Sect. 6 concludes the findings of the paper.

## 2 Home Load Energy Management System Model

The work considers a smart home with renewables with different user-specific smart appliances. The renewables only considered integrated home solar panels. It is expected that the home loads are under the home energy management controller (EMC) and smart meters are installed in residences. The purpose of the smart meter is to share load demand and price signals between the utility and the electricity consumer. Whereas the Energy management controller (EMC) is set up to schedule appliances in accordance with price signals and electricity produced by the micro-grid. RTP tariff is adopted for electricity cost calculation. The duration of scheduling is considered as one day in this work. Additionally, the whole duration is divided into equal-sized sub-intervals.

**Table 1** Home appliances with types and rating [12]

| Category | Appliance name | Power rating(kw) | Optimal start time (h) | Optimal end time(h) | LOT(h) |
|---|---|---|---|---|---|
| Deferrable | Cooker | 3 | 6 | 10 | 1 |
| Deferrable | Oven | 4 | 15 | 20 | 1 |
| Deferrable | Microwave | 1.7 | 6 | 10 | 1 |
| Deferrable | Laptop | 0.1 | 18 | 24 | 2 |
| Deferrable | Desktop | 0.3 | 18 | 24 | 3 |
| Deferrable | Vacuum cleaner | 1.2 | 9 | 17 | 1 |
| Deferrable | Electric Vehicle | 3.5 | 18 | 8 | 3 |
| Deferrable | Dish Washer | 3.5 | 18 | 8 | 3 |
| Non-Shiftable | Lighting | 0.84 | 16 | 24 | 5 |
| Non-Shiftable | Refrigerator | 0.3 | 1 | 24 | 20 |

## *2.1 Home Appliances*

The home consists of a set of smart loads categorized with their power ratings with information on lengths of operational time (LOT), optimal starting times, and optimal ending times. The appliances should start at their optimal starting time or later, but they must finish the operation before the optimal end time. Each appliance either operates or does nothing during the time slots. Furthermore, all the home appliances are pulled into two categories based on their operational time: shiftable appliances and non-shiftable appliances, and is depicted in Table 1.

**Shiftable Appliances**: The appliances that are shifted during their execution. These deferrable loads help with energy management. Examples of this kind of appliance include vacuum cleaners, dishwashers, and cooker hubs.

**Non-Shiftable Appliances**: These are the loads that once started must have to complete their execution without any interruption. The load schedule may or may not involve these appliances. Examples of this kind of appliance include lighting and refrigerators.

## 2.2 *Solar Photo-Voltaic*

The PV is considered in this work for electricity generation in the home environment. Electricity generation from solar PV in the form of direct current (DC), which is then changed into alternating current (AC) with the help of an inverter. The PV generation is explained with the mathematical formulation explained in Eq. 1 [13].

$$P_{pv}(t) = \eta_{pv} * A_{pv} * Irr(t) * [1 - 0.005(Temp(t) - 25)] \quad (1)$$

where $P_{pv}(t)$ is the total electricity generation in KW from the solar panel in t interval, $\eta_{pv}$ is the solar panel efficiency, $A_{pv}$ is the area of PV panel in $m^2$. $Irr(t)$ is the amount of solar radiation in $Kw/m^2$, and $Temp(t)$ is the temperature in °C at t interval.

# 3 Problem Formulation

Reducing electricity consumption costs and imported load for the residence with the incorporation of Solar PV is the aim of this work. The RTP pricing strategy is utilized in this work to calculate costs; this strategy provides a distinct price for energy for each hour. When a load is switched from a high-priced to a low-priced hour, the total cost automatically decreases. In this work, for each hour the algorithm first determines whether energy is accessible from the solar panel or not, if it is, that energy is used, and the remaining energy requirement is purchased from an external grid. Otherwise, the whole energy requirement for that hour is satisfied using energy purchased from the external grid. The objective is defined mathematically in Eq. 2.

$$f_1 = Minimize \sum_{t=1}^{T} (E_s(t) + E_{ns}(t) - E_{pv}(t))E_{pp}(t) \quad (2)$$

Subject to

$$\sum_{i=1}^{N} \gamma_{i,j} = LOT(j)\ \forall j$$

$$\alpha_j \leq \gamma_j \leq \beta_j\ \forall j$$

$$0 < Irr(t) < Kc$$

where $E_s(t)$ is the energy used by all the appliances scheduled at t interval.$E_{ns}(t)$ is the energy used by all the non-shiftable appliance scheduled at t interval. $E_{pv}(t)$ PV energy utilized in t interval. $E_{pp}(t)$ price offered by the utility in tth slot which is the RTP signal. $\gamma_{i,j}$ is ith slot of jth appliance and maximum no of slots is N. Therefore $\sum_{i=1}^{N} \gamma_{i,j}$ is the total slots allotted to jth appliance which should always be equal to the length of operation time (LOT) of jth appliance and this condition should be satisfied for all appliances in the home. $\alpha_j$, $\beta_j$ are the operational start and end time of jth appliance respectively and $\gamma_j$ is the operational time of the jth load, this condition implies that the operation of an appliance should always lie between the start time and end time. $LOT(j)$ is the predefined length of operational time. $Irr(t)$ and $Kc$ are the solar radiation during t interval and clear sky condition respectively.

## 4 Teaching Learning Based Optimization (TLBO) for HLEM

The paper implements the Teaching Learning based optimization for load scheduling, and the cost minimization in home load energy management.

Teaching Learning based optimization is a nature-inspired population-based meta-heuristic optimization technique that simulates the environment of a classroom to optimize a given objective function [14]. The most appealing aspect of this strategy is that it requires just two parameters: population size and the number of iterations for its operation. Aside from that, this method is devoid of derivatives. When compared to other evolutionary algorithms, it is simple to implement as any other algorithm-specific parameters are not required in this algorithm. Lower computation and the simplicity associated with this algorithm is what make it avourable for Home Load Energy Management. Furthermore, there has been less work published on this algorithm for Home Load Energy Management, and the works have not considered factors such as appliance comfort, which is considered in this work. This algorithm works in two-fold scenarios (i) Teacher structure phase, (ii) Learner Phase structure.

The TLBO defines the relationship of teacher and student to improve the quality of knowledge. Firstly, the teacher tries to bring all students to the same knowledge level by utilizing the teaching phase and students' learning capabilities in the teaching phase. In the second learning step, the interaction among themselves increases their knowledge. The pseudocode of the TLBO algorithm is as follows [15].

**Pseudocode**:

Input: Objective function, number of iterations (T), number of populations (N)

Output: Objective function values(F)

Step-1: Initialize random population.

Step-2: Evaluate objective function values of the population.

Step-3:

For t= 1 to T

For k= 1 to N

{Teacher phase}

Identify best solution as teacher $X_{best}$

Find the mean of each decision variable $X_{mean}$

Calculate $TF = round[1 + rand(0,1)]$

Calculate $X_{new} = X_k + rand(0,1)[X_{best} - TF * X_{mean}]$

Based on $X_{new}$ calculate $F_{new}$

Accept $X_{new}$ if it better than $X_k$

{End of Teacher phase}

{Learner phase}

Choose any solution randomly, $X_p$ such that $p \neq k$

Determine $X_{new}$ as

If $F_k < F_p$

$X_{new} = X_k + rand(0,1) * (X_k - X_p)$

else

$X_{new} = X_k - rand(0,1) * (X_k - X_p)$

Based on $X_{new}$ calculate $F_{new}$

Accept $X_{new}$ if it better than $X_k$

{End of Learner phase}

End

End

# 5 Results and Discussion

Simulations have been run to demonstrate the best scheduling of microgrid and household appliances. The usefulness of the proposed work is assessed by reducing energy costs with the best possible use of electricity produced by the own solar generation. A smart house is considered with ten home loads connected to the Energy Management Controller via the WIFI, and the controller decides whether to turn on or off each appliance based on the electricity tariff. As per the operational nature, the devices are categorized as deferrable which can be shiftable and others as non-shiftable. As non-shiftable appliances cannot be scheduled and must operate at predetermined intervals set by energy users, they have little impact on load management and electricity cost reduction. By shifting from expensive to inexpensive hours, shiftable appliances may significantly lower energy costs. Additionally, 24 h are considered for simulation. The data considered for the research work like hourly solar generation, RTP signal, length of operation time of appliances, Ratings of appliances are considered from [1]. The proposed scheme is optimized with TLBO algorithm and compared with the cost when consumption is considered in the traditional way and when scheduled considering Particle Swarm Optimization (PSO) algorithm.

## 5.1 *Scheduling with Traditional Structure*

In this section, the home loads are considered in a conventional way. The Scheduling strategy of home loads harnesses energy from the conventional grid with the structured tariff plan. The results shown here are explained without optimization techniques.

Figure 1. shows the RTP pricing signal [1] as well as the hourly electricity consumption costs. By effectively shifting the load towards low-priced electricity hours, the suggested strategy reduces the amount of electricity used during high-priced electricity hours, or from 8 to 13, as is evident. Only the load demanded by the deferrable appliances are scheduled by the proposed plan.

The hourly electricity uses from the external grid are shown in Fig. 2 for a day. As discussed previously, this graph shows how the suggested plan reduces the energy consumption during periods of high electricity prices. No electricity is purchased from the external grid between the hours of 8 and 14. In this instance, the expense of the electricity was calculated to be 415.1520 Rupees.

## 5.2 *Scheduling with TLBO Optimization*

The hourly energy utilization prices with TLBO optimization along with the RTP pricing signals are depicted in Fig. 3. Most appliances are observed to be used during

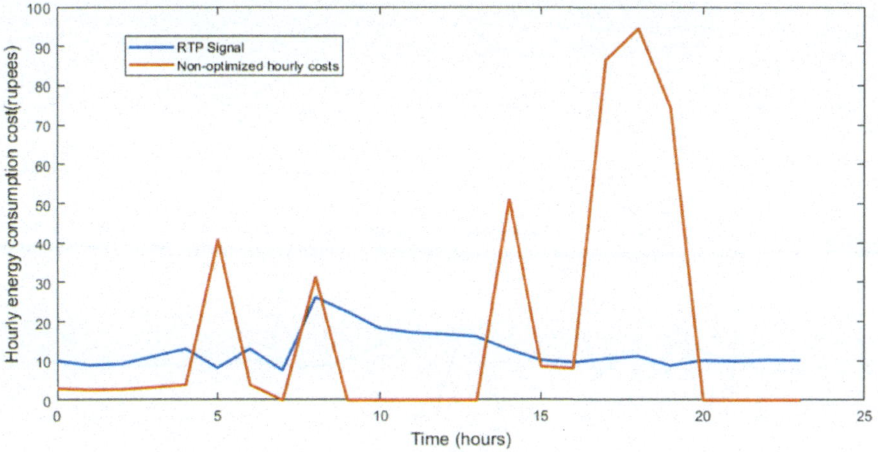


**Fig. 1** Non optimized Hourly energy consumption costs for home loads in comparison with real time pricing signal

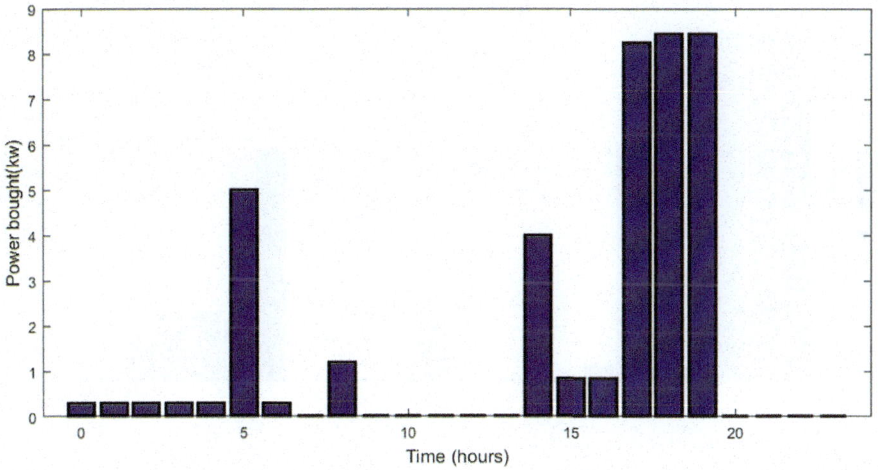


**Fig. 2** Electricity utilisation from external grid for 24 hours in a day

the 1-to-5-h window, which has the lowest energy price range of the entire day. In contrast to the prior instance, where most appliances were scheduled during the 15–20 h, which is also an area with cheap energy prices but higher than the 1–5 h.

The hourly electricity imports from the external grid for the TLBO optimized case are shown in Fig. 4. for the entire time frame. In accordance with the previous figure, this figure shows that a large amount of electricity is drawn during the early morning hours. In this instance, the expense of electricity was determined to be

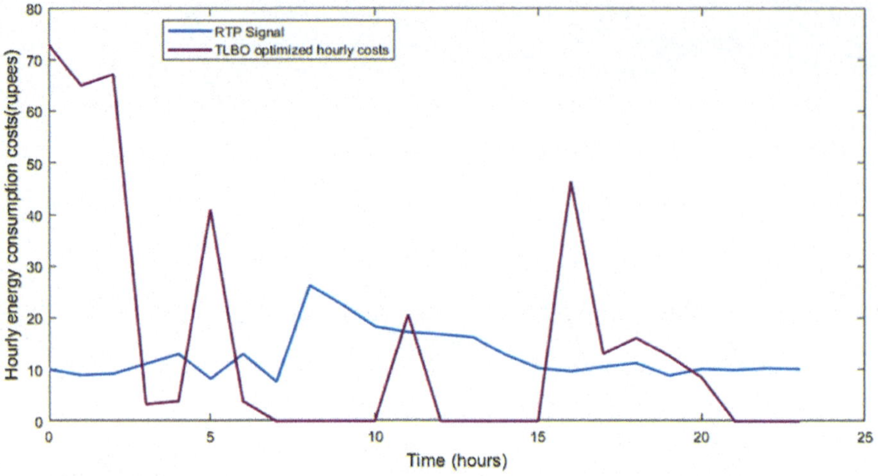


**Fig. 3** Hourly energy consumption costs for home loads with optimal Scheduling

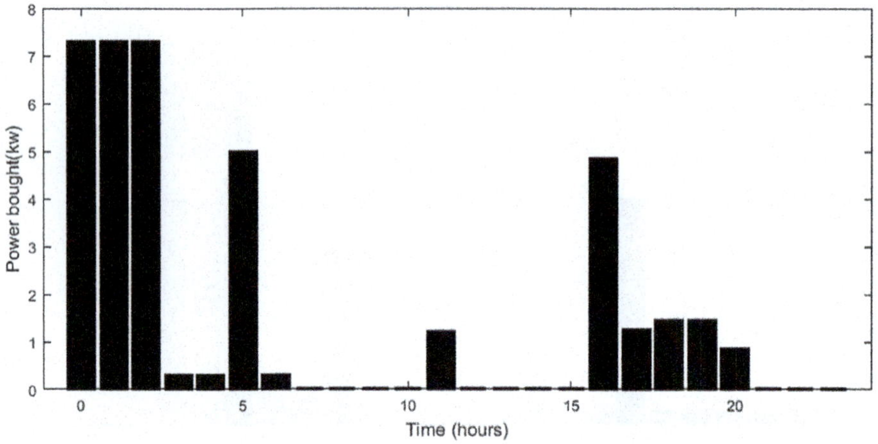


**Fig. 4** Hourly imported electricity from external grid with optimized schedule slots for home loads

374.6680 rupees. Table 2 shows the time of operation of appliances during both the cases along with LOT, optimal start time and optimal end time data.

Figure 5 shows the hourly energy production from the microgrid in the smart home, which is solely made up of solar panels. The figure makes it abundantly obvious that the electricity production varies with the amount of sun that is available throughout the day. This graph shows that energy production from solar panels is greatest during the day. On the other hand, during periods of low solar irradiation or at night, the solar panel produces zero electricity. Figure 6. shows the hourly energy

**Table 2** Scheduled results of home appliances

| Category | Home Appliance | Optimal start time (h) | Optimal end time(h) | LOT(h) | Time of operation during non-optimized case | Time of operation during PSO optimized case | Time of operation during TLBO optimized case |
|---|---|---|---|---|---|---|---|
| deferrable | Cooker | 6 | 10 | 1 | 6 | 6 | 6 |
| deferrable | Oven | 15 | 20 | 1 | 15 | 16 | 17 |
| deferrable | Microwave | 6 | 10 | 1 | 6 | 6 | 6 |
| deferrable | Laptop | 18 | 24 | 2 | 18,19 | 18,19 | 18,19 |
| deferrable | Desktop | 18 | 24 | 3 | 18,19,20 | 18,19,20 | 18,19,20 |
| deferrable | Vacuum cleaner | 9 | 17 | 1 | 9 | 9 | 12 |
| deferrable | Electric vehicle | 18 | 8 | 3 | 18,19,20 | 1,2,3 | 1,2,3 |
| deferrable | Dish Washer | 18 | 8 | 3 | 18,19,20 | 1,2,3 | 1,2,3 |
| Non-Shiftable | Lighting | 16 | 24 | 5 | 16,17,18,19,20 | 16,17,18,19,20 | 17,18,19,20,21 |
| Non-Shiftable | Refrigerator | 1 | 24 | 20 | 1 to 20 | 1 to 20 | 1 to 20 |

**Table 3** Objective function values

| Cases | Non-optimized | PSO optimized | TLBO optimized |
|---|---|---|---|
| Objective function values | 415.1520 | 387.9520 | 374.6680 |

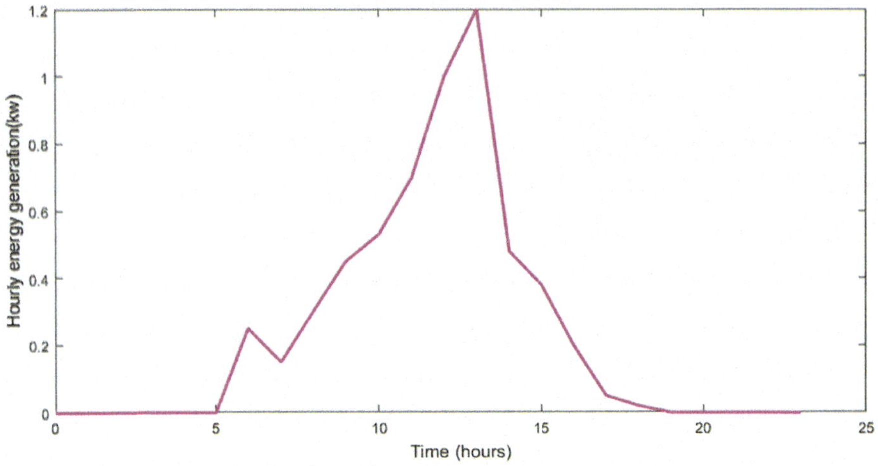


**Fig. 5** Hourly electricity generation of solar PV

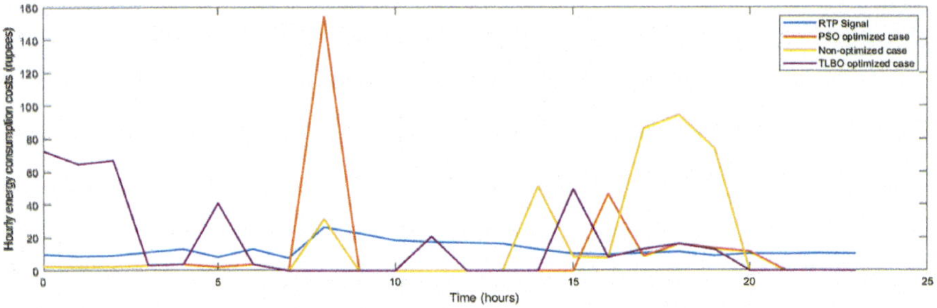


**Fig. 6** Hourly energy consumption costs comparisons for home loads

utilization costs of both cases along with the RTP pricing signal. The numerical cost comparison has been given in Table 3.

Figure 7 shows the comparison between the three cases. Without optimization total electricity cost was 415.1520 rupees whereas, with PSO optimization it was 387.9520 and with TLBO optimization it was 374.6680 rupees which is around 6.55% and 9.75% decrease in the total electricity cost.

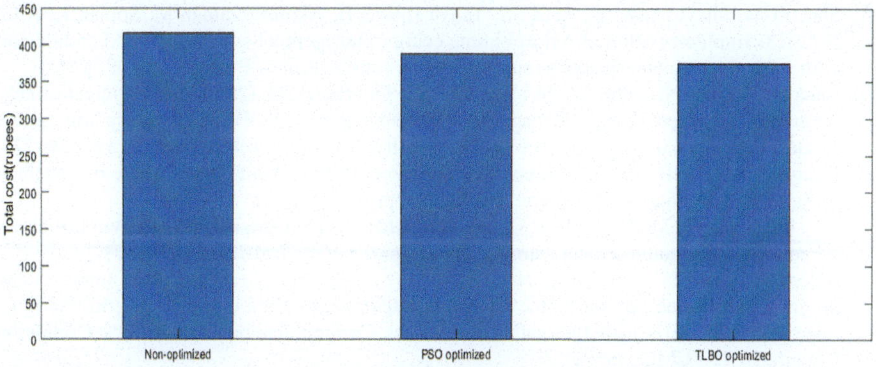


**Fig. 7** Total Consumption cost comparison with optimized scheduled load Scenario

## 6 Conclusion

In this work, appliance scheduling is performed in a single residence by taking total electricity cost and imported electricity from grid minimization with RTP pricing strategy as objectives. The results show; the proposed TLBO scheme schedules the smart appliances in accordance with the real-time pricing. The analysis shows a distinct proficiency of around 10% in cost reduction as compared to the unscheduled traditional case. This also clearly shows the peak demand reduction and wise use of renewables beneficial to consumers and utility. The work can be extended to the analysis of impact in micro grid with more dynamic load integration and DSM technology.

## References

1. Aslam S, Khalid A, Javaid N (2020) Towards efficient energy management in smart grids considering microgrids with day-ahead energy forecasting. Electr Power Syst Res 182:106232
2. Rout S, Biswal GR, Sahoo RM (2020) Optimal scheduling of deferrable loads in home energy management system using elephant herding optimization. In 2020 IEEE International Conference on Computing, Power and Communication Technologies (GUCON), pp 698–702
3. Rahim S, Javaid N, Ahmad A, Khan SA, Khan ZA, Alrajeh N, Qasim U (2016) Exploiting heuristic algorithms to efficiently utilize energy management controllers with renewable energy sources. Energy Build 129:452–470
4. Huang Y, Tian H, Wang L (2015) Demand response for home energy management system. Int J Electr Power Energy Syst 73:448–455
5. Ma K, Yao T, Yang J, Guan X (2016) Residential power scheduling for demand response in smart grid. Int J Electr Power Energy Syst 78:320–325
6. Shakouri H, Kazemi A (2017) Multi-objective cost-load optimization for demand side management of a residential area in smart grids. Sustain Cities Soc 32:171–180
7. Javadi MS, Gough M, Lotfi M, Nezhad AE, Santos SF, Catalão JP (2020) Optimal self-scheduling of home energy management system in the presence of photovoltaic power generation and batteries. Energy 210:118568

8. Shakeri M, Shayestegan M, Reza SS, Yahya I, Bais B, Akhtaruzzaman M, Sopian K, Amin N (2018) Implementation of a novel home energy management system (HEMS) architecture with solar photovoltaic system as supplementary source. Renew Energy 125:108–120
9. Luna AC, Diaz NL, Graells M, Vasquez JC, Guerrero JM (2016) Mixed-integer-linear-programming-based energy management system for hybrid PV-wind-battery microgrids: Modeling, design, and experimental verification. IEEE Trans Power Electron 32(4):2769–2783
10. Rout S, Biswal GR (2022) Bidirectional EV integration in home load energy management using swarm intelligence. Int J Ambient Energy, pp1–8
11. Kim BG, Zhang Y, Van Der Schaar M, Lee JW (2015) Dynamic pricing and energy consumption scheduling with reinforcement learning. IEEE Trans Smart Grid 7(5):2187–2198
12. Aslam S, Javaid N, Asif M, Iqbal U, Iqbal Z, Sarwar MA (2018) A mixed integer linear programming based optimal home energy management scheme considering grid-connected microgrids. In 2018 14th International Wireless Communications and Mobile Computing Conference (IWCMC), pp 993–998
13. Shirazi E, Jadid S (2015) Optimal residential appliance scheduling under dynamic pricing scheme via HEMDAS. Energy Build 93:40–49
14. Rao R (2016) Review of applications of TLBO algorithm and a tutorial for beginners to solve the unconstrained and constrained optimization problems. Decis Sci Lett 5(1):1–30
15. Keraliya D, Ashalata K (2017) Minimizing the detection error in cooperative spectrum sensing using Teaching Learning Based Optimization (TLBO). Int J Eng Res and Technol (IJERT) 6(2):495–500

# Automatic Generation Control of Multi Area Power Systems Using BELBIC

**J. Shankar and G. Mallesham**

**Abstract** The goal of this review article on automatic generation control studies is to offer both a thorough analysis of the literature and a sizable bibliography. It has been addressed to use various methods for controlling frequency and power. Different power generating sources, including conventional, renewable energy sources, and realistic Renewable Energy Sources (RES), are taken into account in both regulated and uncontrolled situations when discussing various linear and non-linear power system models. Several secondary controllers with integer order, fractional order, intelligence, cascade, and a few recently used controllers are also briefly explained in the AGC literature. In this record, the subject of load frequency regulation received considerable attention. A multi-area, connected power system still has a serious LFC issue. For LFC in power systems, numerous control strategies have been put forth in recent years. This essay offers a comprehensive review of the LFC literature. Excellent control action, disturbance handling, and robustness of system parameters are all characteristics of the proposed controller Brain Emotional Learning Based Intelligent Controller (BELBIC). It has been used in the field of control because of its exceptional effectiveness and simplicity of application. A power system that is subjected to disturbance variation change and system-inherent nonlinearity will be the subject of the simulation results, which will be obtained using MATLAB. To prove that the suggested designed controller works as intended.

**Keywords** Load Frequency Control (LFC) · Automatic Generation Control (AGC) · Multi Area Power System · Area Control Error (ACE). Brain Emotional Learning Based Intelligent Controller BELBIC

---

J. Shankar · G. Mallesham (✉)
Electrical Engineering Department, University College of Engineering (Autonomous), Osmania University Hyderabad, Hyderabad, Telangana, India
e-mail: drgm@osmania.ac.in



O. H. Gupta et al. (eds.), *Soft Computing Applications in Modern Power and Energy Systems*, Lecture Notes in Electrical Engineering 1107,
https://doi.org/10.1007/978-981-99-8007-9_13

## 1 Introduction

Maintaining the proper operating level of frequency, voltage control, and load flow conditions is the major duty of a power producing system. To deliver customers with dependable, consistent, high-quality electric power, the relationship between power requirements and power production needs to be rebalanced. The system's easily controllable sources' parameters can be changed to regulate both reactive and actual electricity generation. The fundamental goal of AGC is to maintain a reasonable spacing between the switching frequency and tie-line power [1, 2]. Variations in power demand have an impact on both the frequency of the power system and the flow of electricity via the tie-lines. By attempting to manage the power generation with area control, the major goal of frequency control for load is to maintain the operating frequency at the predetermined value. Maintaining the tie-line changeover schedules, distributing the power among the generators, and preserving the typical frequency of 50 Hz or 60 Hz are the key goals of LFC. Many conventional controllers, including PID, PI, and I, are employed in manipulating machines because they are simple to implement, simple to understand, and inexpensive. Nature of their control approach is dependable and mentioned as sturdy for a few operating situations. However, the reaction of the device with those controllers is gradual and bad in evaluation of the Intelligent controller [3]. A variation in frequency (f) and tie-line actual power (Ptie), 2 separate assessments of the change in rotor angle as illustrated in Fig. 1, can both be recognized using LFC as its primary approach. Together, they can be used to identify this tiny difference in accuracy. Several methods were employed to address the LFC issue, including robust control [4, 5], particle swarm optimization [6], pole placement strategic plan [7], and state input [8]. To control the frequency and tie-line power in LFC systems, a variety of control techniques, including the adaptive neuro-fuzzy inference system (ANFIS), the NARMAL-2 controller, etc., have been described in the literature [9, 10]. Area Control Error (ACE) and area control error reduction using continuous active power adjustment. Historically, the input for the LFC controller used to control the area control error (ACE) and the interoperability frequency deviation has been the area control error (ACE), also known as the sum of the area frequency bias, frequency deviation, net power interchange error, or net tie-line flow error.

The industry-standard conventional type LFC controllers are tuned online through trial-and-error procedures. It has been suggested to simulate the entire system instead of just the control region under consideration so as to use a variety of optimization techniques to fine-tune the control parameters [11, 12]. Tie-lines connect the various locations to one another. The tie-lines are used to exchange energy between

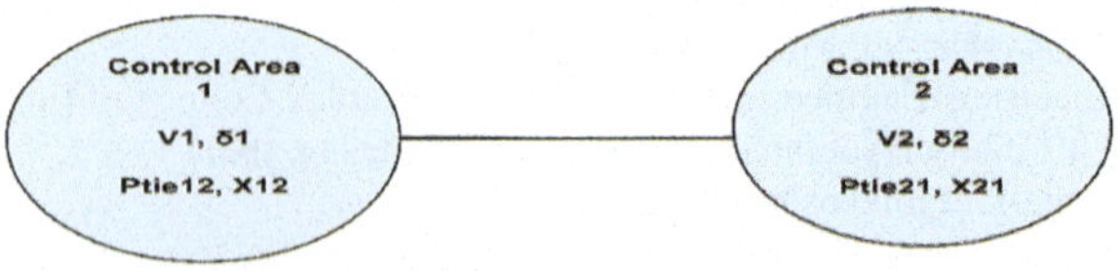


**Fig. 1** Load frequency control in a two-area system

successive two areas and to offer inter-area assistance in the event that the power system experiences abnormal conditions [13]. Some of them mistakenly think that all sub-systems are identical, which is untrue for actual power systems. Numerous decentralized LFC controllers were developed to address the aforementioned issue. The system develops a mismatch in frequency and scheduled power exchange across locations when a load change occurs. Load frequency control (LFC), which is defined as the management of generator power output within a tolerable limit is required to fix this mismatch [14]. The length of time required for frequencies to stabilize in a system with Control scheme, PID, and PID with a fuzzy logic-based controller was measured [15, 16]. Of all the controllers, fuzzy with PID generated the most effective results. The controllers in use are PID with fuzzy logic-based AGC controllers [12, 17–21].

In the following paper, it is suggested to build and develop a brain emotional controller with the primary goal of controlling load frequency. Although the design of the brain's emotional controller is similar for all purposes, the choice of sensory signal and emotional cue or reward functions varies. The variable gain employed to obtain better performances is geared for the brain's emotional controller. The proposed technique's structure, which uses the brain's emotional controller to modulate the load frequency, is depicted in Fig. 4. The design of the brain's emotional controller is distinctive and straightforward, which defeats the complexity found in other intelligent controllers. This essay is structured as follows: The section on "Development of Brain Emotional Controller" discusses the structure of the brain, with an emphasis on the limbic system and its related regions to develop as a controller. Section 'Results and comments' explains the findings, while Section 'Conclusions' presents the conclusions.

## 2 Mathematical Modelling of LFC

The frequency of power systems is dependent on active power, and voltage is dependent on the reactive power limit. The control power system is divided into two distinct problems. Load frequency control (LFC) refers to the control of frequency by active power [18, 22]. An important task of LFC is to keep the frequency deviation constant in the face of continuous load variation, also known as unknown external load disturbance. Power exchange error is a critical task for LFC [23, 24]. A power system is typically composed of several generating units linked together; these generating units are interconnected via tie-lines to become fault tolerant. The block diagram can be made as given below of a two-area interconnected power system is shown in Fig. 2. The use of tie-line power is represented in Eq. (1) introduces a new error into the control problem, which is known as the tie-line power exchange error. Area controller error (ACE) represented in Eq. (1) plays a significant role in interconnected power systems, as well as in minimizing error functions of the given system. The BELBIC controller is proposed in this paper, as is the performance of load frequency control on a two-area power system.

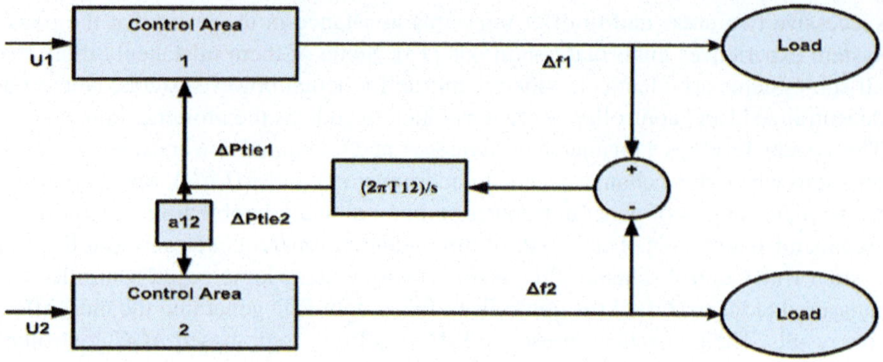


**Fig. 2** Block diagram of two area interconnected power system with controller

Power is transferred from area 1 to area 2.

$$P_{tie12} = \frac{|V_1||V_2|}{X_{12}} \sin(\delta_1 - \delta_2) \quad (1)$$

If the load demands of two areas change, the power angle will change incrementally. $\Delta\boldsymbol{\delta}_1$ and $\Delta\boldsymbol{\delta}_2$ are the incremental changes in $\boldsymbol{\delta}_1$ and $\boldsymbol{\delta}_2$ are the power changes. Because controller input is Area control error, it plays a significant role in interconnected power systems.

$$ACE_i = \sum_{j=1}^{N} \Delta P_{ij} + B_i \Delta\omega_i \quad (2)$$

where N is the no. of areas interconnected areas *i*, $\boldsymbol{P_{ij}}$ is the power deviation between areas i and j from the scheduled values.

$\Delta\omega$ is the speed deviation

$$B_i = \frac{1}{R_i} + D_i \quad (3)$$

$\boldsymbol{B_i}$ is the frequency bias factor represented in Eq. (3).

Where $\Delta f_1$, $\Delta f_2$ are change in frequency in area-1 and area-2.

## 3 Concepts for Designing a Control System

The control system designs were used to discuss LFC advancements in power systems that are interconnected. The vast majority of LFC design features are based on technologies employed in the control system design field. Techniques were the focus of the first era of traditional control theory. Root Locus, Bode, Nyquist, and Routh-Hurwitz are examples of control methods. These methods commonly employ complex frequency domain transfer functions, feedback, and trying to simplify assumptions to evaluate the time response. One of the major drawbacks of traditional control strategies was the use of single-input, single-output (SISO) methods. The frequency domain and transfer functions are also limited to structure linear time-invariant systems.

The use of state-space model-based techniques is a hallmark of the second era of control systems, also referred to as the modern control era. Standard control evaluation and development are performed in the time domain. State space model-based methods removed classical control's limitations and provided insight into system structure and properties, but they obstructed other important control properties that could be studied and manipulated using classical control. During the third era, it offered assistance to the systems' uncertainty problem. As well as optimal and adaptive control, which are also expressed in state space, robust control, which combines contemporary state space and traditional frequency domain techniques, is a type of control.

## 4 Research Gaps and Challenges

Load frequency control's (LFC) primary goal, also known as unidentified explicit disturbance, is to maintain a steady frequency over arbitrarily changing active power loads. Controlling the tie-line power exchange error is another objective of LFC [25–27]. The tie-line power exchange error is a new error that has been added to the control problem as an outcome of the use of tie-line power. The energy gained within the area is obtained via TL from preceding areas when an unanticipated change in active power happens over the area. The balance must ultimately be managed among the areas that are processed to load change without aid from other sources. If not, the regions' economies will diverge to some extent. Since each location needs its own LFC to control the tie-line power exchange error, various set points can be defined for each area within an interconnected power system. The LFC requirement is therefore more resistant to fluctuations in system parameter uncertainty and system model uncertainties. In conclusion, the LFC is in charge of two key duties: keep up with the tie-line power exchange and keep up with the frequency benchmark value in the event of fluctuating load variations.

### *4.1 Improvement of AI Optimization Techniques Over Conventional PI Controllers*

The conventional controller, one of many techniques used in the design of LFC, produces subpar results as shown by pronounced overshoots. When determining and modifying the PID controller parameters, AI optimization techniques such as fuzzy control, particle swarm optimization, also known as (PSO), and genetic programming (GA) will significantly enhance the system outputs. These control methods, which rely on smooth computer technology for LFC parameters, guarantee the system frequency and total eradication of steady state error.

## 5 Proposed Control Strategy

In this work, a novel use of artificial intelligence was discussed. Brain Emotional Learning Based on Intelligent Controller (BELBIC), is the name of the new controller. Its two main components are the medial brain concept and emotional processes. The newly proposed controller is meant to control load frequency in two thermal zones in response to various operating conditions and disturbances, including changes in load demand, parameter uncertainty, and parameter. A direct adaptive controller, an easy structure, strong robustness and performance properties, and quick auto-learning are just a few of the benefits of BELBIC.

### *5.1 BELBIC Controller*

Nevertheless, there are two different kinds of cognitive and intelligent control techniques. The intelligent system modifies the controller's variables mostly through indirect means. We chose the second option since it is a more favourable strategy. In this situation, the BELBIC computer model serves as the intelligent system's controller block. A model representation is shown in Figure 3. A system called BELBIC creates actions in response to sensory input and emotional cues. However, for demonstration purposes, just one Sensory-input and one Emotional-Cue signal (stress) were taken into account in the benchmarks presented in this study. These can normally be vector valued. The amygdala is the main location for emotional learning.

A BELBIC is a Highly intelligent Controller Based on Brain Emotional Learning. It is reliant on mammals' brains' limbic system. The BELBIC controller is originally introduced by Lucas, and the first straightforward computer model of the emotional learning process is created by Moren and Balkenius. There are a total of four parts. The first component is the Thalamus (TH), and its source of signal is stimuli (SI). The reward signal (Rew), which is received as an input by the amygdala (A), is the second component. The remaining third and fourth overall components, respectively,

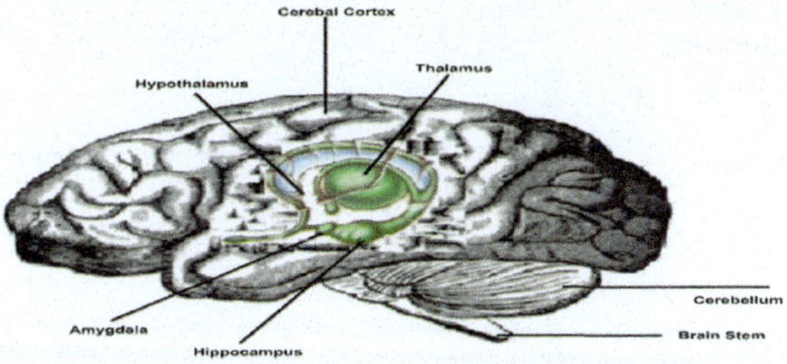


**Fig. 3** Block diagram of BELBIC

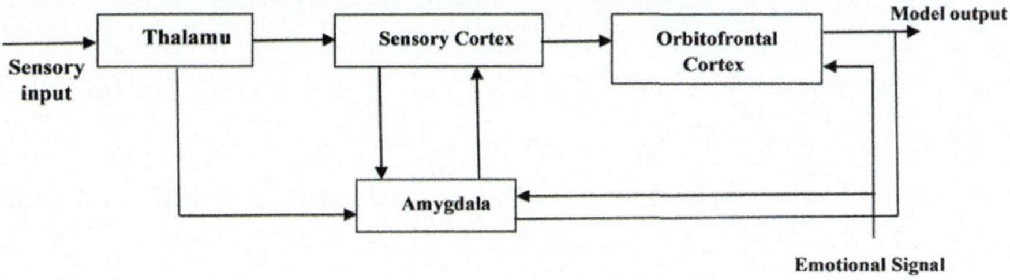


**Fig. 4** A Schematic structure of BELBIC

**Fig. 5** Change in frequency AREA1 with PI & BELBIC for 0.01p.u. Load Change

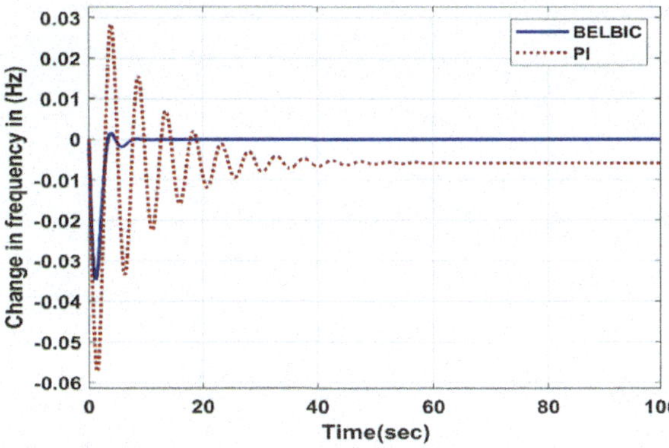


are the sensory cortex of the brain (CX) and the brain's orbital frontal cortex (OFC). As seen in Fig. 4, the system receives a sensory cortex input as well as the major reward signal REW.

This paper describes a controller with a single output and two inputs: the sensory input (SI) and the reward signal (Rew) where kp and ki are the (PI) controller gains and (ACE) is the input. The appropriate controller is created by tuning these gains in order to increase performance. The BELBIC controller's main benefits are its

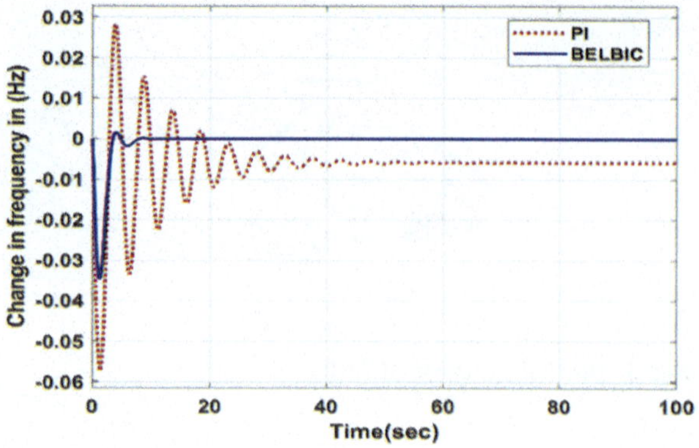


**Fig. 6** Change in frequency AREA2 with PI & BELBIC for 0.01p.u. Load Change

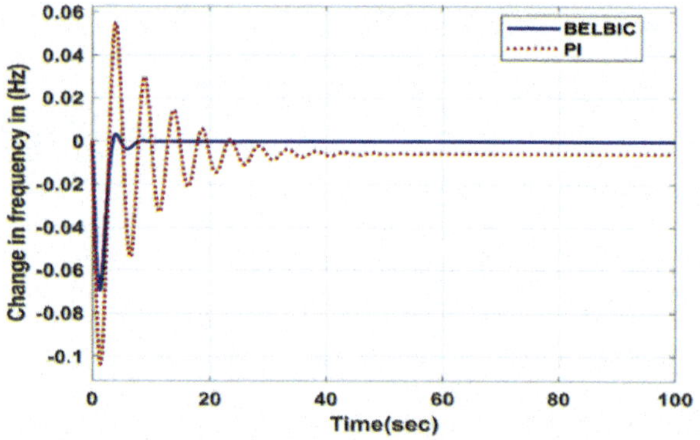


**Fig. 7** Change in frequency AREA1 with PI & BELBIC for 0.02p.u. Load Change

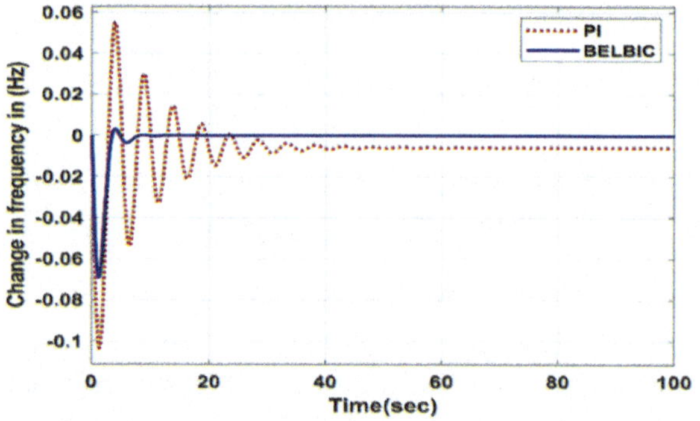


**Fig. 8** Change in frequency AREA2 with PI & BELBIC for 0.02p.u. Load Change

straightforward structure, quick autolearning, good robustness and dynamic performance, and low online computation during controller adaptation. However, for the same application, the BELBIC has large peak overshoots because it reaches a steady state with any random initial values in a shorter amount of time. The sensory signal (***Si***) coming from the thalamus is the only one sent to the sensory cortex (***SCi***) in

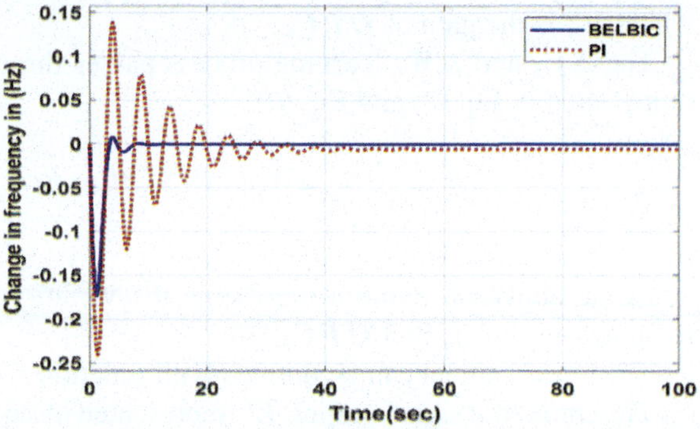


**Fig. 9** Change in frequency AREA1 with PI & BELBIC for 0.05p.u. Load Change

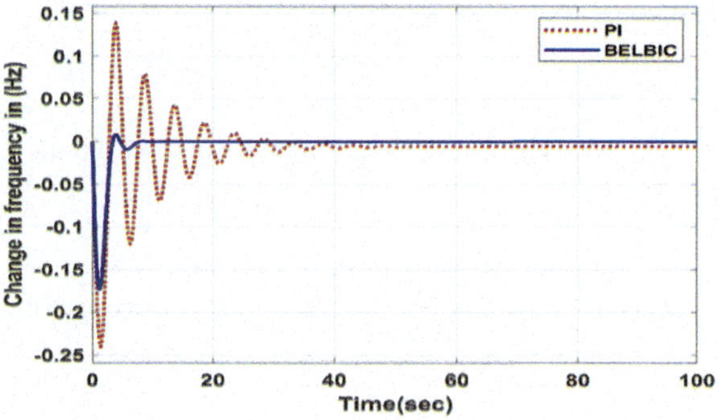


**Fig. 10** Change in frequency AREA2 with PI & BELBIC for 0.05p.u. Load Change

Eq. (4) and Eq. (5) and Amygdala.

$$SC_i = g(S_i) \tag{4}$$

$$g(S_i) = e^{S_i} \tag{5}$$

$A_i$ and $\boldsymbol{O}_i$, Amygdala and Orbitofrontal cortex outputs are represented in Eq. (6) and Eq. (8)

$$A_i = V_i S_i \tag{6}$$

$$\Delta V_i = \propto (S_i \max(0, Rew - \sum_i A_i)) \tag{7}$$

where $\boldsymbol{V}_i$ is the gain of $\boldsymbol{A}_i$.

The orbitofrontal cortex output is expressed as

$$O_i = W_i S_i \tag{8}$$

where $\boldsymbol{W_i}$ is the gain of $\boldsymbol{OFC_i}$.

The $\Delta \boldsymbol{V_i}$ and $\Delta \boldsymbol{W_i}$ learning process passes through the internal weight update represented in Eq. (7) and Eq. (9)

$$\Delta W_i = \beta(S_i(E^i - Rew)) \tag{9}$$

.

In the above, Δ symbol represents the variations in weights. α and β are the learning rates of $\boldsymbol{A_i}$ and $\boldsymbol{OFC_i}$.

The reward signal can be derived with function $J$.

The emotional signal output, $\boldsymbol{E^i}$, represented in Eq. (10) is obtained by combining inhibitory Orbitofrontal Cortex and excitatory Amygdala nodal outputs, as shown below

$$E^i = A_i - O_i \tag{10}$$

The total derived Reward signal is expressed as

$$E = A - O \tag{11}$$

$E$ is the resultant emotional signal of the controller represented in Eq. (11).

## 6 Simulation Results and Discussion

The Interconnected two are power system is simulated in MATLAB with the proposed BELBIC controller under a load demand change of 0.01, 0.02, and 0.05 p.u. in both areas. The frequency deviation Δf1 of the thermal area-1 and the frequency deviation Δf2 of the thermal area-2, for the two areas for the system dynamic responses are shown in Figs. 5, 6, 7, 8, 9, 10. The dynamic response of the system is given in Figs. 7, 8, 9, 10 these performances represent the robustness of the BELBIC Controller where the frequency is better than the PI Controller. The settling time of the BELBIC controller is faster than that of the PI controller. The peak over shoot and peak under shoot of the BELBIC controller are less than that of the PI controller, and steady state error is zero. Finally, the designed controller outperforms the PI controller in terms of dynamic response. The following Tables 1, 2, 3, 4 describe the simulation results of the two controllers BELBIC and PI.

**Table 1** Peak Undershoot (Hz) with different methods

| Controller | Area-I | Area-II |
|---|---|---|
| PI | 0.058 | 0.058 |
| BELBIC | 0.035 | 0.035 |

**Table 2** Peak Overshoot (Hz) with different methods

| Controller | Area-I | Area-II |
|---|---|---|
| PI | 0.028 | 0.028 |
| BELBIC | 0.001 | 0.001 |

**Table 3** Steady State Error with different methods

| Controller | Area-I | Area-II |
|---|---|---|
| PI | −0.003 | −0.003 |
| BELBIC | 00 | 00 |

**Table 4** Settling time with different methods

| Controller | Area-I | Area-II |
|---|---|---|
| PI | 55 | 55 |
| BELBIC | 08 | 08 |

## 7 Conclusion

This paper presents the study on load frequency control (LFC) that was conducted, providing an overview of problems with LFC and approaches for control system design. discussion of various load frequency control techniques, including traditional models and contemporary control principles, will then follow. Modern concepts discuss several optimal LFC schemes as well as intelligent LFC schemes. In this, many models described in the literature that are linked to this research effort were discussed. Intelligent control systems that use various optimization algorithms to manage load frequency may produce better results. The structure of a suitably analyzed BELBIC control algorithm for power grid Control of connected load frequency is discussed. in this article. The Load disturbance in both areas verifies the BELBIC system's rigidity. MATLAB was used to conduct simulation studies. The suggested controller is connected to the load disturbance in both areas more dynamically.

## References


1. Balamurugan CR (2018) Three area power system load frequency control using fuzzy logic controller
2. Tridipta Kumar Pati, Jyoti Ranjan Nayak, Binod Kumar Sahu, Bibek Gantayat (2015) PID controller shows that the proposed approach strategy can reduce the amplitude of oscillation and effectively enhance system stability
3. Khodabakhshian A, Ezatabadi Pour M, Hooshmand R (2012) Design of a robust load frequency control using sequential quadratic programming technique. Int J Electr Power & Energy Syst, 40:1, pp 1–8, ISSN 0142–0615
4. Prakash Ayyappan B, Kanimozhi R (2022). Load frequency control of three area power system using optimal tuning of fractional order proportional integral derivative controller with multi objective grey wolf optimization
5. Arya Y (2019) A new optimized fuzzy FOPI-FOPD controller for automatic generation control of electric power systems. J Franklin Inst 356(11):5611–5629
6. Iroshani Jayawardene, Yawei Wei, Ganesh Kumar (2018) Optimized automatic generation control in a Multi-area power system with particle swarm optimization
7. Sat Aunga, Zaw Min Htikeb (2016) Modeling and simulation of load frequency control for three area power system using proportional Integral Derivative (PID) controller
8. Haiming Niu, Zhuoran Ma, Chaohui Wu (2022) Research on AGC and AVC control technology of photovoltaic power station
9. Wen Tan (2009) Tuning of PID load frequency controller for power systems. Energy Convers Manag, 50:6, pp 1465–1472, ISSN 0196–8904
10. Vijaya Santhi R, Sudha KR, Prameela Devi S (2013) Robust load frequency control of multi-area interconnected system including SMES units using Type-2 Fuzzy controller In: 2013 IEEE International Conference on Fuzzy Systems (FUZZ-IEEE), Hyderabad, India, pp 1–7, https://doi.org/10.1109/FUZZ-IEEE.2013.6622324
11. Rasolomampionona D, Klos M, Cirit C, Montegiglio P, De Tuglie EE (2022) A new method for optimization of load frequency control parameters in multi-area power systems using Genetic Algorithms. In: IEEE International Conference on Environment and Electrical Engineering and 2022 IEEE Industrial and Commercial Power Systems Europe (EEEIC/I&CPS Europe), Prague, Czech Republic, 2022, pp 1–9, https://doi.org/10.1109/EEEIC/ICPSEurope54979.2022.9854535
12. Naresh Kumar, Shri M.G Soni (2018) Automatic generation control for three area system using improved bacteria foraging optimization algorithm (BFOA)
13. Nanda J, Mangla A (2004) Automatic generation control of an interconnected hydro-thermal system using conventional integral and fuzzy logic controller. In: 2004 IEEE International Conference on Electric Utility Deregulation, Restructuring and Power Technologies. Proceedings, Hong Kong, China, pp 372–377 1, https://doi.org/10.1109/DRPT.2004.1338524
14. Farahani M, Ganjefar S, Alizadeh M (2012) PID controller adjustment using chaotic optimization algorithm for multi-area load frequency control. IET Control Theory Appl 6(13):1984–1992. https://doi.org/10.1049/iet-cta.2011.0405,PrintISSN1751-8644,OnlineISSN1751-8652
15. Zaid Hamodat, Galip Cansever. (2021) Automated generation control of Multiple-Area electrical system with an Availability-Based tariff pricing scheme regulated by whale optimized fuzzy PID controller 35
16. Tridipta Kumar Pati, Jyoti Ranjan Nayak, Binod Kumar Sahu, Bibek Gantayat (2015) PID controller shows that the proposed approach strategy can reduce the amplitude of oscillation and effectively enhance system stability
17. Diem-Vuong Doan, Khoat Nguyen, Quang-Vinh Tha (2022) Load-Frequency control of Three-Area interconnected power systems with renewable energy sources using novel PSO~PID-Like fuzzy logic controllers
18. Vasadi Srinivasa Rao PD, Srinivas CH, Ravi Kumar (2015) Load Frequency Control (LFC) of three area interconnected power system using adaptive neuro fuzzy interface system

19. RuiXiang Fan, Jingtao Zhao, Benren Pan, Na Chen, Taoyuan Wang and Hongzhong Ma (2014) Automatic generation control of three-area small hydro system based on fuzzy PID control
20. Surya Prakash, Sinha SK (2011) Load frequency control of three area interconnected hydro-thermal reheat power system using artificial intelligence and PI controllers
21. Emre Ozkop, Ismail H Altas, Adel M Sharaf (2010) Load frequency control in four area power systems using fuzzy logic PI control
22. Snehal Gondaliya, Krishan Arora (2015) Automatic generation control of multi area power plants with the help of advanced controller
23. Electrical Power System Analysis. Front Cover Sivanagaraju, Rami Reddy BV (2007) Firewall Media p 345
24. Power Generation, Operation and Control (2013) 3rd Edition Allen J Wood, Bruce F Wollenberg, Gerald B Sheble. ISBN: 978–0–471–79055–6 p 656
25. Moren J, Balkenius C (2000) A Computational model of emotional learning in the amygdala: from animals to animals. In Proc. 2000 International Conference on the Simulation of Adaptive Behavior, Cambridge, Mass., The MIT Press, pp 383–391
26. Lucas C, Shahmirzadi D, Sheikholeslami N (2004) Introducing BELBIC: brain emotional learning based intelligent controller. Int J Intell Autom Soft Comput 10:11–22
27. João Paulo Coelho, Tatiana M Pinho, José Boaventura-Cunha, Josenalde B de Oliveira (2017) A new brain emotional learning Simulink® toolbox for control systems design, IFAC-Papers Online, 50: 1, pp 16009–16014, ISSN 2405–8963, https://doi.org/10.1016/j.ifacol.2017.08.1912

# Chaotic Quasi-Oppositional Differential Search Algorithm for Transient Stability Constraint Optimal Power Flow Problem

**Sourav Paul, Sneha Sultana, Provas Kumar Roy, Nirmalya Dey, Pravin Kumar Burnwal, and Devjeet Sengupta**

**Abstract** In this research work, the concept of Chaotic Quasi-Oppositional Differential Search Algorithm (CQODSA) has been successfully applied to address the transient stability constraint optimal power flow problem. The effectiveness of the suggested algorithm has been evaluated on WSCC 3-generator, 9-bus system and New England 10-generator, 39-bus system. The recommended algorithm's implementation has been evaluated for different fault conditions with the purpose of demonstrating CQODSA's applicability in this versatile scenario. By contrasting the findings with those of other well-known algorithms, the superiority of the established method has been proven.

**Keywords** Optimal power flow · Transient stability constraint · Chaotic quasi-oppositional based learning · Differential search algorithm

## 1 Introduction

A network used to transfer energy from generators to loads is called a power system. The electric power transmission network is used to convert energy from generators to loads. The complexity of the current power system network is gradually increasing for planning and operation purposes due to large power transfers across greater distances, complicated coordination, the challenging connection between multiple system controllers, and low power reserves. It is considered that a power system is secure when it can resist unexpected interruptions with little degradation in service quality. Whenever there is a disruption, the system experiences the following transient before entering an acceptable stable state in which all operational restraints are kept

S. Paul (✉) · S. Sultana · N. Dey · P. K. Burnwal · D. Sengupta
Dr. B. C. Roy Engineering College, Durgapur, India
e-mail: sourav.p01@gmail.com
URL: http://www.bcrec.ac.in

P. K. Roy
Kalyani Government Engineering College, Kalyani, India



O. H. Gupta et al. (eds.), *Soft Computing Applications in Modern Power and Energy Systems*, Lecture Notes in Electrical Engineering 1107,
https://doi.org/10.1007/978-981-99-8007-9_14

under the permitted boundaries. The true and reactive powers of the generators in an interlinked power system are to be flexible within practical bounds based on using the least amount of fuel possible while still meeting the specific load requirements.

The concept of optimal power flow underpins the operation, control, and planning of power systems. (OPF). Dommel et al. [1] was the one who initially presented it. After that, the community of power system researchers proposed a number of OPF techniques. Currently, OPF is a standard power system analysis technique and has proven to be trustworthy enough for practical application. As a result, the core objective of the OPF solution is to create a power production system that reduces the cost of fuel while meeting all equity and inequality criteria. The OPF problem is typically a wide-ranging, multi-objective, highly nonlinear, non-convex optimization problem. For the purpose of minimizing the desired objective function while still satisfying the power balancing equations and some inequality constraints in the system, the ideal steady-state operation of a power system needs to be determined. This is done using a nonlinear programming problem. The limits of the independent and dependent variables are the restrictions on inequality, as well as the load flow equations, which are the power balancing equations. The generator real powers, omitting slack bus power, generator bus voltages, transformer tap settings, and reactive power injection, are independent variables. Load bus voltages, slack bus power, line flows, and generator reactive powers are the dependent variables. Though reducing fuel costs is the major goal of the OPF, however, because of the continuous rise in energy demand and incomparable expansions in production and transmission capacity, power system construction and operation are being faced with a new challenge: voltage instability. Reactive power sources in electrical systems that are insufficient might cause a significant transmission loss. In these situations, it could be required to make the objective of the OPF issue take into account transmission loss and voltage stability margin. By lowering losses in transmission and generator reactive margins in the Spanish power system, Lobato et al. [2] proposed an OPF-based on linear programming (LP). This method simulates the discrete characteristics of capacitors and shunt reactors by using integer variables. It was discovered that the method, which linearized both the constraints and the goal function in each iteration, was superior to one that only linearized the objective function once. By extending the basic Kuhn-Tucker requirements, Momoh [3] made use of a generalized quadratic-based model for the OPF. The proposed OPF technique took into account the feasibility, convergence, and optimality requirements. In order to include various goal functions and selectable constraints, it was also capable of using hierarchical structures as well. The generalized method combines the sensitivity of a goal function to changes in constraints with optimal performance, leading to a globally optimal solution. Requirements for algorithmic memory and execution speed were reduced. The Newton-based techniques [4, 5], the reduced gradient method, Lagrangian relaxation (LR), interior point methods (IPM) [6, 7], and linear programming (LP) [8]. Roy et al. [9] and Abou [10] discovered in the literary [11–13] are some of the traditional methods for solving OPF problems that have evolved over the past 20 years. These strategies use the starting point and convexity to locate an overall ideal solution. Techniques, however, were unable to resolve complex objective functions that are not differentiable. The nonlinear objective function and constraint problems may be addressed using nonlinear programming (NLP) [7]. This method, however, also has a number of disadvantages, including unstable convergence features and excessive numerical iterations that lead to enormous computations, large execution times, and algorithmic complexity. Some of these methods are effective at finding the best solution, but they frequently converge slowly and have high computing costs. The limitation of Newton-based approaches is that their merging traits are sensitive to the starting circumstances, and due to an improper initial condition, they may not even converge. Heuristic approaches have recently achieved significant advancements in the optimization of power supply network. A significant quantity of heuristic algorithms seemed suggested over the recent years to find the best solution or to be highly effective in addressing nonlinear OPF issues, including GA [14, 15], SA [16], TS [17], PSO [17, 18], Chaotic Ant Swarm Optimization (CASO) [19], BFO [20], DE [21]; Sivasubramani and Swarup [22], Biogeography Based Optimization (BBO) [23], Harmony Search Algorithm (HSA) [24], ABC [25], and GSA [26] are all examples of hybrid evolutionary programming (HEP). They have been effectively used to solve a variety of issues where finding a global solution is preferable to finding a local one or where the issue contains non-differentiable regions. Basu [27] applied the multifaceted discrete evolution method to the OPF problem using FACTS instruments in IEEE 30-bus and IEEE 57-bus systems, and the outcomes were compared to prior studies. In order to update the gravitational acceleration, Hadi [28] presented a non-dominated sorting gravitational search algorithm (NSGSA) in 2012. This Sivasubramani and Swarup [22] introduced the multi-objective Harmony Search Algorithm (HSA) to assess the OPF issue. The outcomes of the suggested method were contrasted with NSGA-II in this study, and numerous objectives including fuel expenditure, L-index, and transmission loss were taken into account. Voltage deviation, total real power losses, total emission, and the overall fuel expense of thermal units were contrasted with that of other algorithms. Abido and Bakhashwain [29] proposed a technique for solving the restricted optimal VAR dispatch problem with objective functions that included real power outages and varying voltages as objective functions. The suggested technique's durability and precision were tested using the IEEE 30-bus system.

One of the numerous OPF issues is the OPF with transient stability limitations, and it makes for an intriguing research topic since it considers not only certain optimum solutions but also all security and stability requirements. The boundaries that system dynamics must meet to be stable under transient conditions are referred to as transient stability constraints. In contemporary power systems, it is extremely expensive to lose synchronous due to transient instability. Employing transient stability-constrained optimal power flow is a more effective way to strike a balance between the security and economics of the power system. (TSC-OPF). The main aim of the TSC-OPF is to combine the economic objectives of stability constraints and steady-state goals into a single, original formulation. One approach is to treat the TSC-OPF as an algebraic optimization problem that may be dealt with using Gan et al. [30] by using standard optimization techniques. The TSC-OPF issue is then interpreted through a transformation of differential equations as inequality constraints into comparable algebraic equations. Based on this, Yuan et al. [31], who developed a fresh method for incorporating a model of transient stability into conventional OPF which significantly decreases computational load, is the source of the multi-contingency concept. However, it has significant limitations brought on by the enormous number of variables, which also causes computation errors for approximations in addition to convergence issues. In order to change the TSC-OPF issue's infinitesimal dimension into a finite-dimensional issue, Chen et al. [32] propose a new technique called functional transformation. But due to the high sensitivity issue of the beginning conditions, these solutions are unfortunately not applicable to real-world systems. Because of the computing challenges, types of limits in OPF concerns which are the transient stability limit and the flexibility of the objective functions are both limited. Because of this, it is essential to develop optimization methods that can get around these limitations and manage these challenges. The artificial bee colony (ABC) by Karaboga et al. [33], the improved genetic algorithm (GA) by Chan et al. [34], and differential evolution by Cai et al. [35] are a number of population-based optimization techniques that have been effectively used to solve TSC-OPF issues. In this chapter, the researcher suggested chaotic quasi-oppositional based differential search algorithm (CQODSA), for addressing TSC-OPF issues by taking into consideration the New England 10-generator, 39-bus system, and the WSCC 3-generator, 9-bus system.

## 2 Formulation of Mathematical Problems

### 2.1 Objective Function

The purpose of OPF is to reduce the objective function while preserving all equality and inequality constraints. The OPF issue might be stated as follows:

$$\text{Minimize OF}(x, y) \tag{1}$$

$$\text{Subject to: } \begin{Bmatrix} e(x, y) = 0 \\ ie_i \leq ie(x, y) \leq ie_u \end{Bmatrix} \tag{2}$$

in which, $\text{OF}(x, y)$ and $e(x, y)$ are the objective function and equality constraints respectively; $ie(x, y)$ is the inequality constraints; $ie_i, ie_u$ are the minimum and maximum limits of the inequality constraints; x is the vector of dependent variables composed of slack bus active power, load voltages, reactive power of generators, and loadings of transmission lines; y is an independent variable vector comprised of reactive power injections, alternator voltages, transformer tap settings, and active power injections from generators other than the slack bus. Thus, the expression for $x$ and $y$ is

$$x = [P_{g1}, V_{l_1} \ldots V_{l_{NL}}, Q_{g_1} \ldots Q_{g_{NG}}, S_{l_1} \ldots S_{l_{NTL}}] \tag{3}$$

$$y = [P_{g2}...P_{gNG}, V_{g1}...V_{gNG}, T_1...T_{NT}, Q_{c_1}...Q_{cNC}] \tag{4}$$

where *NG* and *NL* are the number of generator and load buses; *NTL* is transmission line's number; the numbers *NT* and *NC* represent the amount of transformers and shunt compensators.

The overall generation cost is provided with an optimal remedy in this study. The total fuel expenditure of the generating units is given for the exponential expense function by

$$F_c(u, v) = \sum_{p=1}^{NG} (a_p P_{g_p}^2 + b_p P_{g_p} + c_p) \tag{5}$$

where $a_P$, $b_p$, $c_P$ are the gasoline expenditure multipliers, and $P_{g_p}$ is the $p$th generation.

Because many valve steam turbines are used in real generators to provide more precise and adaptive operation when determining an acceptable cost function, the valve spot effect should be considered by generating units with valve point loading in order to meet the overall fuel expense:

$$F_c(u, v) = \sum_{p=1}^{NG} (a_p P_{g_p}^2 + b_p P_{g_p} + c_p + d_p \sin[e_p(P_{g_p}^{\min} - P_{g_p})]) \tag{6}$$

where $d_p$, $e_p$ are the energy expense factors with inlet point transferring impact and $P_{g_p}^{\min}$ is the $p$th generator's minimum generation.

### 2.2 *Transient Stability Assessment*

To characterize the fluctuating behaviour of a system containing unstable moments, the angle of rotor angles in relation to the inertial centres for each alternator are given here. According to Athay et al. [36], the centre of inertia is located as follows:

$$\delta_{\mathrm{COI}} = \frac{\sum_{k=1}^{N_{NG}} M_k \delta_k}{\sum_{k=1}^{N_{NG}} M_k} \tag{7}$$

where $M_k$ is the $k$th generator's inertia constant and $k$ is the $k$th generator's rotating angle. The rotor angle limitation can be expressed in terms of limitations of inequality as follows:

$$\delta' \leq \delta_k - \delta_{\mathrm{COI}} \leq \delta'' \tag{8}$$

$$K \subseteq S_G \tag{9}$$

where the set of generators is $S_G$ and the lower and higher power source rotor angle parameters are $\delta', \delta''$.

## 2.3 Constraints

**Equality Constraints**: The equality constraints applied to the load flow equations are as follows:

$$\begin{cases} P_{gi} - P_{Li} = \sum_{j=1}^{N} |V_i| \left|V_j\right| G_{ij}\cos\delta_{ij} + B_{ij}\sin\delta_{ij} \\ Q_{gi} - Q_{Li} = \sum_{j=1}^{N} |V_i| \left|V_j\right| G_{ij}\sin\delta_{ij} - B_{ij}\cos\delta_{ij} \end{cases} \quad (10)$$

where $V_i$, $V_j$ are the voltages of the $i$th and the $j$th bus; $P_{gi}$, $Q_{gi}$ are the active and reactive powers of the $i_{th}$ generator; $P_{Li}$, $Q_{Li}$ are active and reactive power of the $i_{th}$ load bus; $G_{ij}$, $B_{ij}$, $\delta_{ij}$ are the conductance, admittance, and phase difference of voltages between the $i$th and the $j$th bus; $N$ is the number of buses.

**Inequality Constraints**

(i) **Generator constraints**: The $i$th bus's generator voltage, reactive power, and active power are all between their lower and upper limits, as represented by

$$\begin{cases} V_{gi}^{\min} \le V_{gi} \le V_{gi}^{\max} & i = 1, 2, \ldots., NG \\ P_{gi}^{\min} \le P_{gi} \le P_{gi}^{\max} & i = 1, 2, \ldots., NG \\ Q_{gi}^{\min} \le Q_{gi} \le Q_{gi}^{\max} & i = 1, 2, \ldots., NG \end{cases} \quad (11)$$

in which, $V_{gi}^{\min}$, $V_{gi}^{\max}$ are the minimum and maximum generator voltages of the $i$th generating unit; $P_{gi}^{\min}$, $P_{gi}^{\max}$ are the minimum and maximum active powers of the $i$th generating unit; $Q_{gi}^{\min}$, $Q_{gi}^{\max}$ are the minimum and maximum reactive powers, of the $i$th generating unit, respectively.

(ii) **Load bus constraints**:

$$V_{l_i}^{\min} \le V_{l_i} \le V_{l_i}^{\max} \quad i = 1, 2, \ldots., NL \quad (12)$$

where $V_{l_i}^{\min}$, $V_{l_i}^{\max}$ are the minimum and maximum load voltages respectively of the $i$th generating unit.

(iii) **Transmission cable limitations**:

$$S_{li} \le S_{l_i}^{\max} \qquad i = 1, 2, \ldots.NTL \quad (13)$$

where $S_{l_i}$ is the evident electricity flow of $i$th branch; $S_{l_i}^{\max}$ is the perceived high-powered flow limit of the $i$th branch.

(iv) **Transformer outlet restrictions**: The upper and lower boundaries of the transformer input settings are mentioned below

$$TC_i^{\min} \leq TC_i \leq TC_i^{\max} \quad i = 1, 2, ....NT \tag{14}$$

where $TC_i^{\min}$, $TC_i^{\max}$ are the minimal and most prominent maximum setting tap, respectively, of the $i$th transformer.

(v) **Constraints on shunt compensators**: The following are the boundaries of shunt compensation:

$$Q_{C_i}^{\min} \leq Q_{C_i} \leq Q_{C_i}^{\max} \quad i = 1, 2, ...., NC \tag{15}$$

where $Q_{C_i}^{\min}$, $Q_{C_i}^{\max}$ are the lower and highest VAR injection limits, respectively, of the $i$th shunt capacitor.

# 3 Differential Search Algorithm (DSA)

## 3.1 *Inspiration of DSA*

This section discusses Civicioglu's differential search algorithm (DSA), a novel metaheuristic optimization method [68]. DSA takes its cues from the natural migration patterns of living things. Due to changes in the seasons throughout the year, there are differences in the availability and effectiveness of food areas that exist in nature. Living things exhibit migration behaviour as a result throughout the year. This makes it possible for living things to transition from one habitat to a different habitat, where natural resources are more plentiful and effectively used. This serves as the primary drive for calculating DSA in the literature [37].

## 3.2 *Modelling of DSA Mathematically*

**Initialization** T' *artificial organism* DSA is generated through a chance procedure that points to the difficulty of optimization. DSA synthetic creature initialization is characterized by Eq. (2.8) [37]:

$$X(i, j) = \text{rand} \times (ub(j) - lb(j)) + lb(j) \tag{16}$$

where $i = 1, 2, 3, ......, n_p$, $j == 1, 2, 3, ......, \dim$, $X(i, j)$ is described as the initial population of the $j$th variable of $i$th agent, rand is a number chosen randomly in between [0, 1], $n_p$ and dim are interpreted as the number of search agents (superorganism) and a number of control variables of the optimization problem, and also $ub(j)$ and $ib(j)$ are the upper and lower limits of $j$th variable, respectively.

**Superorganism Reshuffle at Random**. Choose a *superorganism* at random from the beginning populace and move it towards the target *donor* using random shuffling in search of the *stopover site*, which plays a crucial role in an effective. *scale* Gamma randomly generate establishes scale value to regulate the functions of synthetic Eq. (2.9) [37]:

$$\text{scale} = \text{randg}(2 \times r_1) \times (r_2 - r_3) \tag{17}$$

where randg is a gamma distribution value selected at random and $r_1$, $r_2$, and $r_3$ are numbers that are chosen at random through (0,1). In DSA, the *stopoversite* of *superorganism* is identified as [37].

**Boundary Condition**. In DSA, there are two additional variables at most, i.e., $p_1$ and $p_2$ are members of that group chosen at random *superorganism* to interact in *stopoversite*. The values that are typically created between [0, 0.3] whenever a part of the *stopoversite* exceeds the restrictions of the environment, then a boundary condition is put on the components of *stopoversite* them into the ecosystem through forcibly [37].

## 4 Quasi-Oppositional Based Learning (Q-OBL)

OBL was first introduced by Tizhoosh [38] as an revolutionary notion for intelligence-or soft virtualization tactics to progress diverse simulated procedures. It appears to be among the most successful theories in algorithmic wisdom for tackling non-linear optimization problems, enhancing the searching capabilities of traditional population-based optimization strategies. The establishment of the first estimate, which is based on some prior knowledge about the answer or random, is where the OBL process begins. The best course of action may be in any way, or at least the opposite. The opposing set of estimates is taken into account for convergence, replacing the starting estimates repeatedly for a better answer in the way of optimal solutions.

### 4.1 Opposite Numbers

Let the real number be represented by $P \in [y, z]$ and $P^0$ be the meaning of a numerical value for its reverse by

$$P^0 = y + z - P. \tag{18}$$

### 4.2 Opposite Pointing

Say $R = (X_1, X_2, \ldots\ldots, X_n)$ is a point in n-dimensional space, where $P_r \in [y_r, z_r]$, $r \in 1, 2, \ldots, n$. The $R^0 = ({X_1}^0, {X_2}^0, \ldots\ldots {X_n}^0)$. Its constituents determine the opposite case:

$$ {P_r}^0 = y_r + z_r - P_0. \tag{19} $$

### 4.3 Quasi-Opposite Number and Quasi-Opposite Point

Let $P$ be a real number between $[y, z]$. The quasi-opposite number is defined as

$$ P^{Q_0} = \text{rand}(C, \tilde{P}) \tag{20} $$

where $C$ is given by $C = \frac{y+z}{2}$. Let $P$ be a real number between $[a, b]$. The quasi-opposite point ${P_r}^{Q_0}$ is defined as

$$ {P_r}^{Q_0} = \text{rand}(C_r, \tilde{P_r}) \tag{21} $$

$$ \text{where} \quad C_r = \frac{y_r + z_r}{2}. $$

### 4.4 Chaotic DSA

Chaotic DSA (CDSA), which combines chaotic action with differential search optimization (DSA), was created to lessen the shortcomings of DSA. Chaos, which is stochastic and non-repeating by nature, conducts general searches at higher speeds, which is essential for quickening a deify algorithm's convergence rate. A chaotic set is composed of a total of 10 chaotic maps, each of which exhibits a different action. This set's starting point has been determined to be 0.7 between 0 and 1. The many chaotic map actions help to solve issues with local optimum and convergence speed.

## 5 CQODSA Steps for OPF Problem

Different steps of the CQODSA for solving the OPF problem are presented below

**Step 1**: Define *superorganism* and create the first wave of people (P) of each individual at random. Also, generate the PSS control parameters with compute the feature subset in the specified subspace.

**Step 2**: Generate chaotic quasi-opposite population and calculate the fitness function.

**Step 3**: Arrange the group from best to worst according to their fitness value. Thus far, the best answer achieved is represented by *PE*. An initial *KE* is given to every particle.

**Step 4**: According to value, only a select handful of the best answers are privileged status-preserved solutions.

**Step 5**: Those not belonging to the elite solutions are once more subjected to the varied chaotic map conduct of the CDSA in order to change the independent variables. In CDSA, the whales target the focuses particularly and the location of the fixates is regarded as the best location.

**Step 6**: Randon shuffling of *superorganism* takes place and evaluate the scale using the gamma distribution.

**Step 7**: The method is finished and the outcomes can be seen if the stopping requirement is met; otherwise, proceed to **Step 4**.

## 6 Simulation Results and Discussions

Two test systems, the Western Systems Co-ordinating Council (WSCC) 3-generator, 9-bus system and the New England 10-generator, 39-bus system, were used to evaluate the applicability and validity of the various optimization strategies. Each time, the synchronous generators are all represented by a classical generator model, while the loads are all represented by constant impedance models. All simulations are run on a personal computer with an Intel Core i5 3.1 GHz processor and 8GB of RAM using MATLAB 2022b. The entire simulation time is set at 3.0 s, with an integration time step of 0.01 s. 100 iterations are chosen as the terminating condition.

### *6.1 Test System-I (WSCC 3-Machine 9-Bus System)*

**Case I: Base loading conditions (OPF without transient stability constraint)** As shown in Fig. 1, The infrastructure is made up of 3 generating units, 9 buses, 2 regulatory transformers and 2 shunt converters, and 11 transmission lines. The data for the system bus, the lines, the capacities of the generators, and the cost coefficients are extracted from Nguyen and Pai [39]. In contrast, the VAR and the regulatory transformers follow the prompt injections of the shunt capacitors, which are both active powers and voltages regarded as parameters of a generator (apart from slack bus-1) are considered continuous variables. The span of [0.95 p.u., 1.1 p.u.] and [0.95 p.u., 1.05 p.u.] [39] for the load buses' and generators' respective voltage amplitudes is taken into consideration. Regulating transformer tap options are bound to a range of [0.9 p.u. and 1.1 p.u.]. For the shunt capacitors' Var injection restrictions, the range of [0 Mvar, 5 Mvar] with a distinct initial value taken into

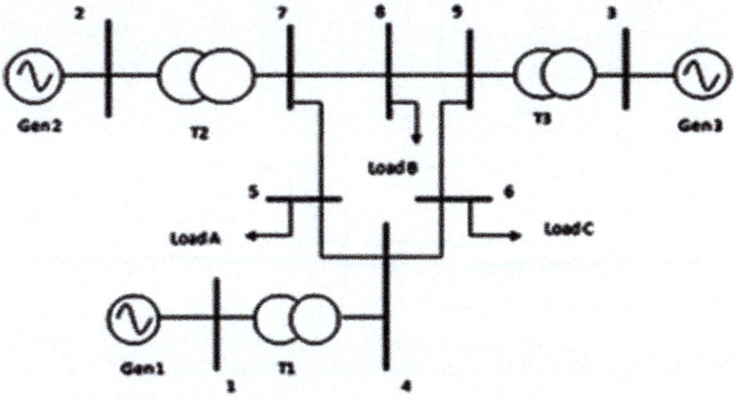


**Fig. 1** The WSCC 3-machine 9-bus test system

**Table 1** Simulation outcomes achieved by different methodologies for Case I (Test system-I)

| Algorithms | BPD | ABC | DE | TS | TDS | TLBO | QODSA | CQODSA |
|---|---|---|---|---|---|---|---|---|
| Pg1 (MW) | 105.94 | 107.15 | 105.94 | 106.19 | 105.94 | 105.3200 | 104.7600 | **104.1235** |
| Pg2 (MW) | 113.04 | 114.18 | 113.04 | 112.96 | 113.04 | 113.1800 | 112.6751 | **112.6512** |
| Pg3 (MW) | 99.23 | 96.74 | 99.29 | 99.20 | 99.24 | 99.3700 | 97.9095 | **98.1218** |
| Vg1 (p.u.) | 1.05 | 1.033 | 1.050 | 1.000 | 1.05 | 1.0998 | 1.0321 | **1.0054** |
| Vg2 (p.u.) | 1.05 | 1.024 | 1.050 | 1.000 | 1.05 | 1.1000 | 1.0980 | **1.0457** |
| Vg3 (p.u.) | 1.04 | 1.028 | 1.040 | 1.000 | 1.04 | 1.0999 | 1.0323 | **1.0012** |
| FC ($/hr) | 1132.17 | 1131.87 | 1132.30 | 1132.59 | 1132.18 | 1131.0340 | 1131.0112 | **1131.0067** |

consideration of 0.5 Mvar, the simulation outcomes for the fuel price fitness function are shown in Table 1. The computed results have been compared with MDE [40], BBO [41], GSA [42], and TLBO [43]. The minimum objective function values for CQODSA and DSA are 1131.0112 and 1131.0067, respectively, when contrasted with the additional strategies in Table 1. Hence, it can be inferred from Table 1 that, in comparison to other algorithms, CQODSA and DSA algorithms provide more appropriate controller parameters for Case I.

**Case II: OPF with transient stability constraint (3-phase to ground fault at bus 7)** To be precise, an asymmetric specifically 3-phase to ground at bus 7 and among lines 7–5 as illustrated of Fig. 2 in a network, in order to discuss if the recommended algorithms are suitable for handling the power systems. By opening the circuit breaker contact for 0.35 s, the aforementioned problem was fixed. This case's outcome complies with the transient stability limit. In Table 2, the minimal full cost values are compared: Artificial Bee Colony Simplification (ABC) [44], Dynamic Variation (DV) [35], Sensitive to Trend line (TS) [35], Simulations in Spatial Domain [35], and TLBO [43]. The outcome demonstrates the superiority of the suggested CQODSA in finding the overall ideal solution. Using the suggested CQODSA algorithm, the solution is reliable and safe.

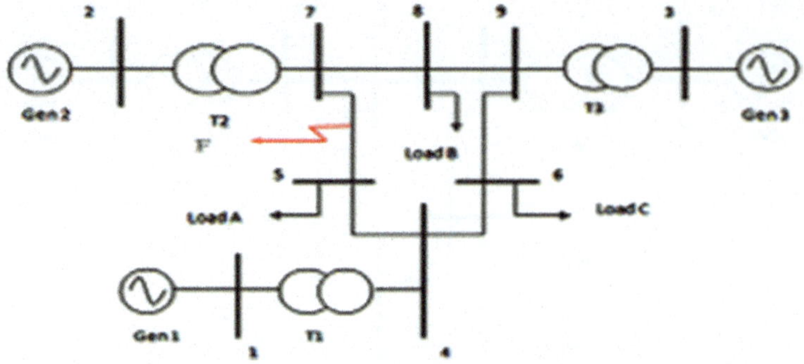


**Fig. 2** The 3-machine, 9-bus WSCC test system has a problem at bus 7

**Table 2** Simulations outcomes achieved by various methodologies for Case II (WSCC 9-bus system)

| Algorithms | ABC | DE | TS | TDS | TLBO | QODSA | CQODSA |
|---|---|---|---|---|---|---|---|
| Pg1 (MW) | 117.69 | 130.94 | 170.20 | 117.85 | 105.3100 | 104.1209 | **103.7840** |
| Pg2 (MW) | 105.89 | 94.46 | 48.94 | 103.50 | 113.3300 | 101.8956 | **101.8765** |
| Pg3 (MW) | 94.23 | 93.09 | 98.74 | 96.66 | 99.2400 | 95.99 | **97.8970** |
| Vg1 (p.u.) | 1.025 | 0.9590 | 1.000 | 1.05 | 1.0988 | 1.04 | **1.0789** |
| Vg2 (p.u.) | 1.070 | 1.0139 | 1.000 | 1.05 | 1.0987 | 1.05 | **1.0459** |
| Vg3 (p.u.) | 1.070 | 1.0467 | 1.000 | 1.04 | 1.0999 | 1.03 | **1.0897** |
| FC ($/hr) | 1133.18 | 1140.06 | 1179.95 | 1134.01 | 1131.1908 | 1132.12 | **1130.1289** |

## *6.2 Test System II (New England 10-Machine 39-Bus System)*

The New England 10-machine 39-bus test system is also used to perform the suggested techniques as depicted in Fig. 3. There are 10 generator buses and 19 load buses in the network. Bus 39 is going to act as the load bus. The load buses' and the generators' respective voltage magnitude spans are [0.95 p.u., 1.1 p.u.] and [0.95 p.u., 1.05 p.u.]. Limitations are established among [0.9 p.u. and 1.1 p.u.], [0.9 p.u., 1.1 p.u.], [0.9 p.u., 1.1 p.u.], [0.9 p.u., 1.1 p.u.], [0.9 p.u., 1.1 p.u.], and [0.9 p.u., 1.1 p.u.]. The [0 Mvar, 5 Mvar] span with a discrete value of 0.5 Mvar moves is taken into consideration for using shunt capacitors' Var insertion limitations. The primary objective is to lower the total gasoline prices. The simulation results for the wellness feature (the price of fuel), 3–4 indicate active production and voltage. The system features a 3-phase asymmetrical fault across lines 28 and 29, as well as at bus 29. By briefly opening the connections on the circuit breakers, the aforementioned issue can be resolved. The actual power generation and voltage of various generating units, as well as the fuel price and transmission loss, were calculated using the Classical Model (CM) [45] and the Dynamic model (DM [45], Variables simulation method (DSA) [46] prototype and TLBO [43] (strategies described in the literature

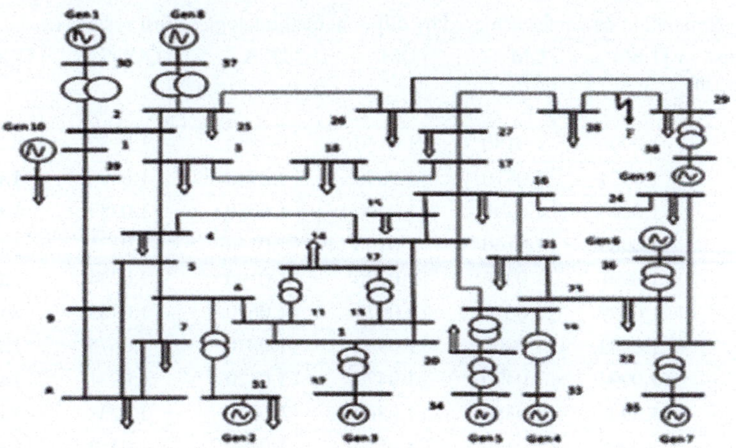

**Fig. 3** The 10-machine, 39-bus New England test system

**Table 3** Simulation outcomes achieved by different methodologies (Test system-II)

| Active power generation (in MW) | TLBO | CM | DM | DSA | QODSA | CQODSA |
|---|---|---|---|---|---|---|
| PG30 | 2.4422 | 248.7300 | 249.4470 | 247.8300 | 249.4470 | **247.8300** |
| PG31 | 5.6007 | 577.8400 | 578.3590 | 577.2300 | 578.3590 | **577.2300** |
| PG32 | 6.4363 | 654.4700 | 654.3560 | 653.4100 | 654.3560 | **653.4100** |
| PG33 | 6.3298 | 645.0000 | 641.7600 | 643.2800 | 641.7600 | **643.2800** |
| PG34 | 5.1098 | 518.8200 | 517.4100 | 517.7800 | 517.4100 | **517.7800** |
| PG35 | 6.5306 | 664.3200 | 660.7310 | 662.4600 | 660.7310 | **662.4600** |
| PG36 | 5.6103 | 571.3700 | 568.1810 | 569.5900 | 568.1810 | **569.5900** |
| PG37 | 5.3286 | 547.8100 | 547.6950 | 543.8800 | 547.6950 | **543.8800** |
| PG38 | 8.4015 | 752.0200 | 754.6180 | 774.5400 | 754.6180 | **774.5400** |
| PG39 | 9.5845 | 995.6000 | 1003.1810 | 1000.3500 | 1003.1810 | **1000.3500** |
| Cost ($/hr) | 60873.5574 | 61600.7600 | 61597.7600 | 61799.6800 | 61597.7600 | **61799.6800** |

are compared with the wellness feature values). An analysis of Table 4 reveals that, when compared to the other algorithms listed there, CQODSA gives the smallest minimum objective function value, 41.1290. Accordingly, it can be inferred from Table 4 that CQODSA and QODSA algorithms as shown in Tables 3 and 4 give more appropriate controller settings than alternative algorithms. Table 4 includes information on the voltage and actual power generation of different generating units, the fuel usage, and the energy losses as determined by various methods. We could conclude from this that the gasoline cost acquired using the provided CQODSA and QODSA approaches is significantly less than that gathered utilizing other methods.

**Table 4** Simulation outcomes achieved by different methodologies (Test system-II)

| Generated voltages (in p.u.) | TLBO | CM | DM | DSA | QODSA | CQODSA |
|---|---|---|---|---|---|---|
| V30 | 1.0996 | 1.0150 | 1.0150 | 0.9840 | 1.0132 | **0.7867** |
| V31 | 1.0924 | 1.0870 | 1.0870 | 1.0740 | 1.0675 | **1.0345** |
| V32 | 1.0863 | 1.0290 | 1.0290 | 1.0080 | 1.0289 | **1.0066** |
| V33 | 1.0985 | 1.0160 | 1.0160 | 1.0140 | 1.0131 | **1.0091** |
| V34 | 1.0997 | 1.0220 | 1.0220 | 1.0190 | 1.0234 | **1.0237** |
| V35 | 1.0999 | 1.0620 | 1.0620 | 1.0670 | 1.0453 | **1.0453** |
| V36 | 1.0999 | 1.0900 | 1.0900 | 1.0870 | 1.0675 | **1.0651** |
| V37 | 1.0994 | 1.0470 | 1.0470 | 1.0120 | 1.0231 | **1.0089** |
| V38 | 1.0995 | 1.0380 | 1.0380 | 1.0510 | 1.0231 | **1.0432** |
| V39 | 1.0998 | 1.0530 | 1.0530 | 1.0190 | 1.0319 | **1.025** |
| Loss (MW) | 43.15359 | NA | NA | NA | 42.1231 | **41.1290** |

## 7 Conclusion

Various TSC-OPF problem-solving strategies have been introduced, developed, and effectively employed in this chapter. The WSCC 3-generator 9-bus system and the New England 10-generator 39-bus system are The described method is used to evaluate the realistic generator operating on two test devices. The verdict drawn from the simulation results produced by various algorithms is that the suggested CQODSA and QODSA algorithms are capable of discovering the best solutions in every scenario without violating any operating limits. To further demonstrate the superiority, it can be stated that CQODSA and QODSA significantly beat all other classifiers presented in this chapter when it comes to verifying the finished method's efficiency, rated, and continuity. Furthermore, the mentioned techniques have high convergence properties. The recommended CQODSA and QODSA algorithms are thus highly intriguing evolutionary optimization approaches for the goal of overall optimization of any TSC-OPF difficulties based on originality. Even if the presented algorithms give superior results, these algorithms can be further tuned to produce the likely best response in a fair time period.

## References

1. Dommel HW, Tinney WF (1968) Optimal power flow solutions. IEEE Trans Power Apparatus Syst (10):1866–1876
2. Lobato E, Rouco L, Navarrete MI, Casanova R, Lopez G (2001) An LP-based optimal power flow for transmission losses and generator reactive margins minimization. In: 2001 IEEE Porto power tech proceedings (Cat. No. 01EX502), vol 3. IEEE, p 5

3. Momoh JA (1989) A generalized quadratic-based model for optimal power flow. In: Conference proceedings, IEEE international conference on systems, man and cybernetics. IEEE, pp 261–271
4. Momoh JA, Zhu JZ (1999) Improved interior point method for OPF problems. IEEE Trans Power Syst 14(3):1114–1120
5. Sun DI, Ashley B, Brewer B, Hughes A, Tinney WF (1984) Optimal power flow by Newton approach. IEEE Trans Power Apparatus Syst (10):2864–2880
6. Yan X, Quintana VH (1999) Improving an interior-point-based OPF by dynamic adjustments of step sizes and tolerances. IEEE Trans Power Syst 14(2):709–717
7. Momoh JA, El-Hawary ME, Adapa R (1999) A review of selected optimal power flow literature to 1993. II. Newton, linear programming and interior point methods. IEEE Trans Power Syst 14(1):105–111
8. Pudjianto D, Ahmed S, Strbac G (2002) Allocation of VAR support using LP and NLP based optimal power flows. IEE Proc-Gen Trans Distrib 149(4):377–383
9. Roy PK, Mandal D (2011) Quasi-oppositional biogeography-based optimization for multi-objective optimal power flow. Electr Power Comp Syst 40(2):236–256
10. Abou El-Ela AA, Abido MA (1992) Optimal operation strategy for reactive power control. Model Simul Control Gen Phys Matter Waves Electr Electron Eng 41:19–19
11. Jr Santos A, Da Costa GRM (1995) Optimal-power-flow solution by newton's method applied to an augmented lagrangian function. IEE Proc-Gen Trans Distrib 142(1):33–36
12. De Carvalho EP, dos Santos Júnior A, Ma TF (2008) Reduced gradient method combined with augmented lagrangian and barrier for the optimal power flow problem. Appl Math Comput 200(2):529–536
13. Capitanescu F, Wehenkel L (2013) Experiments with the interior-point method for solving large scale optimal power flow problems. Electr Power Syst Res 95:276–283
14. Gaing Z-L, Chang R-F (2006) Security-constrained optimal power flow by mixed-integer genetic algorithm with arithmetic operators. In: 2006 IEEE power engineering society general meeting. IEEE, p 8
15. Lai LL, Ma JT, Yokoyama R, Zhao M (1997) Improved genetic algorithms for optimal power flow under both normal and contingent operation states. Int J Electr Power Energy Syst 19(5):287–292
16. Roa-Sepulveda CA, Pavez-Lazo BJ (2003) A solution to the optimal power flow using simulated annealing. Int J Electri Power Energy Syst 25(1):47–57
17. Abido MA (2002) Optimal power flow using Tabu search algorithm. Electr Power Comp Syst 30(5):469–483
18. Swain AK, Morris AS (2000) A novel hybrid evolutionary programming method for function optimization. In: Proceedings of the 2000 congress on evolutionary computation. CEC00 (Cat. No. 00TH8512), vol 1. IEEE, pp 699–705
19. Cai J, Ma X, Li L, Yang Y, Peng H, Wang X (2007) Chaotic ant swarm optimization to economic dispatch. Electr Power Syst Res 77(10):1373–1380
20. Tripathy M, Mishra S (2007) Bacteria foraging-based solution to optimize both real power loss and voltage stability limit. IEEE Trans Power Syst 22(1):240–248
21. Abou El Ela AA, Abido MA, Spea SR (2010) Optimal power flow using differential evolution algorithm. Electr Power Syst Res 80(7):878–885
22. Sivasubramani S, Swarup KS (2012) Multiagent based differential evolution approach to optimal power flow. Appl Soft Comput 12(2):735–740
23. Roy PK, Ghoshal SP, Thakur SS (2010) Biogeography based optimization for multi-constraint optimal power flow with emission and non-smooth cost function. Exp Syst Appl 37(12):8221–8228
24. Khazali AH, Kalantar M (2011) Optimal reactive power dispatch based on harmony search algorithm. Int J Electr Power Energy Syst 33(3):684–692
25. Ayan K, Kılıç U (2012) Artificial bee colony algorithm solution for optimal reactive power flow. Appl Soft Comput 12(5):1477–1482

26. Roy PK, Mandal B, Bhattacharya K (2012) Gravitational search algorithm based optimal reactive power dispatch for voltage stability enhancement. Electr Power Syst Res 40(9):956–976
27. Basu M (2011) Multi-objective optimal power flow with facts devices. Energy Conv Manag 52(2):903–910
28. Nobahari H, Nikusokhan M, Siarry P (2012) A multi-objective gravitational search algorithm based on non-dominated sorting. Int J Swarm Intell Res (IJSIR) 3(3):32–49
29. Abido MA, Bakhashwain JM (2005) Optimal VAR dispatch using a multiobjective evolutionary algorithm. Int J Electr Power Energy Syst 27(1):13–20
30. Gan D, Thomas RJ, Zimmerman RD (2000) Stability-constrained optimal power flow. IEEE Trans Power Syst 15(2):535–540
31. Yuan Y, Kubokawa J, Sasaki H (2003) A solution of optimal power flow with multicontingency transient stability constraints. IEEE Trans Power Syst 18(3):1094–1102
32. Chen L, Taka Y, Okamoto H, Tanabe R, Ono A (2001) Optimal operation solutions of power systems with transient stability constraints. IEEE Transa Circ Syst I: Fundamental Theory Appl 48(3):327–339
33. Karaboga D et al (2005) An idea based on honey bee swarm for numerical optimization. Technical report, Technical report-tr06, Erciyes University, Engineering Faculty, Computer...,
34. Chan KY, Chan KW, Ling SH, Iu HHC, Pong GTY (2007) Solving multi-contingency transient stability constrained optimal power flow problems with an improved GA. In: 2007 IEEE Congress on Evolutionary Computation. IEEE, pp 2901–2908
35. Cai HR, Chung CY, Wong KP (2008) Application of differential evolution algorithm for transient stability constrained optimal power flow. IEEE Trans Power Syst 23(2):719–728
36. Athay T, Podmore R, Virmani S (1979) A practical method for the direct analysis of transient stability. IEEE Trans Power Apparatus Syst (2):573–584
37. Civicioglu P (2012) Transforming geocentric cartesian coordinates to geodetic coordinates by using differential search algorithm. Comput Geosci 46:229–247
38. Tizhoosh HR (2005) Opposition-based learning: a new scheme for machine intelligence. In: International conference on computational intelligence for modelling, control and automation and international conference on intelligent agents, web technologies and internet commerce (CIMCA-IAWTIC'06), vol 1. IEEE, pp 695–701
39. Nguyen TB, Pai MA (2003) Dynamic security-constrained rescheduling of power systems using trajectory sensitivities. IEEE Trans Power Syst 18(2):848–854
40. Vickers NJ (2017) Animal communication: when I'm calling you, will you answer too? Curr Biol 27(14):R713–R715
41. Bhattacharya A, Chattopadhyay PK (2011) Application of biogeography-based optimisation to solve different optimal power flow problems. IET Gen Trans Distrib 5(1):70–80
42. Duman S, Güvenç U, Sönmez Y, Yörükeren N (2012) Optimal power flow using gravitational search algorithm. Energy Conv Manag 59:86–95
43. Mukherjee A, Paul S, Roy PK (2014) Transient stability constrained optimal power flow using teaching learning-based optimization. Int J Energy Optim Eng (IJEOE) 3(4):55–71
44. Kürşat Ayan and ULAŞ KILIÇ (2013) Solution of transient stability-constrained optimal power flow using artificial bee colony algorithm. Turk J Electr Eng Comput Sci 21(2):360–372
45. Ahmadi H, Ghasemi H, Haddadi AM, Lesani H (2013) Two approaches to transient stability-constrained optimal power flow. Int J Electr Power Energy Syst 47:181–192
46. Xiaoping T, Dessaint L-A, Nguyen-Duc H (2013) Transient stability constrained optimal power flow using independent dynamic simulation. IET Gen Trans Distrib 7(3):244–253

# Chaotic Quasi-Oppositional Moth Flame Optimization for Radial Distribution Network Reconfiguration with DG Allocation

**Sneha Sultana, Sourav Paul, Poulami Acharya, Provas Kumar Roy, Devjeet Sengupta, and Nirmalya Dey**

**Abstract** This research aimed to reconfigure radial distribution networks in the presence of distributed generators (DGs) using the Chaotic Quasi-Oppositional Moth Flame Optimization (CQOMFO) method so as to minimize power losses in the power system network and keep the voltage profile consistent throughout the power system network, which will aid in increasing system efficiency. The primary goal is to demonstrate the proper placement of Distributed Generators (DGs) in the radial distribution network, as well as the reconfiguration and installation of DGs in the radial distribution network. The main advantage of this algorithm is continuous guiding search with changing goals, which can be used for real-time applications with only minor adjustments because the power from distributed generation is constantly changing. This algorithm's efficiency and suitability for real-time applications have been determined by testing for loss minimization on typical 33- and 69-bus radial distribution systems.

**Keywords** Distribution network reconfiguration · Distributed generation · Optimal DG allocation problem · Quasi-oppositional moth flame optimization · Power loss minimization

## 1 Introduction

Network reconfiguration is a systematic and effective way for reducing power loss in power distribution networks. The goal of Distribution Network Reconfiguration (DNR) is to improve voltage profile and load balance, reduce power losses, and enhance network reliability. This can be done by using tie switches and section-

S. Sultana (✉) · S. Paul · P. Acharya · D. Sengupta · N. Dey
Dr. B. C. Roy Engineering College, Durgapur, India
e-mail: sneha.sultana@gmail.com
URL: http://www.bcrec.ac.in

P. K. Roy
Kalyani Government Engineering College, Kalyani, India

 
O. H. Gupta et al. (eds.), *Soft Computing Applications in Modern Power and Energy Systems*, Lecture Notes in Electrical Engineering 1107,
https://doi.org/10.1007/978-981-99-8007-9_15

alizing. The goal of the DNR issue is to find the best method of operation for the radial distribution network while minimizing power losses and satisfying all network constraints.

If the optimal location and size of DG resources are not selected correctly, network efficiency will decline, the network will become unstable, voltage levels will exceed operational norms, and power losses will rise. As a result, the primary problems that can be addressed by implementing DNR and strategically arranging DGs are lowering power losses and improving the voltage profile. The purpose of this paper is to show how to reconstruct and position capacitors on a Radial Distribution System.

In order to correctly determine the most efficient network for distribution configuration, traditional model-based techniques necessitate network parameters. The functioning of distributed generators (DG) can have both positive and negative effects on distribution networks (DN). According to research, incorrect DG location and size may result in higher system losses than losses without DG [1]. The incorporation of distributed generators (DGs) into distribution networks is increasing as a result of energy market liberalization, environmental concerns, and technical advancement. Reference [2] investigates a distribution network reconfiguration process using the model-free reinforcement learning (RL) strategy and the NoisyNet Deep Q-learning Network (DQN). The paper successfully demonstrated that the exploration can be carried out automatically with no changes to the parameters, resulting in improved performance and reduced losses with minimal voltage deviation. Paper [3] explores a scenario-based convergent programming model to increase distribution network reconfiguration potential through optimal reserve branch placement, despite the fact that only a few branches are switchable. Two test networks have been set up for this reason. Reference [4] is based on the data batch-constrained RL algorithm that addresses the flaws of physical model-based control methods while also resolving the DNR problem. Three networks were investigated, and it was determined that this technique enhances behaviour control policy, is extremely scalable, and can deliver a workable resolution immediately. In order to achieve a superior operating condition during critical landing, Cao et al. [5] have improved and generalized the traditional reconfiguration strategy using dynamic microgrids. The research is supported by a two-stage mixed-integer conic programme model with risk aversion. DGs are small sources of electrical energy with variable capacities of up to 10 MW, with either a renewable or non-renewable primary energy source, and are connected to the distribution network immediately or at the site of consumption [6]. According to the study, Particle Swarm Optimization outperformed the Genetic Algorithm in terms of discovering the optimal answer, speeding convergence, and reducing running time. Another study looked at various optimal and maximal models for utility-based dispersed generation penetration. The main aim was to increase the penetration of distributed generations into rural distribution networks. Paper [7] addressed simultaneous radial distribution system remodelling and changed condenser allotment to improve distribution network efficiency. The primary objectives were to reduce active and reactive power losses and improve the voltage profile. To boost a system's power generation capability and meet increasing electricity demand, two alternative techniques—distribution system reconfiguration (DSR) and optimal capacitor placement (OCP)—are explored [8]. It was better to use both tactics simultaneously than one at a time. The individual OCP mode and the dual DSR mode after the OCP were the two operating modes that were used. The artificial ecosystem optimizer (AEO) approach is employed to distribute distributed generators (DGs) and capacitors while accounting for the reorganization of power distribution networks. (RPDS). This was inspired by three energy-transfer procedures in an ecosystem that concerned production, consumption, and decomposition [9]. For instance, Tan et al. [10], Srinivasa Rao et al. [11], Mohamed Imran et al. [12], novel integrates the DNR and DG positioning issues to boost the distribution network's efficiency. In Srinivasa Rao et al. [11] the presence of DG, the DNR issue is addressed to lower real power loss and improve the voltage profile of the distribution network (HSA). In an attempt to minimize electricity loss and enhance voltage consistency, Mohamed Imran et al. [12] suggested an approach for handling DNR and DG placement that relies on the fireworks optimization algorithm (FWA). Both studies employed a variety of methods, including the voltage stability index (VSI) and the loss sensitivity factor, to pre-identify the suitable bus sites for DG installation (LSF). The earliest techniques for network reconfiguration were built on heuristic methods. Merlin et al. published a pioneering work on network reconfiguration for loss reduction [13]. They proposed a heuristic method that starts with a network that resembles a mesh, which can be achieved by turning off all the switches. The switches are subsequently opened one at a time, according to the least current criterion, to restore the radial structure. This approach does not always result in global optimization. Shirmohammadi et al. [14] used a compensation-based power flow strategy in accordance with the methodology described in Shirmohammadi et al. [14] to accurately model the weakly meshed networks. To reduce network losses, Civanlar et al. [15] suggested a straightforward heuristic technique. However, the network setup can have an impact on the positioning and sizing of the DG. Both reconfiguration of the network and DG installation problems have to be addressed in order to benefit the entire distribution network [16]. Sultana et al. [17] narrated the oppositional krill herd (OKH) algorithm, and this was successfully integrated to perform optimal reconstruction of distribution network difficulties. In the future, this could encourage researchers to concentrate on difficult power system efficiency problems such as self-regulating generation control, optimal energy flow, economic discharge of load dispatch, hydro-thermal planning, power supply stability, and so on. Reference [18] suggested the first-ever application of the suggested quasi-reflection-based slime mould algorithm (QRSMA) for optimal capacitor bank positioning and size, as well as radial distribution network reconfiguration. In addition, the authors of Barnwal et al. [19] provided a novel approach that, by meticulously balancing the positioning of DGs, DNR, and PVQ bus voltage control, enhances voltage stability, reduces power losses while maintaining the desired voltage profile of radial distribution networks, and takes into account the presence of fluctuating reactive power at the P bus. This leads to the definition of a multi-objective function. The grey wolf optimization (GWO) method is suggested. The simultaneous network reconstruction and DG placement in radial distribution systems utilizing a new quasi-oppositional chaotic neural network method are covered in Ref. [20] (QOCNNA). It targeted to reduce active power losses and maintain stable voltage in distribution networks like 33-, 69-, and 118-bus systems. To minimize the accumulative cost for dispatchable DER operation and load reduction, an optimal distributed energy resource (DER) scheduling problem is solved in Ref. [21]. Finally, the topology with the lowest accumulative cost is chosen from among all radial topologies. It was tested on IEEE 69 and IEEE 123 bus systems.

In this study, the author created a novel algorithm, the chaotic quasi-oppositional moth flame optimization, to solve the reconfiguration issue with optimal DG placement in RDN, taking into account the profile of the voltage and the stability of the voltage index. CQOMFO is a relatively new optimization algorithm that is considerably simpler and more robust than other optimization issues. This study used the optimization of a CQOMFO to minimize losses and optimize the voltage profile and stability index in the distribution system. A comparison of simulation results obtained using MATLAB software and other approaches recommended by others is given.

To organize the paper, use the following technique. The second section discusses mathematical reasoning. Section 3 describes the chaotic quasi-oppositional moth flame optimization method. Section 4 addresses the use of CQOMFO for DNR as well as the ODGA problem. Section 5 discusses the findings and research. The conclusion is addressed in Sect. 6.

## 2 Mathematical Formulation

### *2.1 Intended Purpose*

The goal of this study is to reduce distribution network losses by reconfiguring and allocating (location and size) DGs. The proposed approach is used to solve the problem here. In this context, power loss ($OF_{Ploss}$) is used as an objective function along with DG:

$$P_{LOSS} = \sum_{M=1}^{N}\sum_{P=1}^{N} \frac{R_{M,P}}{V_M V_P} Cos(\mu_M - \mu_P)(A_M A_P + B_M B_P) + \frac{R_{M,P}}{V_M V_P} Sin(\mu_M - \mu_P)(B_M A_P - A_M B_P). \quad (1)$$

Here, $P_{LOSS}$ is the real power loss; $R_{M,P}$ and $V_M V_P$ are resistance and voltage of the branch connected between $M$th and $P$th buses; $\mu_M$, $\mu_P$ are the voltage angle of $M$th and $P$th buses; $A_M B_M$ and $A_P B_P$ are real and reactive power of $M$th and $P$th buses, respectively.

## 2.2 Modelling of Load

We have now taken into consideration various forms of load. The following details are provided for the active and reactive loads:

$$PL_p = PL_o(p)(s_1 + s_2|v(p)| + s_3|v(p)|^2) \tag{2}$$

$$QL_p = QL_o(p)(t_1 + t_2|v(p)| + t_3|v(p)|^2) \tag{3}$$

where $(s_1, t_1)$, $(s_2, t_2)$, and $(s_3, t_3)$ are considered as the arrangement of fixed power, fixed current, and fixed impedance, respectively. Here, three factors have been taken into account: constant power, constant current, and constant impedance. The constant power load circumstances $s_1 = t_1 = 1$ and $s_2 = t_2 = s_3 = t_3 = 0$; and constant current load $s_2 = t_2 = 1$ and $s_1 = t_1 = s_3 = t_3 = 0$.

## 2.3 Constraints

**Power Balanced Constraints** To achieve load-balanced conditions, the total demand for bus and system losses must be satisfied by the total electricity generated by DG at a specific bus and that provided by the substation. The following are the list of load equilibrium restrictions:

$$p_{\text{sub-station}} + \sum_{i=1}^{N} p_{DG,i} = \sum_{i=1}^{N} p_{TD,i} + P_{\text{loss}} \tag{4}$$

$$q_{\text{sub-station}} + \sum_{i=1}^{N} q_{DG,i} = \sum_{i=1}^{N} q_{TD,i} + q_{\text{loss}}. \tag{5}$$

**Voltage Limit**: In order to ensure system stability and power quality from the following sources, the bus voltage must be between its maximum and lowest voltage limits:

$$V_{M,\text{MIN}} \leq V_M \leq V_{M,\text{MAX}}. \tag{6}$$

**Range of Voltage Angle**: At the bus, the voltage angles must fall within the range of permissible angles, including minimum and maximum:

$$\delta_{M,\text{MIN}} \leq \delta_M \leq \delta_{M,\text{MAX}}. \tag{7}$$

# 3 Optimization Technique

## 3.1 Moth Flame Optimization

In the year 2015, Seyadali Mirjali developed the MFO [22] a brand-new population-based meta-heuristic technique dubbed Moth Flame optimization. Elegant insects called moths resemble butterflies quite a little. Over the course of their existence, they typically go through two phases: the larval phase and the adult phase. The distinctive night-time behaviour of the flying moth organism served as the inspiration for this technique. Moths are thought to have a distinctive night-time navigation system called a transverse mechanism.

Moths fly using the aforementioned process by keeping a steady angle at night with respect to the moon. The following strategy makes sure that the moths go in a straight line even when the moon is quite far from them. Nevertheless, when the moths are placed close to a man-made light source, they have a tendency to move in a dangerous spiral pattern. This specific conduct is helpful for resolving issues in everyday life. In basic MFO, individual moths represent potential solutions, and each position is expressed as a matrix of choice variables, as shown below

$$Y = \begin{bmatrix} Y_1 \\ Y_2 \\ \vdots \\ Y_1 \end{bmatrix} = \begin{bmatrix} Y_{1,1} & Y_{1,2} \ldots & Y_{1,n-1} & Y_{1,n} \\ Y_{2,1} & \ddots \quad \ldots & \ldots & Y_{2,n} \\ \vdots & \ldots \quad \ddots & \ldots & \ldots \\ Y_{N-1,1} & \ldots \quad \ldots & \ddots & Y_{N,n-1} \\ Y_{N,1} & Y_{N,2} \ldots & Y_{N-1,n} & Y_{N,n} \end{bmatrix} \tag{8}$$

where $Y_i = \left[Y_{i,1}, Y_{i,2}, \ldots, Y_{i,n}\right], i \in 1, 2, \ldots, n$. The moth and the flame are the two main characters of MFO. To get the desired consequences, the moth must pass across the flame. The logarithmic spiral function, which is described in the equation below, is used to model the spiral movement of the moth:

$$Y_i^{k+1} = \begin{cases} \mu_i.e^{an}.cos\,(2\pi\ t) + f_{k_i}(k),\ i \leq n.fm \\ \mu_i.e^{an}.cos\,(2\pi\ t) + f_{kn.fm}(k),\ i \geq n.fm \end{cases} \tag{9}$$

where $\mu_i = |Y_{i^k} - f_{ki}|$ represents the distance between a moth at point $x_i$ from its corresponding flame $f_{ki}$.

In comparison to other meta-heuristic algorithms, the MFO algorithm is seen to have a greater convergence rate in its results which provides better quality solutions in a very less amount of time. But after this above-mentioned process is fulfilled, there is one more concern to be thought about which is that the position refreshing of the moths with respect to various locations in the particular search space that may reduce the chances of achieving the best solutions. This concern can be remedied using the below-stated mathematical formulation where the number of flames is reduced with every successful iteration:

$$N_g = \text{round}(\gamma - \delta * \frac{\gamma - \delta}{\sigma}) \tag{10}$$

where $N_g$ is the flame number, $\gamma$ is then the maximum number of flames at the present time, $\delta$ is the current cycle number, and $\sigma$ is the maximum amount of iterations.

Several real-world optimization problems have been solved using the moth flame optimization algorithm (MFO), which has the benefits of being quick to converge, having few setting parameters, and being easy to understand and apply. However, the MFO struggles to strike a good balance between exploration and exploitation, and there is little information sharing among people, especially when it comes to working out some challenging mathematical issues.

## *3.2 Quasi-Oppositional Based Learning*

Tizhoosh [23] initially released OBL which is a cutting-edge idea in intelligence-based problem solving or soft computing that can be utilized to enhance several optimization methodologies. It looks to be one of the most effective theories in computational intelligence, which can handle nonlinear optimization problems and enhance the search performance of conventional population-based optimization procedures. OBL's primary objective is to compare an estimate or assumption with its opposite or reciprocal in order to increase the possibility that a solution will be found more rapidly. The OBL approach begins with initializing the initial estimate, which is done either randomly or based on prior knowledge about the solution. Any direction, or at the very least the opposite, may be the best course of action. The opposing set of estimates for a superior solution is taken into consideration for convergence after iteratively replacing the initial estimates in the direction of optimality. Let the real number be denoted by $B \in [k, l]$ and $B^0$ be its corresponding opposite number is defined real number by

$$B_0 = k + l - B. \tag{11}$$

Say $R = (Z_1, Z_2, \ldots\ldots., Z_n)$ is a number in n-dimensional space with upper and lower bounds, where $B_u \in [k_u, l_u], u \in 1, 2, \ldots., n$. The $R^0 = (Z_1{}^0, Z_2{}^0, \ldots\ldots.Z_n{}^0)$ opposite point is defined by its components:

$$B_u{}^0 = k_u + l_u - B_0. \tag{12}$$

Let $B$ be a real number defined between upper and lower limits $[k, l]$. The quasi oppositional number is described by

$$B^{Q_0} = \text{rand}(C, \tilde{B}) \tag{13}$$

where $C$ is given by $C = \frac{k+l}{2}$. Suppose $B$ is a real integer $[e, f]$. Define the quasi -oppositional point $B_r{}^{Q_0}$ as

$$B_r{}^{Q_0} = \text{rand}(C_u, \tilde{B}_u) \tag{14}$$

$$\text{where} \quad C_r = \frac{k_u + l_u}{2}.$$

## 4 Chaotic Moth Flame

In order to get even closer to the ideal global solution, the current researchers expanded the QOBL to the CQOBL. A key factor in accelerating the convergence speed of a meta-heuristic algorithm is the stochastic and non-repeating nature of chaos, which performs the overall searches at greater speeds. Different chaotic maps are taken into account to control the MFO's parameters in the CQOMFO optimization issue. There is an aggregate of 10 chaotic maps chosen for a chaotic set, and each one has a unique behaviour. This set's starting point has been determined to be 0.7 ranging from 0 to 1. The many chaotic map behaviours help to solve the issues of local optimal and speed of convergence.

## 5 CQOMFO Applied to Reconfiguration Problem Along with DG

The reconfiguration of the distribution network with the simultaneous allocation of DGs is implemented using the CQOMFO as follows:

**Step 1**: Initialize the size of population ($P_n$), and read data for the system with constraints and the maximal iteration.

**Step 2**: Determine the DG size following the maximum and minimum limits.

**Step 3**: Calculate the objective function (1) by performing power flow calculations and determine the minimum power losses.

**Step 4**: Depending upon the present candidate solution in the search space, a logarithmic spiral defining a little better solution is retained as noble solutions.

**Step 5**: Use functions (9)–(12) to produce opposite population.

**Step 6**: If the obtained objective function is weak, replace the best solution with the previous best solution; otherwise, return to Step 3.

**Step 7**: Increase the number of iterations and return to Step 3. Print the results and stop the algorithm.

# 6 Results

The above-proposed approach CQOMFO is evaluated using two different systems for testing that consist of 33 and 69 buses, with an operating voltage of 12.66 kV, to assess its usability and superiority in solving the MFO problem for reconfiguration with DG installation identifies the best location and DG size to diminish active power loss. Simulations are executed using MATLAB software on a PC with an Intel i3-7020U @ 2.30 GHz and 8 GB of RAM. The algorithm's population size is set at 50, and its iteration count is set at 100.

## *6.1 Radial Distribution Network 33-Bus Demonstration*

As seen in Fig. 1, 33-bus demonstration RDN, which has 33 buses and 32 branches, is used to test the CQOMFO methodology. Line values and the load values for the systems are taken from Ref. [24]. The results of the 33-bus test system are shown in Table 1. For a 33-bus test system by application of CQOMFO and MFO, the real power and losses are obtained at 119.01 and 151.21 kW respectively after placement of DG. But after reconfiguration of the radial distribution network, power losses decreased to 49.03 and 56.13 kW, respectively. After reconfiguration, the 33-bus system for constant power type load and constant current type load is shown in Figs. 2 and 3, respectively. The obtained results are compared with Tran et al. [25] for constant power type load which shows power loss reduction is more significant

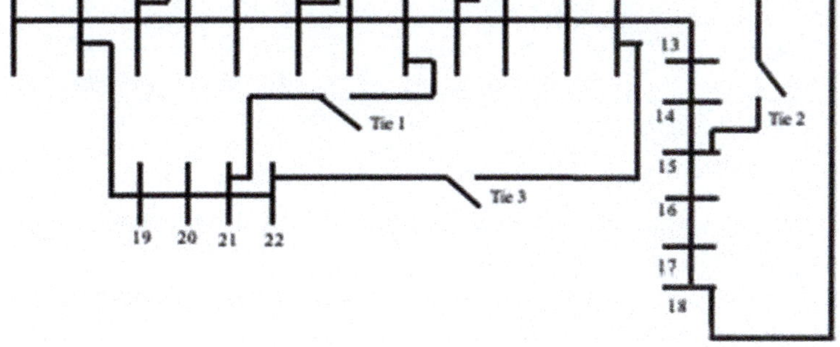


**Fig. 1** Topology of 33-bus RDN

**Table 1** Summary of the results for 33-bus RDN before and after reconfiguration with DG for constant power type load

| *Without reconfiguration* | | | | |
|---|---|---|---|---|
| Parameters | Without DG placement | With DG placement | | |
| | | SFS [25] | MFO | CQOBL-MFO |
| Power loss (KW) | 200.23 | NA | 151.21 | **119.01** |
| Optimal position of DG | NA | NA | 29 | **29** |
| | | | 26 | **26** |
| | | | 24 | **24** |
| Optimal size of DG | NA | NA | 992.35 | **876.12** |
| | | | 900.31 | **731.02** |
| | | | 733.02 | **655.17** |
| Opening branches | NA | NA | NA | **NA** |
| Closing branches | NA | NA | NA | **NA** |
| *With reconfiguration* | | | | |
| Power loss (KW) | | 53.01 | 56.13 | **49.03** |
| Optimal position of DG | | 22 | 12 | **12** |
| | | 25 | 17 | **17** |
| | | 33 | 27 | **27** |
| Optimal size of DG | | 775.3 | 856.56 | **997.14** |
| | | 1285.8 | 602.43 | **645.8** |
| | | 735.6 | 600.01 | **601.12** |
| Opening branches | | 7–8 | 7–8 | **7–8** |
| | | 9–10 | 9–10 | **9–10** |
| | | 14–15 | 14–15 | **14–15** |
| | | 27–65 | 28–29 | **28–29** |
| | | 30–31 | 32–33 | **32–33** |
| Closing branches | | NA | 21–8 | **21–8** |
| | | | 9–15 | **9–15** |
| | | | 22–12 | **22–12** |
| | | | 18–33 | **18–33** |

by the CQOMFO technique than other techniques. And also for constant current type load, CQOMFO is superior to MFO.

Figure 3 shows the building layout of a 33-bus radial distribution network for a continuous current type load multiplied by 1.0.

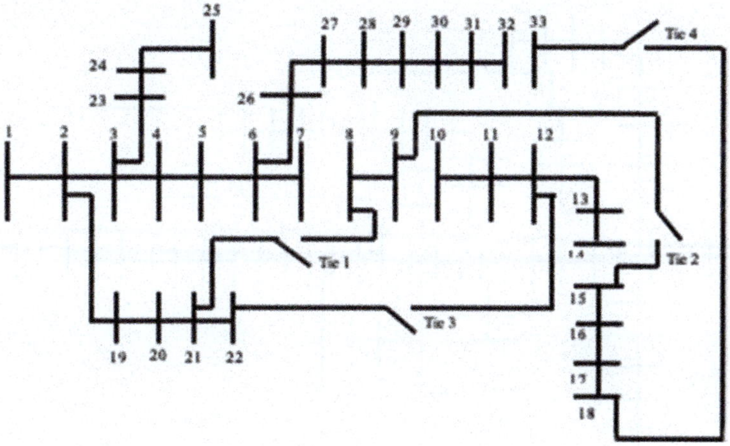


**Fig. 2** The design of a 33-bus circumferential distribution network for a constant power type load with a load multiplication factor of 1.0

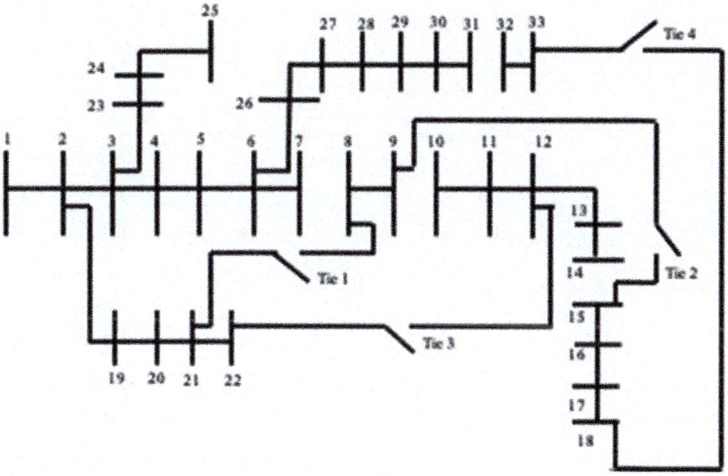


**Fig. 3** The construction layout of a 33-bus radial distribution network for a continuous current type load for load multiplication factor 1.0

## 6.2 *69-Bus Test Radial Distribution Network*

As seen in Fig. 4, the 69-bus test RDN, which consists of 69 buses and 68 branches, is used to test the CQOMFO methodology. Line values and the load values for the systems are taken from Ref. [24]. In Table 2, 33-bus test systems are explained. For a 69-bus test system by application of CQOMFO and MFO, the real power losses are obtained as 97.01 and 125.26 kW respectively after placement of DG. But after reconfiguration of the radial distribution network, power losses decreased to 33.03

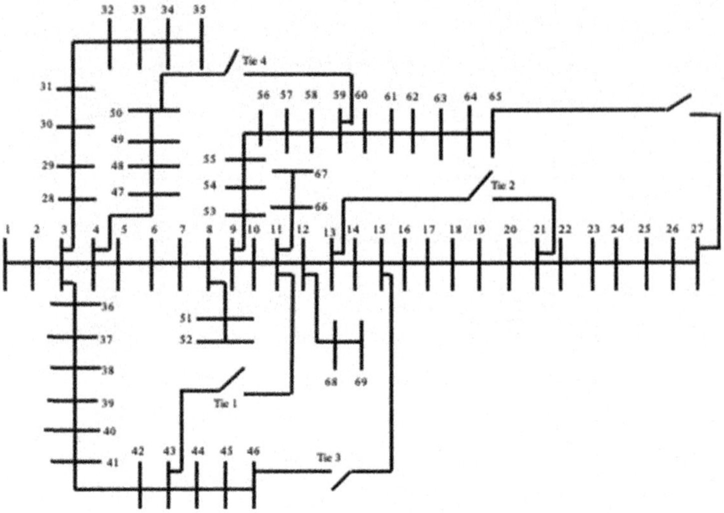


**Fig. 4** Topology of 69-bus RDN

**Table 2** Summary of the results for 33-bus RDN before and after reconfiguration with DG for constant current type load

| *Without reconfiguration* | | | |
|---|---|---|---|
| Parameters | Without DG placement | With DG placement | |
| | | MFO | CQOBL-MFO |
| Power loss (KW) | 179.68 | 117.31 | **102.46** |
| Optimal position of DG | NA | 8 | **8** |
| | | 18 | **18** |
| | | 23 | **23** |
| Optimal size of DG | NA | 923.12 | **908.47** |
| | | 934.5 | **905.1** |
| | | 965.12 | **902.3** |
| Opening branches | NA | NA | **NA** |
| Closing branches | NA | NA | **NA** |
| *With reconfiguration* | | | |
| Power loss (KW) | | 51.69 | **38.12** |
| Optimal position of DG | | 11 | **11** |
| | | 23 | **23** |
| | | 29 | **29** |
| Optimal size of DG | | 943.23 | **912.12** |
| | | 911.12 | **898.35** |
| | | 878.21 | **945.3** |
| Opening branches | | 7–8 | **7–8** |
| | | 9–10 | **9–10** |
| | | 14–15 | **14–15** |
| | | 31–32 | **31–32** |
| Closing branches | | 21–8 | **21–8** |
| | | 9–15 | **9–15** |
| | | 22–12 | **22–12** |
| | | 18–33 | **18–33** |

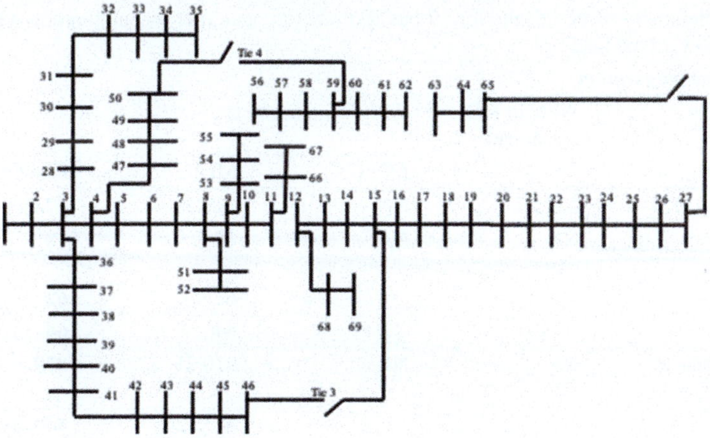


**Fig. 5** The construction layout of 69-bus radial distribution network for continuous power type load for load multiplication factor 1.0

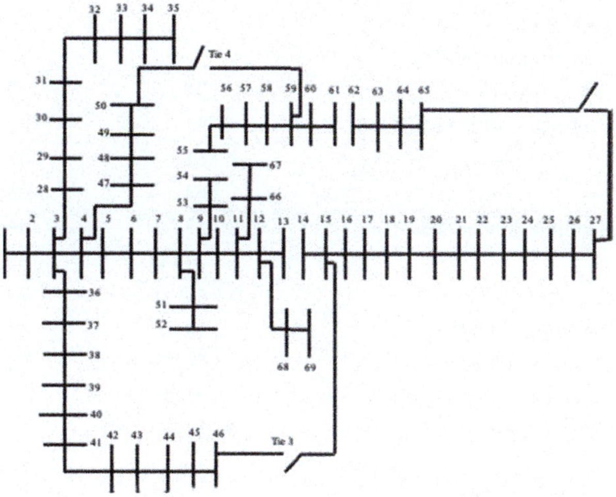


**Fig. 6** The construction layout of 69-bus radial distribution network for continuous current type load for load multiplication factor 1.0

and 37.21 kW, respectively. After reconfiguration, 33-bus systems for constant power type load and constant current type load are shown in Figs. 5 and 6 each. The obtained results are compared with Tran et al. [25] which shows power loss reduction is more significant by CQOMFO technique than other techniques (Tables 3 and 4).

**Table 3** Summary of the results for 69 bus RDN before and after reconfiguration with DG for constant power type load

| *Without reconfiguration* | | | | |
|---|---|---|---|---|
| Parameters | Without DG placement | With DG placement | | |
| | | SFS [25] | MFO | CQOBL-MFO |
| Power loss (KW) | 223.20 | NA | 125.26 | **97.01** |
| Optimal position of DG | NA | NA | 18 | **18** |
| | | | 62 | **62** |
| | | | 64 | **64** |
| Optimal size of DG | NA | NA | 910.21 | **879** |
| | | | 883.12 | **833.20** |
| | | | 874.02 | **690.47** |
| Opening branches | NA | NA | NA | **NA** |
| Closing branches | NA | NA | NA | **NA** |
| *With reconfiguration* | | | | |
| Power loss (KW) | | 35.16 | 37.21 | **33.03** |
| Optimal position of DG | | 11 | 21 | **21** |
| | | 61 | 47 | **47** |
| | | 64 | 58 | **58** |
| Optimal size of DG | | 537.6 | 868.43 | **764.01** |
| | | 1434.0 | 901.12 | **866.32** |
| | | 490.3 | 933.14 | **910.03** |
| Opening branches | | 14–15 | 55–56 | **55–56** |
| | | 56–57 | 62–63 | **62–63** |
| | | 61–62 | 11–12 | **11–12** |
| Optimal size of DG | | 11–43 | – | – |
| | | 13–21 | – | – |
| Closing branches | | NA | 50–59 | **50–59** |
| | | | 27–65 | **27–65** |
| | | | 15–46 | **15–46** |

**Table 4** Summary of the results for 69-bus RDN before and after reconfiguration with DG for constant current type load

| *Without reconfiguration* | | | |
|---|---|---|---|
| Parameters | Without DG placement | With DG placement | |
| | | MFO | CQOBL-MFO |
| Power loss (KW) | 198.24 | 105.09 | **83.21** |
| Optimal position of DG | NA | 25 | **25** |
| | | 34 | **34** |
| | | 45 | **45** |
| Optimal size of DG | NA | 923.01 | **837.54** |
| | | 867.42 | **867.32** |
| | | 801.56 | **754.75** |
| Opening branches | NA | NA | **NA** |
| Closing branches | NA | NA | **NA** |
| *With reconfiguration* | | | |
| Power loss (KW) | | 49.5 | **35.01** |
| Optimal position of DG | | 18 | **18** |
| | | 39 | **39** |
| | | 47 | **47** |
| Optimal size of DG | | 812.03 | **743.11** |
| | | 899.43 | **797.52** |
| | | 921.03 | **932.33** |
| Opening branches | | 13–14 | **13–14** |
| | | 54–55 | **54–55** |
| | | 61–62 | **61–62** |
| Closing branches | | 50–59 | **50–59** |
| | | 27–65 | **27–65** |
| | | 15–46 | **15–46** |

## 7 Conclusion

In this research, the CQOMFO method was effectively implemented for the simultaneous location and magnitude of the DG problem and distribution network reconfiguration. The goal is to reduce active power loss while increasing the voltage stability index of power distribution networks. Additionally, various approaches for voltage stability improvement and loss reduction, including only DG installation and network reconfiguration after the exact placement of DG, are simulated in order to demonstrate the superiority of the proposed method. The main goal was to improve the voltage profile while lowering active power loss in the distribution network. The CQOMFO was a reliable and effective method that quickly converged in all circumstances taken into account. On 33- and 69-bus test systems, the suggested procedure is put to the test. According to the test results, CQOMFO can manage extremely intricate and sizable distribution networks. Moreover, CQOMFO outperformed other approaches in terms of improving voltage profiles and reducing power loss for all applications. As a result, the suggested CQOMFO method may be a very promising approach for resolving the DNR problem in conjunction with the ideal location of DGs.

## References

1. Mithulananthan N, Oo T et al (2004) Distributed generator placement in power distribution system using genetic algorithm to reduce losses. Sci Technol Asia 55–62
2. Wang B, Zhu H, Honghua X, Bao Y, Di H (2021) Distribution network reconfiguration based on Noisynet deep q-learning network. IEEE Access 9:90358–90365
3. Tabares A, Puerta GF, Franco JF, Romero RA (2021) Planning of reserve branches to increase reconfiguration capability in distribution systems: a scenario-based convex programming approach. IEEE Access 9:104707–104721
4. Gao Y, Wang W, Shi J, Nanpeng Yu (2020) Batch-constrained reinforcement learning for dynamic distribution network reconfiguration. IEEE Trans Smart Grid 11(6):5357–5369
5. Cao X, Wang J, Wang J, Zeng B (2019) A risk-averse conic model for networked microgrids planning with reconfiguration and reorganizations. IEEE Trans Smart Grid 11(1):696–709
6. Purlu M, Turkay BE (2022) Optimal allocation of renewable distributed generations using heuristic methods to minimize annual energy losses and voltage deviation index. IEEE Access 10:21455–21474
7. Gebru Y, Bitew D, Aberie H, Gizaw K (2021) Performance enhancement of radial distribution system using simultaneous network reconfiguration and switched capacitor bank placement. Cogent Eng 8(1):1897929
8. Hussain AN, Al-Jubori WKS, Kadom HF (2019) Hybrid design of optimal capacitor placement and reconfiguration for performance improvement in a radial distribution system. J Eng 2019:1–15
9. Shaheen A, Elsayed A, Ginidi A, El-Sehiemy R, Elattar E (2022) Reconfiguration of electrical distribution network-based dg and capacitors allocations using artificial ecosystem optimizer: Practical case study. Alexandria Eng J 61(8):6105–6118
10. Tan S, Xu J-X, Panda SK (2013) Optimization of distribution network incorporating distributed generators: an integrated approach. IEEE Trans Power Syst 28(3):2421–2432
11. Srinivasa Rao R, Ravindra K, Satish K, Narasimham, SVL: Power loss minimization in distribution system using network reconfiguration in the presence of distributed generation. IEEE Trans Power Syst 28(1):317–325
12. Mohamed Imran A, Kowsalya M, Kothari DP (2014) A novel integration technique for optimal network reconfiguration and distributed generation placement in power distribution networks. Int J Electr Power Energy Syst 63:461–472
13. Merlin A (1975) Search for a minimum-loss operating spanning tree configuration for an urban power distribution system. In: Proceedings of 5th PSCC, vol 1, pp 1–18
14. Shirmohammadi D, Wayne Hong H (1989) Reconfiguration of electric distribution networks for resistive line losses reduction. IEEE Trans Power Del 4(2):1492–1498
15. Civanlar S, Grainger JJ, Yin H, Lee SSH (1988) Distribution feeder reconfiguration for loss reduction. IEEE Trans Power Del 3(3):1217–1223
16. Georgilakis PS, Hatziargyriou ND (2013) Optimal distributed generation placement in power distribution networks: models, methods, and future research. IEEE Trans Power Syst 28(3):3420–3428

17. Sultana S, Roy PK (2016) Oppositional krill herd algorithm for optimal location of capacitor with reconfiguration in radial distribution system. Int J Electr Power Energy Syst 74:78–90
18. Biswal SR, Shankar G, Elavarasan RM, Mihet-Popa L (2021) Optimal allocation/sizing of DGS/capacitors in reconfigured radial distribution system using quasi-reflected slime mould algorithm. IEEE Access 9:125658–125677
19. Barnwal AK, Yadav LK, Verma MK (2022) A multi-objective approach for voltage stability enhancement and loss reduction under PQV and p buses through reconfiguration and distributed generation allocation. IEEE Access 10:16609–16623
20. Tran TV, Truong B-H, Nguyen TP, Nguyen TA, Duong TL, Vo DN (2021) Reconfiguration of distribution networks with distributed generations using an improved neural network algorithm. IEEE Access 9:165618–165647
21. Shi Q, Li F, Olama M, Dong J, Xue Y, Starke M, Winstead C, Kuruganti T (2021) Network reconfiguration and distributed energy resource scheduling for improved distribution system resilience. Int J Electrl Power Energy Syst 124:106355
22. Mirjalili S (2015) Moth-flame optimization algorithm: a novel nature-inspired heuristic paradigm. Knowl-Based Syst 89:228–249
23. Tizhoosh HR (2005) Opposition-based learning: a new scheme for machine intelligence. In: International conference on computational intelligence for modelling, control and automation and international conference on intelligent agents, web technologies and internet commerce (CIMCA-IAWTIC'06), vol 1. IEEE, pp 695–701
24. Kashem MA, Ganapathy V, Jasmon GB, Buhari MI (2000) A novel method for loss minimization in distribution networks. In: DRPT2000. International conference on electric utility deregulation and restructuring and power technologies. Proceedings (Cat. No. 00EX382). IEEE, pp 251–256
25. Tran TT, Truong KH, Vo DN (2020) Stochastic fractal search algorithm for reconfiguration of distribution networks with distributed generations. Ain Shams Eng J 11(2):389–407

# Global Horizontal Irradiance Prediction Using Clustering and Artificial Neural Network

**Deep Rodge, Janavi Popat, and Akanksha Shukla**

**Abstract** Due to the limited supply of fossil fuels and their negative effects on the ecosystem, renewable energy sources, in particular solar energy, are becoming more and more significant. In order to maximize the use of solar energy and increase the effectiveness of solar energy facilities, solar energy forecasting is an essential area of research. In this study, we investigate the application of machine learning methods, particularly linear regression, polynomial regression, and artificial neural networks, for Global Horizontal Irradiance (GHI) prediction which would be important for effective solar energy facilities. Before applying the machine learning methods, we also examine the effects of clustering the data using the K-means algorithm. The accuracy of the different methods is evaluated using mean absolute error (MAE), root mean squared error (RMSE), and R-squared ($R^2$) values. According to the results, the accuracy of ANN along with K-Means performed better than the other methods, with an MAE of 53.941971, RMSE of 25.018180, and $R^2$ of 0.966465. Our results indicate that machine learning methods, in particular ANN, can be helpful for precise forecasting of GHI, and clustering the data can further enhance the models' accuracy. The renewable energy sector may be significantly impacted by these findings, especially in terms of optimizing solar energy use and improving the effectiveness of solar energy facilities.

**Keywords** Artificial neural network · GHI · K-means clustering · Regression · Solar energy

D. Rodge (✉) · J. Popat · A. Shukla
Sardar Vallabhbhai National Institute of Technology, Surat, India

e-mail: deeprodge14@gmail.com

A. Shukla
e-mail: ashukla@eed.svnit.ac.in

 
O. H. Gupta et al. (eds.), *Soft Computing Applications in Modern Power and Energy Systems*, Lecture Notes in Electrical Engineering 1107,
https://doi.org/10.1007/978-981-99-8007-9_16

# 1 Introduction

The conventional sources of electricity not only contribute to greenhouse emissions but also mandates centralized bulk power generation, which leads to additional cost of transmission and reduced efficiency. Whereas Renewable Energy Sources (RES) will enable decentralized, sustainable, and reliable power generation. It is estimated that due to industrialization, urbanization, and the inclusion of electric vehicles (EVs), the demand for electricity could rise by 3% YOY by 2030. On the other end, sources like coal, natural gas, and oil are getting depleted. To meet the increased energy demand, RES will play a vital role [1].

According to the COP27 report, India is responsible for 7% of the world's carbon dioxide emissions. India's 4% of total energy demand is met by renewable and nuclear power, to which, the contribution of solar is 42.5%, i.e., 1.7% of the total power demand. In the first quarter of 2022, solar power generation has saved 19.4 million tons of coal and 4.2 billion dollars in fuel costs, which is equivalent to 9% of total fuel costs. India is among the top economies when it comes to solar generation capacity and is aiming to increase the overall capacity to 500 GW by 2030, which will reduce the cumulative emissions by one billion tons [2].

Increasing the solar generation capacity isn't as simple as increasing the thermal or hydro generation capacity, as it depends on the natural resource, which is intermittent in nature. The duration of energy availability and its intensity varies widely with location. Eg., Yuma, USA, has 4127 sunlight hours, whereas Tórshavn has only 840. It also differs in nearby regions based on weather conditions such as temperature, pressure, relative humidity, wind speed, wind direction, etc.

Due to the intermittency of power generation, solar power planning and development are much more complex than that for thermal and hydropower plants. When we are concerned about conventional generation, the load demand is dynamic, but generation will follow certain rules and combinations which can be controlled. In the case of solar, both load and generation are dynamic, and none of them are easily controllable. Hence, it needs complex mathematical methods to plan accurately. Hourly forecasting of both load and solar generation are needed to analyze how much power can be generated by solar and what should be the energy storage or conventional backups' capacity to make the system reliable [3].

Solar energy is directly related to solar irradiance and temperature. There are three types of solar irradiance, namely, Global Horizontal Irradiance (GHI), Direct Normal Irradiance (DNI), and Diffused Horizontal Irradiance (DHI). As far as solar-to-electrical conversion is concerned, GHI is the only parameter that impacts the generation. Solar cell current increases significantly with an increase in GHI, resulting in increased power. Temperature is directly related to GHI, but inversely to the voltage of the solar cells. The impact of each parameter needs to be considered and correlated to have a better forecasting model.

There are various methods available to predict to process the time series data. The traditional ones, such as ARMA, ARIMA, ARIMAX, SARIMA, SARIMAX, etc., and also the neural network based, such as ANN, LSTM, RNN, CNN, etc. Authors of [4] have used ANN to predict solar energy for Malaysia using the historical data, with 4.8% of Mean Absolute Percentage Error (MAPE). But they have taken only irradiance, temperature, and relative humidity into consideration. While authors of [5] have developed an advanced ANN model, but it establishes the relationship between cloud cover and PV energy output. The energy output is dependent on other parameters as well, it demands an ANN model which establishes a relationship between different weather parameters and GHI. In [6, 7], authors have used ANN and LSTM, respectively, to predict the future load depending upon the numerous weather parameters. A similar approach can be developed to predict global horizontal irradiance. Sharma and Kakkar [8] compare different methods such as, compare the performance of several machine learning models, including decision tree regression, random forest regression, and ANN. Along with the algorithms used in the previous article, authors of [9] also analyzed Support Vector Machine (SVM). In both cases, ANN outperformed other algorithms when compared using MAPE. Researchers of Mishra et al. [10] have proposed an unique ensemble learning approach which combines ANN, SVR, and RF, but it fails to take in account the seasonality and sensitivity of different parameters.

This research article tries to visualize the seasonal data and the correlation by the Pearson method, and polar plots, and the algorithms are further applied to the well-processed clustered data. Section 2 describes the methodologies used, i.e., K-Means Clustering and ANN. Section 3 gives brief insights into data, its seasonality, and how the pre-processing is done. Section 4 provides the comparison of MAE, RMSE, and $R^2$ for different algorithms used on clustered and non-clustered data. It also provides the results of sensitivity analysis and shows how the parameters are strongly correlated with each other.

## 2 Methodology

### 2.1 Problem Formulation

Accurate prediction of Global Horizontal Irradiance (GHI) is essential for planning and designing solar energy systems. Traditional methods such as linear regression and polynomial regression have been commonly used for GHI prediction. However, these methods may not adequately consider the seasonal variations in GHI data. This limitation can lead to inaccurate predictions, which can have significant economic and environmental implications. To overcome these challenges, this research aims to investigate the effectiveness of artificial intelligence (AI) methods, specifically K-means clustering and artificial neural networks (ANN), for predicting GHI. By leveraging the capabilities of AI, the research seeks to develop a more precise and reliable GHI prediction model.

The proposed methodology involves the development of an ANN-based GHI prediction model that incorporates both the time series component and external factors.

The time series component, denoted as $X^t(m(t), d(t), h(t))$, represents the historical GHI data, with $m(t)$ representing the month of the year, $d(t)$ denoting the day of the month, and $h(t)$ indicating the hourly component. The external factors, represented by $X^w(t)$, capture the relevant weather information. The forecasting model aims to establish the relationship between the input variables ($X^t$ and $X^w$) and the GHI output, denoted as $Y$, while accounting for random errors that may be present in the data. The model can be expressed as

$$\mathrm{Y} = f(X^t, X^w) + \epsilon \tag{1}$$

The primary objective of this research is to compare the accuracy of different methods, specifically linear regression, polynomial regression, K-means clustering, and ANN, for GHI prediction. By evaluating the performance of these methods using appropriate metrics, such as mean absolute error (MAE), root mean squared error (RMSE), and R-squared ($R^2$), the research aims to identify the most effective approach for GHI prediction. Furthermore, the research seeks to explore the potential of AI methods, specifically ANN, in enhancing the efficiency and effectiveness of solar energy systems. By providing more accurate GHI predictions, the research aims to facilitate the widespread adoption of solar energy systems and reduce our reliance on fossil fuels.

In summary, this research addresses the limitations of traditional GHI prediction methods by investigating the effectiveness of AI techniques, such as K-means clustering and ANN. The research aims to develop a robust GHI prediction model and evaluate its accuracy compared to traditional methods. The ultimate goal is to contribute to the advancement of solar energy systems and promote sustainable energy practices.

## 2.2 *Artificial Neural Network (ANN)*

An Artificial Neural Network (ANN) is a computational model that borrows features from the structure and operation of biological neurons found within the human brain. ANNs are made up of many levels of interconnected processing units called neurons that are arranged into input, hidden, and output layers. In an ANN, each neuron gets input signals from other neurons or outside sources, processes the data, and then generates an output signal that is sent to other neurons or an output layer. In most cases, mathematical functions known as activation functions control how neurons interpret signals.

A feed-forward multi-layer perception (FFMLP) network was chosen as the neural network, one of the most widely used neural networks that learn from instances. Figure 1 depicts a visual representation of the basic architecture. The input, 5 hidden, and output layers make up the network's multiple layers. Each layer is connected to other layers using strengths called weights [11].

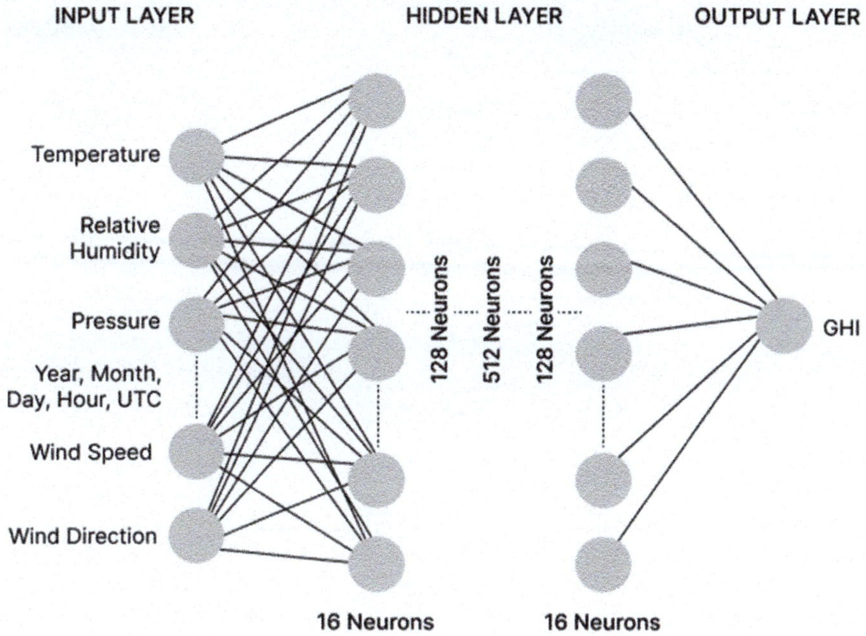


**Fig. 1** ANN architecture

The input factors used for the nodes of the input layers consisted of five temporal and five climatic variables. The year, month, day, hour, and minute were the temporal factors, and temperature, wind speed, wind direction, pressure, and relative humidity were the climatic factors. The output layer had a single node with the calculated global horizontal irradiance as its output. Relu function $f(z_i)$ was used as the neurons' activation function.

$$f(z_i) = max(0, x) \tag{2}$$

$$z_i = \sum_{j=0}^{n} w_{ij} x_j + \beta_i \tag{3}$$

where $w_{ij}$ is the weight on the link directed from neuron $j$ to neuron i, $x_j$ is the signal coming from the $j$th neuron, $\beta_i$ is the bias of neuron $i$, and $z_i$ is the weighted sum of the input signals. Supervised instruction, the approach that is used the most, was employed. GHI measurements were provided, and the network learned by analyzing the measured and estimated data. Using a back-propagation training algorithm, the difference, the error, is transmitted backward from the output layer.

To make the weights more accurate and produce the desired group of outputs, training the model with the data is done. Before training starts, a small random value is assigned to each weight to prevent saturation by large values. The goal is to reduce the squared difference between the produced output and what is desired. An epoch is

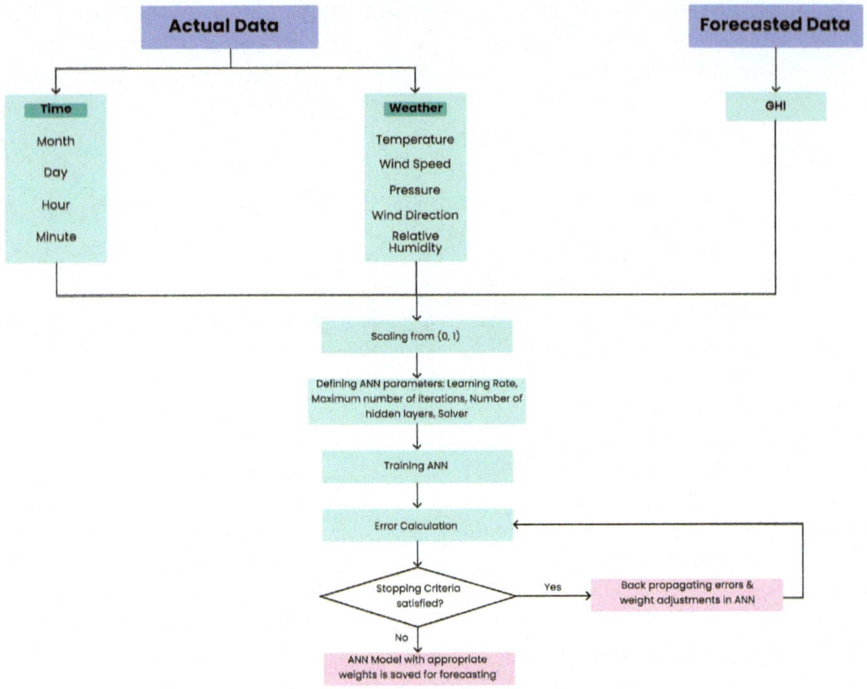


**Fig. 2** ANN training algorithm

completed when a neural network is trained for an entire cycle. The training process is repeated epoch after epoch till a desired error value is achieved.

Figure 2 depicts the flowchart method for training ANN. The raw data are scaled from 0 to 1 prior to training. The ANN's parameters are set, such as learning rate, iterations, hidden layer count and size, and solver. The neural network is then taught, and the error resulting from the output data is determined. The errors are back-propagated and the ANN's weights are revised if the stopping conditions are not met. Retraining of the model is done. We obtain an ANN model with the suitable weights for forecasting when the stopping criteria are satisfied.

## *2.3 Performance Metrics*

$$\text{MAE} = \frac{\sum_{i=0}^{N-1} |y_i - \hat{y}_i|}{N} \tag{4}$$

$$\mathrm{RMSE} = \sqrt{\frac{\sum_{i=0}^{N-1}(y_i - \hat{y}_i)^2}{N}} \tag{5}$$

$$R^2 = 1 - \frac{\sum_{i=0}^{N-1}(y_i - \hat{y}_i)^2}{\sum_{i=0}^{N-1}(y_i - \overline{y}_i)^2} \tag{6}$$

where $n$ represents the number of observations, $y_i$ represents the actual value, $y_i$ represents the predicted value and $\overline{y}_i$ represents the mean value.

We've not used Mean Absolute Percentage Error (MAPE) metric as it can be problematic when applied to prediction problems where real values (for our case, GHI) may be zero. This is because MAPE can result in a division by zero error if the real value is zero, making it challenging to use this metric for evaluation. An alternative to MAPE is the Root Mean Squared Error (RMSE), which we've considered and is a more widely used metric for evaluating regression models.

# 3 Experiment

## *3.1 Data Description*

The time series data of Surat city available on National Renewable Energy Laboratory (NREL) was used for the experiment. Each hourly value from 2017 to 2019 is included. This results in 26280 data points with 11 features (weather parameters) and GHI as the output variable [12]. The statistical description of the data is mentioned in the Table 1.

**Table 1** Statistical data

| Parameter | Mean | Std | Min | Min (25%) | Min (50%) | Min (75%) | Max |
|---|---|---|---|---|---|---|---|
| Temperature | 27.43 | 4.37 | 12.6 | 25.2 | 27.6 | 29.8 | 41.7 |
| Wind speed | 3.37 | 1.58 | 0.3 | 2.1 | 3.1 | 4.4 | 9.7 |
| Pressure | 1007.51 | 4.76 | 992 | 1004 | 1008 | 1011 | 1022 |
| Wind direction | 202.89 | 90.39 | 0 | 120 | 230 | 260 | 360 |
| Relative humidity | 64.74 | 23.75 | 13.14 | 43.2 | 69.28 | 86.56 | 100 |
| GHI | 212.4 | 295.66 | 0 | 0 | 0 | 413 | 1049 |

**Table 2** Pearson coefficient of different parameters with GHI

| Parameter | Correlation with GHI |
|---|---|
| Temperature | 0.630184 |
| Wind speed | −0.022556 |
| Pressure | 00.066675 |
| Wind direction | 00.140302 |
| Relative humidity | −0.515194 |

### 3.2 Correlation Between Parameters

In order to understand the relationship between the weather parameters and the GHI, we used Pearson's correlation coefficient [13]. The correlation coefficient, whose values range from −1 to +1, assesses the strength and direction of the linear connection between two variables. A value of +1 indicates a perfect positive correlation, a value of −1 indicates a perfect negative correlation and a value of 0 indicates no correlation.

The Pearson correlation coefficients between the various weather factors and the GHI are displayed in Table 2. As expected, temperature is highly positively correlated with the GHI, with a coefficient of 0.63. On the other hand, the correlation between wind speed and the GHI is extremely weak, with a coefficient of −0.02. Pressure and wind direction both have very weak positive correlations with the GHI, with coefficients of 0.07 and 0.14, respectively. Interestingly, there is a moderate negative correlation between relative humidity and the GHI, with a coefficient of −0.52.

The correlation coefficients give us a general sense of how strongly the various parameters are related to the GHI, but they do not reveal the nature or pattern of the relationship. We can visualize the correlations using a heat map to better understand how the various parameters link to the GHI. The heat map of the correlation matrix between the various factors and the GHI is displayed in Fig. 3. As anticipated, the greatest positive correlation exists between the GHI and temperature, while the strongest negative correlation exists between the GHI and relative humidity. The relationships between the other variables are weaker, with wind speed and pressure almost completely unrelated to the GHI.

### 3.3 Seasonality

The GHI varies over a year due to changes in the sun's position in the sky and other factors such as atmospheric conditions and cloud cover. Specifically, the GHI changes over time within a year due to Seasonal variation. The amount of solar radiation received at a particular location is affected by the sun's angle in the sky, which changes throughout the year due to the earth's axial tilt. In the Northern Hemisphere, the GHI is typically highest during the summer months when the sun is highest in the sky and lowest during the winter months when the sun is lowest in the sky. The GHI can also be affected by weather conditions such as cloud cover,

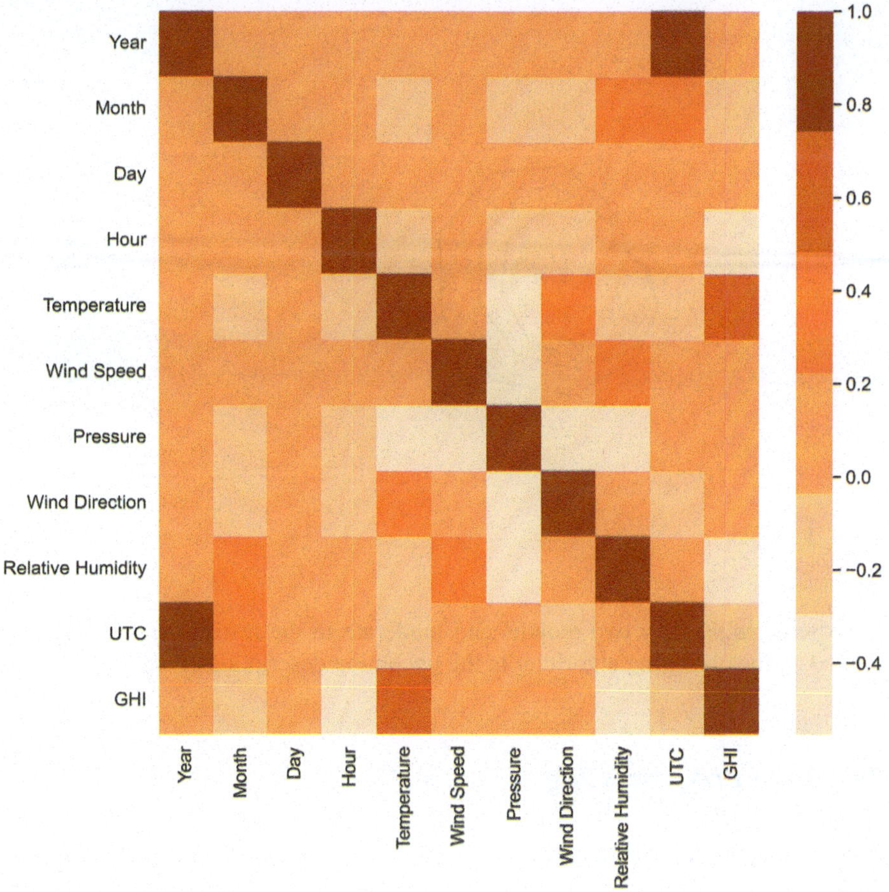


**Fig. 3** Correlation heat map

which can reduce the amount of solar radiation that reaches the surface. In some locations, cloud cover can be more prevalent during certain seasons, affecting the seasonal variation of GHI. Understanding these changes is essential for accurate solar radiation estimates for planning and forecasting purposes [14].

The GHI being highly correlated with weather parameters which are proven to be seasonal [15], it becomes necessary to assess the seasonality of the historical data of GHI in order to develop a better model. To establish the seasonality, a polar plot is used here. It visualizes the time series coordinates as polar and connects the clusters accordingly.

In Fig. 4 of the polar plot, the plots of March, April, and May are far from the origin, and those of December and January are closer, which shows that GHI data has a significant seasonal component. For Surat, the data can be visualized as 6 clusters, each of 2 months, because the weather conditions are such that any season will have

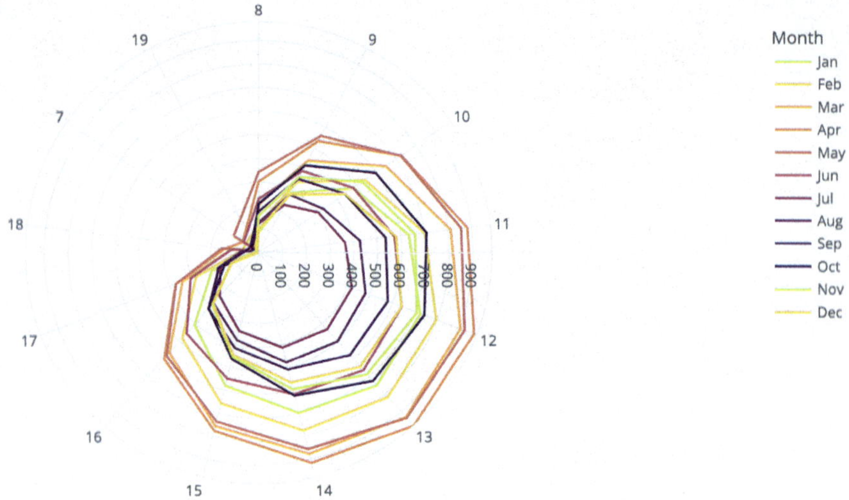


**Fig. 4** Polar plot

its extreme impact for two months and moderate in the rest of the two. (Will be proved by the clustering algorithm in the upcoming description as well)

## *3.4 Data Pre-processing*

Before using the data for the prediction of GHI, it is necessary to normalize it because AI models are highly sensitive to the scale of data points. The GHI data ranges from 0 to 1050 W/m$^2$, values of wind speed are between 0 and 10, but those of pressure are in thousands. Without normalizing it, bigger values will have a greater impact on the prediction model and will lead to poor accuracy. Moreover, neural network-based algorithms are reliant on distance matrix for prediction. Normalizing ensures the consistency of the distance matrix across all features. Hence, it becomes mandatory to scale the data from 0 to 1 for better performance [16].

We have used Min-Max Scalar to normalize the data. Using the Min-Max Scalar, each feature's minimal value is subtracted, and the remaining range—the difference between the minimum and maximum values—is divided. The resultant numbers will range from 0 to 1, with 0 denoting the least value and 1 denoting the greatest value. Min-Max scaling is often used in machine learning algorithms, particularly those that rely on distance calculations, such as k-nearest neighbors and clustering algorithms, where differences in the features' scale can significantly impact the results. These algorithms can be more accurate and efficient by scaling the features to a common range.

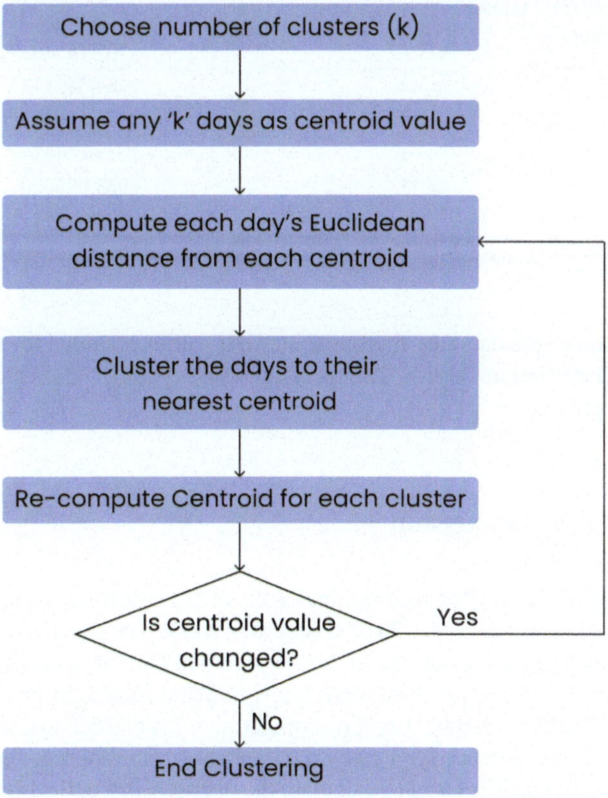


**Fig. 5** K-means clustering algorithm

To take advantage of the seasonality, we performed K-means clustering on the data to group them seasonally. K-means clustering can be a helpful technique for analyzing seasonal data, particularly when we want to identify patterns or groupings within the data. Seasonal data typically exhibit repeating patterns over time, and K-means clustering can be used to identify groups of observations that share similar patterns [17].

The K-Means clustering flowchart method is shown in Fig. 5. To cluster days seasonally; we need to consider all 24 h data points of each day together, so we used customized K-Means. The first step is deciding how many clusters to split our data into. Then any $k$ number of days are selected randomly as centroid values. The Euclidean distances of each node from the centroids are calculated, and the nodes are then grouped together around the centroid that is closest to them. By averaging all the nodes in each cluster, new centroids are obtained. If the centroid values are changed, the process is repeated, or else final clusters are obtained.

**Table 3** Centroid GHI values for different clusters

| No. | Centroid GHI |
|---|---|
| 1 | 707.155405 |
| 2 | 284.586777 |
| 3 | 775.156757 |
| 4 | 739.124424 |
| 5 | 968.085020 |
| 6 | 595.384181 |

After clustering the data into 6 groups, the GHI values obtained for the 7 h of each centroid are displayed in Table 3. The table shows that the GHI data has a significant seasonal component.

## 4 Results and Discussion

By using 80% of dataset for training and 20% of it for testing, Linear Regression (LR), Polynomial Regression (PR) of sixth order, and ANN was implemented on the dataset non-clustered and clustered. It can be observed that the accuracy is maximum by using ANN. In each algorithm, there is a significant reduction in error when it is implemented on clustered data. Error minimization is crucial because the GHI values are in W/m$^2$. They will be multiplied by the total area of the solar generation facility, and the total power will be used to calculate the energy. Hence, the error will also be multiplied (Table 4).

**Table 4** Performance metrics for various algorithms

| Methods | MAE | RMSE | $R^2$ |
|---|---|---|---|
| Linear Regression | 178.14132 | 143.430586 | 0.639508 |
| Linear Regression with K-Means | 169.046487 | 133.542961 | 0.675377 |
| Polynomial Regression | 102.957678 | 77.567908 | 0.879584 |
| Polynomial Regression with K-Means | 69.056983 | 46.638887 | 0.945827 |
| Artificial Neural Network | 65.448092 | 27.941005 | 0.950632 |
| **Artificial Neural Network with K-Means** | **53.941971** | **25.018180** | **0.966465** |

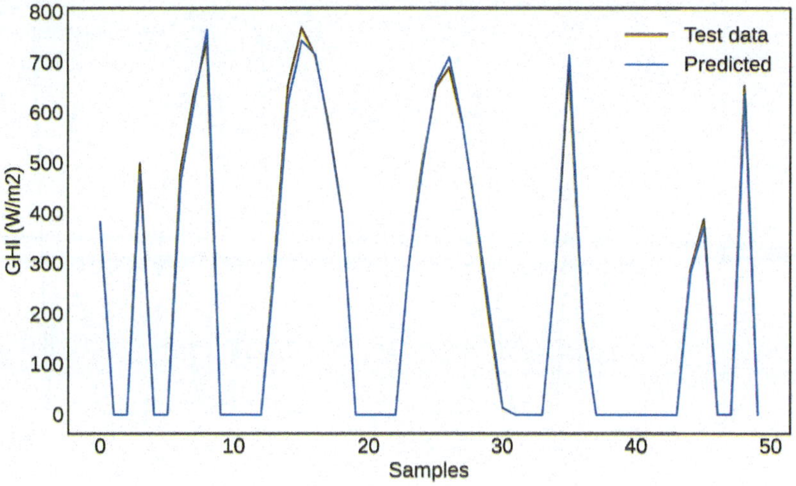


**Fig. 6** Artificial Neural Network

Figures 6 and 7 show that the ANN often provided the closest match between the actual value and the dataset. Its forecasting profile is frequently too clean to include random variations caused by meteorological data since it learns the current dynamics of data sets. Furthermore, it could appear as though the errors are too big for the forecasting model. This results from the significant uncertainty in the solar and weather data. More hidden layers in the model can enhance results, but doing so would prolong computation time and burden. In order to agree on the number of hidden layers, a trade-off between computational burden and minimizing errors is given consideration. Trial and error are used to achieve this.

Studying the effects of various features on system performance is essential because the computational load will increase due to many features. As a result, in this research, various weather inputs are disregarded one at a time, while actual data is fed with the remaining weather inputs. The values of the performance matrices for the various input factors for the ANN-based prediction model with and without K-Means are listed in Tables 5 and 6, respectively. The importance of temperature in the prediction can be expected. When all weather inputs are taken into account, the least is achieved with the observed error values. Hence, emphasizing the significance of taking into account all external variables. When the temperature is disregarded, forecast model performance is at its lowest. The error significantly grows by 8.9% as the MAE rises from 65.45 to 71.28. When the input factors pressure, wind speed, and wind direction are disregarded, there is less variation in the model's performance. Therefore, it can be said temperature and humidity are crucial factors in forecasting and must be considered.

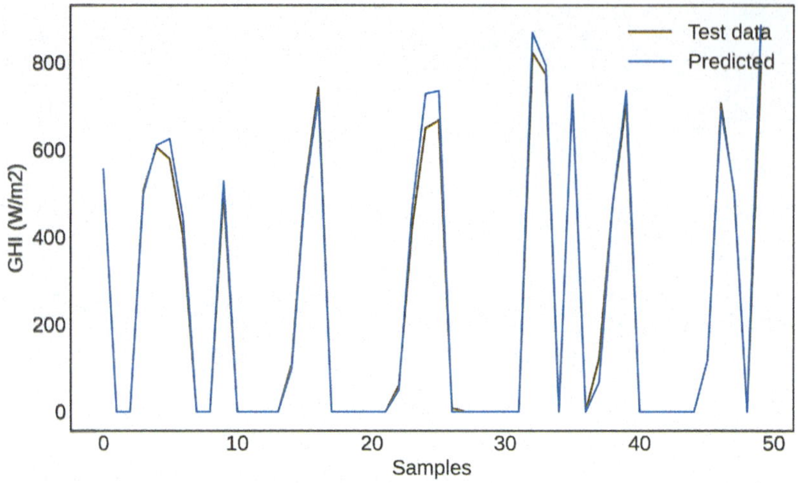


**Fig. 7** Artificial Neural Network with K-means

**Table 5** Sensitivity analysis for ANN

| Parameter | MAE | RMSE | $R^2$ |
|---|---|---|---|
| **All weather inputs** | **65.448092** | **27.941005** | **0.950632** |
| Without temperature | 71.275043 | 34.480865 | 0.941450 |
| Without wind speed | 70.198105 | 30.862293 | 0.943206 |
| Without pressure | 68.878392 | 30.296059 | 0.945322 |
| Without wind direction | 70.615335 | 33.149972 | 0.942529 |
| Without relative humidity | 70.860863 | 32.136967 | 0.942129 |

**Table 6** Sensitivity analysis for ANN with K-means

| Parameter | MAE | RMSE | $R^2$ |
|---|---|---|---|
| **All weather inputs** | **53.941971** | **25.018180** | **0.966465** |
| Without temperature | 60.730970 | 29.425981 | 0.957491 |
| Without wind speed | 58.859790 | 27.685562 | 0.960071 |
| Without pressure | 57.945705 | 27.154220 | 0.961301 |
| Without wind direction | 55.908939 | 26.339465 | 0.963974 |
| Without relative humidity | 59.442174 | 28.325263 | 0.959277 |

## 5 Conclusion

In this study, we have presented a comprehensive analysis of the performance of different algorithms for predicting the global horizontal irradiance (GHI) in a solar power generation system. The dataset used in this study covers a period of 3 years and includes variables such as temperature, wind speed, pressure, wind direction, and relative humidity.

We used linear regression, polynomial regression with and without K-means clustering, and artificial neural network algorithms to predict the GHI values. Three error metrics—mean absolute error (MAE), root mean squared error (RMSE), and R-squared ($R^2$)—were used to assess each algorithm's efficiency. Our results show that the artificial neural network along with K-Means outperforms the other algorithms, achieving the lowest error values for all three metrics.

Furthermore, we have analyzed the impact of seasonality on the prediction accuracy. The analysis of seasonal variations showed that the models' efficacy varied greatly throughout the year. The results suggest that the models must be retrained for each season in order to achieve the highest accuracy in GHI predictions, necessitating the clustering of the data.

Overall, our research shows how effectively machine learning algorithms can be used to forecast GHI values in solar power generation systems. As precise GHI value predictions can help maximize the output and efficiency of solar power plants, our research findings can offer useful insights to the solar power industry. These findings can serve as a guide for similar studies in other locations too. Future research can focus on developing more advanced machine learning algorithms, incorporating more diverse data sources to further improve the accuracy and using these GHI predictions for designing efficient solar power plant and integrating them with the power grid as well.

## References

1. Majid MA et al (2020) Renewable energy for sustainable development in india: current status, future prospects, challenges, employment, and investment opportunities. Energy, Sustainability and Society 10(1):1–36
2. United Nations. COP27: Delivering for people and the planet|United Nations—un.org. https://www.un.org/en/climatechange/cop27
3. Popat J, Kakadiya H, Tak L, Singh NK, Majeed MA, Mahajan V (2021) Reliability of smart grid including cyber impact: a case study. In: Computational methodologies for electrical and electronics engineers. IGI Global, pp 163–174
4. Khatib T, Mohamed A, Sopian K, Mahmoud M (2012) Solar energy prediction for Malaysia using artificial neural networks. Int J Photoenergy 2012
5. Chen Changsong, Duan Shanxu, Cai Tao, Liu Bangyin (2011) Online 24-h solar power forecasting based on weather type classification using artificial neural network. Solar energy 85(11):2856–2870

6. Majeed MA, Mudgal S, Tak L, Popat J, Kakadiya H, Singh NK, Mahajan V (2021) Artificial neural network modelling for short term load forecasting. In: 2021 international conference on technology and policy in energy and electric power (ICT-PEP). IEEE, pp 302–306
7. Majeed MA, Mudgal S, Tak L, Popat J, Kakadiya H, Singh NK, Mahajan V (2022) Hourly load forecasting using sequence-to-sequence LSTM-based deep machine learning model. In: Sustainable technology and advanced computing in electrical engineering: proceedings of ICSTACE 2021. Springer, pp 469–477
8. Sharma Amandeep, Kakkar Ajay (2018) Forecasting daily global solar irradiance generation using machine learning. Renewable and Sustainable Energy Reviews 82:2254–2269
9. Mutavhatsindi Tendani, Sigauke Caston, Mbuvha Rendani (2020) Forecasting hourly global horizontal solar irradiance in south africa using machine learning models. IEEE Access 8:198872–198885
10. Mishra DP, Jena S, Senapati R, Panigrahi A, Salkuti SR (2023) Global solar radiation forecast using an ensemble learning approach. Int J Power Electron Drive Syst 14(1):496
11. Hongjun Lu, Setiono Rudy, Liu Huan (1996) Effective data mining using neural networks. IEEE transactions on knowledge and data engineering 8(6):957–961
12. NSRDB—nsrdb.nrel.gov. https://nsrdb.nrel.gov/data-sets/international-data
13. Aldrich J (1995) Correlations genuine and spurious in Pearson and yule. Stat Sci 364–376
14. Davey AM, Flores BE (1993) Identification of seasonality in time series: A note. Mathematical and computer modelling 18(6):73–81
15. Frimpong K, Oosthuizen J, Van Etten EJ (2014) Recent trends in temperature and relative humidity in Bawku east, northern Ghana
16. Sola Jorge, Sevilla Joaquin (1997) Importance of input data normalization for the application of neural networks to complex industrial problems. IEEE Transactions on nuclear science 44(3):1464–1468
17. Hartigan JA, Wong MA (1979) Algorithm as 136: a k-means clustering algorithm. J Royal Stat Soc. Series C (Appl Stat) 28(1):100–108

# Optimal Co-Ordination of Directional Overcurrent Relays in Distribution Network Using Whale Optimization Algorithm

**Manoharan Premkumar, Ravichandran Sowmya, Jagarapu S. V. Siva Kumar, Pradeep Jangir, Laith Abualigah, and Chandran Ramakrishnan**

**Abstract** Modern power distribution networks are incredibly complex due to the growing incorporation of distributed generators in the past few years. The coordination of Directional Overcurrent Relays (DORs) in interconnected systems with many relays is significantly hindered by this complexity. In a nonlinear and constrained optimization problem, optimal DOR coordination is essential for protecting such

M. Premkumar (✉)
Department of Electrical and Electronics Engineering, Dayananda Sagar College of Engineering, Bengaluru, Karnataka 560078, India
e-mail: mprem.me@gmail.com

R. Sowmya
Department of Electrical & Electronics Engineering, Manipal Institute of Technology, Manipal Academy of Higher Education, Manipal, Udupi, Karnataka, India
e-mail: sowmyanitt@gmail.com

J. S. V. S. Kumar
Department of Electrical and Electronics Engineering, GMR Institute of Technology, Rajam, Andhra Pradesh 532127, India

P. Jangir
Department of Biosciences, Saveetha School of Engineering, Saveetha Institute of Medical and Technical Sciences, Chennai 602105, India

L. Abualigah
Computer Science Department, Faculty for Information Technology, Al Al-Bayt University, Prince Hussein Bin Abdullah, Mafraq 25113, Jordan

Hourani Center for Applied Scientific Research, Al-Ahliyya Amman University, Amman 19328, Jordan

MEU Research Unit, Middle East University, Amman, Jordan

School of Computer Sciences, Universiti Sains Malaysia, Pulau Pinang 11800, Malaysia

C. Ramakrishnan
Department of Electrical and Electronics Engineering, SNS College of Technology, Tamil Nadu, Coimbatore 641035, India



O. H. Gupta et al. (eds.), *Soft Computing Applications in Modern Power and Energy Systems*, Lecture Notes in Electrical Engineering 1107,
https://doi.org/10.1007/978-981-99-8007-9_17

complex systems and necessitates rigorous constraints. In order to address the optimal coordination problems of DORs, this study suggests using the Whale Optimisation Algorithm (WOA), a bio-inspired metaheuristic technique. WOA can optimize the fitness function in electrical engineering applications by taking insights from the humpback whales' hunting strategies. Using various fault data from 3-bus, 9-bus, and 30-bus standard systems, the effectiveness of WOA in promoting optimal DOR coordination is assessed. The main objective is to delineate the implementation of WOA to deal with DOR coordination problems. As a result, we are not comparing WOA's performance against any currently used algorithms. Rather, we use three case studies to test the algorithm's effectiveness with various population sizes and maximum iterations. The outcomes convincingly show that WOA is highly efficient in reducing the total period that primary relays are required to operate.

**Keywords** Bio-inspired algorithm · Operating time · Overcurrent relay coordination · Whale optimization algorithm

## 1 Introduction

When used in combination with fuses, reclosers, and Circuit Breakers (CBs), Overcurrent Relays (OCRs) are a common kind of protection for traditional radial distribution networks. The incorporation of Distributed Generators (DGs) into distribution networks has grown in recent years because of the significant technological, economic, and environmental advantages that DGs provide. The incorporation of DGs, on the other hand, transforms the radial topology of the traditional distribution network into an interconnected framework, leading to bidirectional power flow [1, 2]. A significant influence on the amount of the short-circuit current is exerted by the kind of DG and penetration level of DGs. When it comes to protecting such complex networks quickly and reliably, Directional Overcurrent Relays (DORs) are typically preferred over simple OCRs [3]. In order to improve the overall effectiveness of the protection scheme, DORs must work together in the most efficient manner possible. In transmission and distribution infrastructure, DORs are commonly employed for main protection, and they can also be used to safeguard distance relays in transmission systems as a backup protective device [4, 5].

The DORs might become necessary in order to achieve fault zone differentiation in the ring main, double-end fed, parallel feeder, and multi-looped systems. The primary function of a DOR is to identify a fault as soon as it occurs if it occurs within its

L. Abualigah
Department of Electrical and Computer Engineering, Lebanese American University, Byblos, Lebanon

Applied Science Research Center, Applied Science Private University, Amman, Jordan

School of Engineering and Technology, Sunway University Malaysia, Petaling Jaya, Malaysia

operational zone without any intended delay [6]. Primary protection is defined as the identification of a problem inside the specified zone without the use of any intended delay. Occasionally, primary protection neglects to eliminate a problem due to the failure of relays and/or CBs to function properly. It is necessary to have backup protection in place to correct the problem. Backup protection is an extra layer of security offered to a spot that only triggers after an intended delay if the main security of that part neglects to function properly [7]. In the coordination of DORs, the task of determining the most appropriate relay settings, specifically the Time Multiplier Setting (TMS) and the Plug Setting (PS), in such a way that the main relay reacts quicker than any other relays in the system is referred to as coordination. To minimize the overall working time of all primary relays and prevent miscoordination between primary and backup relay pairs, the coordination of DORs must find the optimal TMS and PS values for each relay while taking into consideration specific constraints. In the event of a failure of the main relay or the related circuit breaker, backup relays must be activated after a certain time interval, ensuring that the Primary/Backup (P/B) relay pairs are operated in the correct sequential order [8, 9].

Several ways to achieve the best possible coordination of DORs utilized for the protection of meshed distribution networks have been described in the scientific research literature. For relay coordination, it was common to practice in the past to use trial-and-error approaches [10]. Trial and error procedures, on the other hand, are hampered by the demand for many iterations and the sluggish rate of convergence. Topological modelling was used to provide optimal coordination of DORs, which were later implemented [11, 12]. When contrasted to trial-and-error procedures, topological analysis-based methods specify a minimum number of iterations to arrive at an appropriate solution. Nevertheless, using the topological analysis approach, it is not assured that the global optimal of TMS and PS of DORs would be obtained. In subsequent phases, numerous optimization algorithms for overcoming the coordination challenges of DORs were suggested, with the most prominent programming. Simplex [13], two-phase simplex [14], Sequential Quadratic Programming (SQP) [15], and dual simplex [16] are examples of classical optimization methods that are based on Linear Programming (LP) methods [17]. These approaches are fast and straightforward. However, because the operating time of the DOR is a linear function of the TMS, the LP-based optimization strategies are only useful for optimizing the TMS in this case. Because DOR management is a nonlinear problem, Nonlinear Programming (NLP) methods such as the gradient search technique [18], random search technique, and sequential quadratic programming [19] have been suggested in previous studies to defeat the shortcomings of LP-based approaches. It is possible to tune both the TMS and PS of DORs simultaneously using an NLP-based optimization technique. When it comes to addressing the DOR coordination problem, NLP-based solutions outperform LP techniques by a wide margin. However, both traditional optimization strategies have the potential to become stuck at the local optima and struggle to reach the global optima as a result. Furthermore, as the scale of the system grows, the pace of convergence of such optimization techniques becomes more and slower. As a result, over the last few decades, heuristic methods have emerged as useful tools for solving the relay coordination problem. There are a variety of methods available, including Genetic Algorithm (GA), Particle Swarm Optimization (PSO), Differential Evolutionary (DE) algorithm, Backtracking Search Algorithm (BSA), Artificial Bee Colony (ABC) algorithm, Ant Colony Optimization (ACO), Biogeography-Based Optimization Algorithm (BBOA), Gravitational Search Algorithm (GSA), Teaching Learning-Based Optimization (TLBO), Grey Wolf Optimizer (GWO), Cuckoo Search (CS), and Harmony Search (HS) algorithm [20–24]. For the coordination problem of DORs, hybrid methods are also used to improve computational speed. These methods include GA-LP, GA-NLP, BBOA-LP, GSA-SQP, PSO-SQP, and PSO-GSA. When compared with trial and error, topological, and traditional LP and NLP algorithms, such heuristic and evolutionary optimization algorithms outperform them in terms of reaching the global optimal solution. Most metaheuristic algorithms, on the other hand, need greater computing time and suffer from premature convergence. Researchers have put forth a great amount of effort to date to solve the optimal relay coordination challenges that arise in DOR networks. It seems that practically every time, the primary aim is to improve the relay settings to reduce the total Operating Time (OT) of the relays. However, the challenge of coordination between relays (P/B relay pairs) has not been properly handled so far. As a result, to close the research gap, this paper addresses both objectives, namely, optimization of coordination between P/B relay pairs and relay settings, at the same time and with better results.

Recently, a bio-inspired metaheuristic optimization technique called Whale Optimization Algorithm (WOA) for solving optimization issues was published in [25] and is described in detail. In comparison to other algorithms, WOA is made to balance the exploration and exploitation of the search area, which can improve the convergence rate and overall optimization. It is simpler to develop and utilize for many optimization problems because it only requires a few tuning parameters. It can rapidly reach an equivalent to an ideal solution in a short period due to its strong convergence rate. Compared to other optimization techniques, it is more computationally efficient and relatively simple to implement. It can solve various optimization problems, including those with one or more objectives. In addition, this algorithm is not applied to optimal coordination of directional overcurrent relay optimization problems. This motivated us to select the whale algorithm for the above-said optimization problem. Therefore, this paper uses WOA to provide the best possible coordination of DORs. The recommended algorithm's performance is tested on standard 3-, 9-, and 30-bus test power systems, with faults introduced at the midway of the lines to evaluate its effectiveness. The recommended algorithm's performance is revealed to be an optimal tool in terms of attaining the shortest overall OT of relays.

# 2 Optimization Problem Formulation

## 2.1 Objective Function

The problem of DOR can be expressed as either a linear function or a nonlinear function. In the scenario of a linear function, plug-setting has remained constant among the upper and lower bounds of the flow of current while TMS is computed; however, in the scenario of a nonlinear function, both TMS and PS are minimized at the same time in both cases. Apart from that, the quantized characterization of relay configurations increases the difficulty in coordinating the operation of the system. When it comes to solving the DOR's coordination challenge, there are primarily two objectives that must be addressed. The initial aim is to reduce the overall running time of all relays put in the network to the bare minimum, allowing the problem to be addressed in the shortest amount of time. Second, the coordination between the backup and primary relays should be retained, i.e., the backup relay would only act after a specified period if the primary relay could not function properly. The mathematical expression of the relay can be stated as follows [26].

$$T_{ik} = \frac{TMS_i * \beta}{\left(\frac{I_f}{PS_i.\,\mathrm{CT}_{Rating,P}}\right)^{\alpha} - \delta} \tag{1}$$

$$I_f = \frac{I_{R_i,k}}{\mathrm{CT}_{Rating,P}} \tag{2}$$

where $T_{ik}$ denotes the ith relay operating time for a fault at the kth site, $I_{R_i,k}$ denotes the fault current observed in the relay $R_i$ for a fault at the the kth site, and $I_\mathrm{f}$ denotes the fault current at the Current Transformer (CT) primary terminal. $PS_i$ denotes the PG of the relay $R_i$ above which it begins to operate, while $TMS_i$ denotes the TMS of the relay $R_i$. $\alpha$, $\beta$, and $\delta$ are coefficients that change depending on the characteristics of the relay. According to [26], the values of the parameters are 0.02, 0.14, and 1, respectively, as per IEC Standard 60,255–151 for Inverse Definite Minimum Time (IDMT) relays. $\mathrm{CT}_{Rating,P}$ denote the primary rating of the respective CT. The relationship between $I_{R_i,k}$ and $PS_i$ denotes the factor that determines the degree of nonlinearity. If the fault occurs nearer to the relay, then the fault is called the near-end fault (or close-in fault), and if the fault occurs at another end of the line is called the far-end fault (or far-bus fault), as depicted in Fig. 1.

The primary objective of coordinating the DORs problem is to find the best TMS and PS values so that the overall weighted sum of all primary relays' Operation Time (OT) at their related zones is as small as possible. As a result, the objective function can be written as follows [27, 28]

$$\mathrm{Min}: f = \sum_{i=1}^{N_{cl}} T^i_{cl_{in},P} + \sum_{j=1}^{N_{far}} T^j_{far_{bus},P} \tag{3}$$

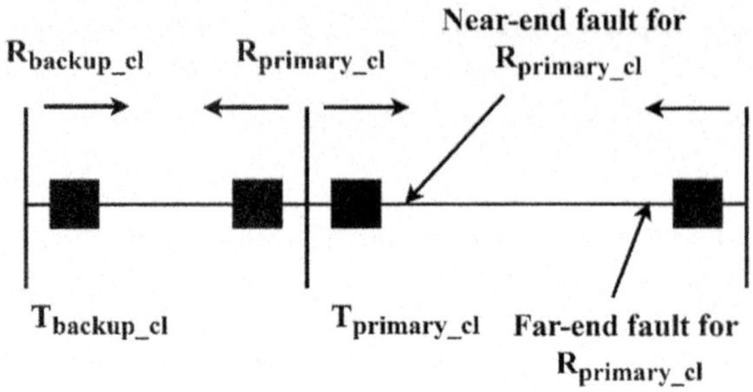


**Fig. 1** Schematic of close-in and far-bus faults for relay

$$T^{i}_{cl_{in},P} = \frac{0.14 \cdot \mathrm{TMS}^{i}}{\left(\frac{I^{i}_{f}}{text PS^{i} \cdot \mathrm{CT}^{i}_{Rating,P}}\right)^{0.02} - 1} \tag{4}$$

$$T^{j}_{far_{bus},P} = \frac{0.14 \cdot \mathrm{TMS}^{j}}{\left(\frac{I^{j}_{f}}{text PS^{j} \cdot \mathrm{CT}^{j}_{Rating,P}}\right)^{0.02} - 1} \tag{5}$$

where $N_{cl}$ denotes the number of relays responding to a near-end fault and $N_{far}$ denotes the number of relays responding to far-end faults.

## 2.2 Constraints

There are specific limits on how long the relay can operate that must be met for the relay to function properly. To function within the constraints, the TMS and PS should be constrained. The TMS has a boundary requirement that should be met for the relay to operate quickly and correctly. The relay must not cross the bounds. It should function within the TMS's upper ($ub$) and lower ($lb$) limits. As a result, the ith relay's TMS limit setting can be written as follows:

$$TMS_{i}^{lb} \leq TMS_{i} \leq TMS_{i}^{ub}, i = 1, 2, 3, \ldots, N_{cl} \tag{6}$$

In this paper, the $lb$ and $ub$ values of TMS are chosen to be 0.05 and 1.1, respectively. The relay's PS should be such that it stays quiet when the feeder is receiving peak load current, but it should function when the feeder is experiencing minimum fault current. To satisfy the above conditions, the bounds of the $i$ th relay's PS can be stated as follows [27]:

$$PS_{j}^{lb} \leq PS_{j} \leq PS_{j}^{ub}, j = 1, 2, 3, \ldots, N_{far} \tag{7}$$

In this paper, the $lb$ and $ub$ values of PS are chosen to be 1.25 and 1.5, respectively. The TMS, PS, and the fault current shown by the relay determine the relay's OT. The OT is calculated using analytic formulas or standard inverse curves based on the characteristics of the relay. Each element of the objective function is constrained to be between 0.1 and 1.1. The backup relay should only be used after the primary relay has failed to prevent maloperation. It is required for the backup and primary relays' selectivity to be maintained. The Coordination Time Interval (CTI) is the summation of the OT of the circuit breaker connected with the primary relay and the overshoot time. The gap between the OT of the backup relay and the primary relay should be greater than the CTI in coordinating two overcurrent relays. The CTI can be characterized as follows [27]:

$$CTI = T_{j,k} - T_{i,k}, i = 1, 2, \ldots, N_{cl} \quad (8)$$

where $T_{j,k}$ denotes the OT of the jh B-relay for a fault at kth location within the protected zone by the ith P-relay. Thus, the constraint for CTI can be represented as follows [27]:

$$CTI \geq CTI_{min} \quad (9)$$

where $CTI_{min}$ denotes a minimum CTI, and it is typically between 0.2 and 0.3 s. For all P/B pairs, the constraint for CTI should be met. Close-in, far-bus, and middle-point fault currents are commonly utilized, and they provide coordination for many fault scenarios.

## 2.3 *Modified Objective Function*

However, the least possible CTI between the backup and primary relays is essential for correct selection; considerably deferred backup relay operation is not preferred from the standpoint of successful relay coordination. Therefore, the objective function is adjusted to improve the CTI between primary and backup relays, as follows [29]:

$$\text{Min}: f = \alpha_1 \sum_{i=1}^{m} \sum_{k} T_{i,k}^2 + \alpha_2 \sum_{p=1}^{m_p} \left[\Delta T_{mp} - \beta\left(\Delta T_{mp} - \left|\Delta T_{mp}\right|\right)\right]^2 \quad (10)$$

$$\Delta T_{mp} = T_{j,k} - T_{i,k} - CTI \quad (11)$$

where $\Delta T_{mp}$ denotes the difference in OT with CTI between pth pair of relays, $m$ denotes the number of relay units, $m_p$ denotes the number of P/B relay pairs, $\alpha_1$ and $\alpha_2$ denote weight factors, and $\beta$ denotes factor to consider the miscoordination. If the value of $\beta$ increases, the level of miscoordination decreases; however, it increases the OT of the relay units. Therefore, it is necessary to select the optimal values of $\beta$. The synchronization between DORs is stated as a highly constrained and nonlinear optimization problem in which the TMS and PS of each relay are considered as design parameters. The major aim is to reduce the OTs of all P-relays, which are supposed to work to clear the faults in their respective regions after the faults are cleared.

## 3 Whale Optimization Algorithm (WOA)

WOA is a recently proposed metaheuristic algorithm based on swarms that have been presented for continual problems. It has been demonstrated to outperform contemporary meta-heuristic approaches in terms of overall performance [25, 30]. For example, it is simple and robust compared to other metaheuristic approaches, making it analogous to diverse nature-inspired algorithms in terms of implementation and robustness. The algorithm requires a smaller number of control parameters; in practice, only one parameter has to be fine-tuned. As depicted in Fig. 2, the humpback whale population in WOA searches for food in a multi-dimensional search space. In this model, the positions of whale populations are depicted as various decision vectors, and the range between whale populations and food relates to the level of fitness values. The three operational procedures described below are used to determine the time-dependent position of a whale population: (i) prey encircling, (ii) bubble-net attacking, and (3) prey search. The WOA is depicted in Fig. 3 in its most fundamental form.

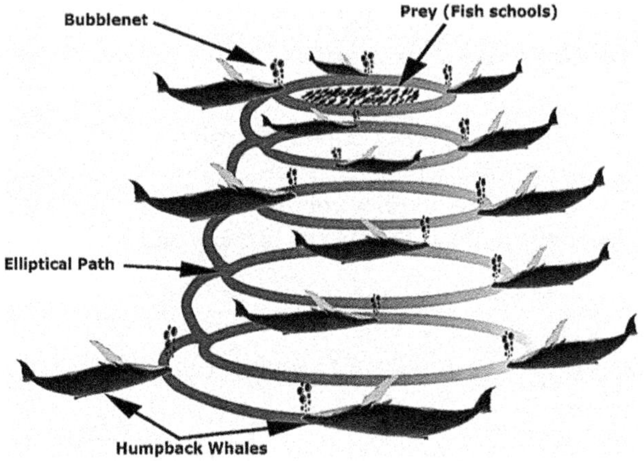


**Fig. 2** Bubble-net attacking of the prey

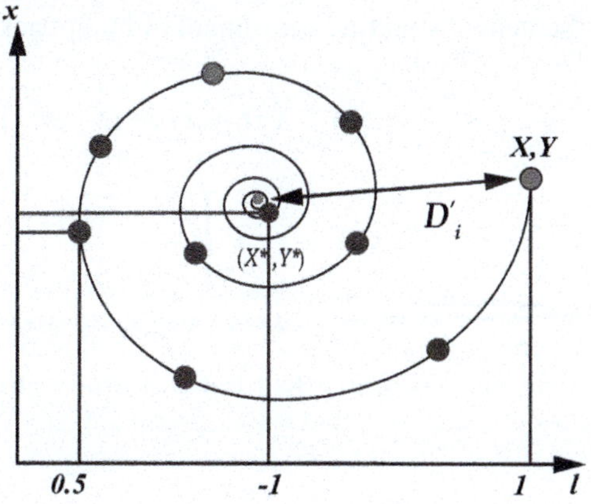


**Fig. 3** Position update of the whale using spiral strategy

## 3.1 *Prey Encircling*

Whales can distinguish and encircle their prey while they are in their natural habitat. Because the location of the design optimization in the search area is not known a priori, the WOA assumes that the current best candidate solution is either the target prey or very close to the best solution in the search space. The top search agent would be identified, and the remaining search agents would adjust their positions to be as close as possible to that of the top search agent. According to [25], the following equations can be used to express the behaviour described above:

$$\vec{D} = \left|\vec{C}.\vec{X^*}(t) - \vec{X}(t)\right| \tag{12}$$

$$\vec{X}(t+1) = \vec{X^*}(t) - \vec{A}.\vec{D} \tag{13}$$

$$\vec{A} = 2\vec{a}\cdot\vec{r} - \vec{a} \tag{14}$$

$$\vec{C} = 2\cdot\vec{r} \tag{15}$$

where $\vec{X^*}(t)$ denote the best position of the agent, $\vec{X}(t)$ denote the current position of the agent, $t$ denote the current iteration, $\vec{a}$ is constant, and it varies linearly from 2 to 0, and $\vec{r}$ denotes a uniform random number between [0, 1].

## 3.2 *Bubble-Net Attacking*

For the whale's bubble-net behaviour to be described mathematically, a spiral mathematical model is used between the positions of the whale and the prey to simulate the helical structure movement of humpback whales [25].

$$\overrightarrow{X}(t+1) = \overrightarrow{D'} \cdot e^{\text{bl}} \cdot \cos(2\pi l) + \overrightarrow{X^*}(t) \quad (16)$$

$$\overrightarrow{X}(t+1) = \begin{cases} \overrightarrow{X^*}(t) - \overrightarrow{A} \cdot \overrightarrow{D} & \text{if } p < 0.5 \\ \overrightarrow{D'} \cdot e^{\text{bl}} \cdot \cos(2\pi l) + \overrightarrow{X^*}(t) & \text{if } p \geq 0.5 \end{cases} \quad (17)$$

where $p$ denotes a constant which describes the logarithmic spiral shape, and $l$ denotes a uniformly distributed arbitrary number between [–1,1].

### 3.3 Prey Search

For global optimization algorithms to function, if $A > 1$ or $A \leq -1$, the population is updated in accordance with the directions provided by a randomly chosen population in the role of the best population [25].

$$\overrightarrow{D} = \left| \overrightarrow{C} \cdot \overrightarrow{X_{\text{rand}}} - \overrightarrow{X} \right| \quad (18)$$

$$\overrightarrow{X}(t+1) = \overrightarrow{X_{\text{rand}}} - \overrightarrow{X} \cdot \overrightarrow{D} \quad (19)$$

where $X_{\text{rand}}$ denotes random population position in the current iteration. Readers should refer to [25] for additional information.

## 4 Results and Discussions

It is verified in this paper that the WOA can be used to find the optimal coordination of directional overcurrent relays in the distribution network is considered. In this paper, there are three test cases, such as 3-bus (Case-1), 9-bus (Case-2), and 30-bus (Case-3) models, are considered. The results were presented for all three test cases. The control parameters of the WOA were taken based on the original paper, literature study, and trial and error method. The population size is 10 times the problem dimension, and the maximum number of iterations is selected as per the problem's complexity. The problem dimensions for the 3-bus test model is 6, for the 9-bus test model is 24, and for the 30-bus test model is 38. Therefore, the best population size for the 3-bus model is 60, for the 9-bus model is 240, and for the 30-bus model is 380. The value of the constant $a$ is linearly decreased from 2 to 0, the value of $l$ varies between [–1,1], and the value of $p$ is a random number between [0,1]. The experimental results are discussed in three sub-sections. The sub-Sect. 1 defines the results attained by the WOA for Case-1 with different population sizes. The sub-Sect. 2 describes the results attained by the WOA for Case-2 with different population sizes. The sub-Sect. 3 defines the results attained by the WOA for Case-3 with different population sizes. A computer with an Intel Core i5 CPU operating at 4.45 GHz and 16 GB of memory is used to execute the experiment through MATLAB software. The lower and upper bounds for all case studies are selected as per the data collected from the different literature.

### *4.1 Case-1: 3-Bus Test Model*

As demonstrated in Fig. 4, a three-bus (B1-B3) test system with one generator (G1) and six DORs (R1-R6) is investigated. The ratings of each component, as well as line data for each component, are taken from [12, 26]. As indicated in Table 1, the fault current at each bus is computed using the provided standard data, with the event of a fault in the centre of the line being taken into consideration. Table 1 also contains the primary rating of the CT. In clearing all near-end and far-end faults, it is necessary to synchronize the settings of each of the six relays that respond. As a result, there are a total of 12 control vectors in the DORs problem, which are designated as TMS1-TMS6 and PS1-PS6, respectively. The TMS values at the lower and upper levels are 0.05 and 1.1, respectively. PS is in the range of 1.25–1.50, depending on the model. The $CTI_{min}$ is set to its default value of 0.3 s.

As shown in Table 3, the optimal TMS and PS parameters derived by the WOA are shown. For near- and far-end faults, the OT of all primary relays is also listed in Table 3. The OTs are within an adequate level of 0.1–1.1 s, which is satisfactory. The overall OT of the P-relays is 4.8613 s. As shown in Table 4, the DORs do not have any miscoordination pairings when they are in operation. To examine the convergence behaviour of WOA, the convergence curve is depicted in Fig. 5.

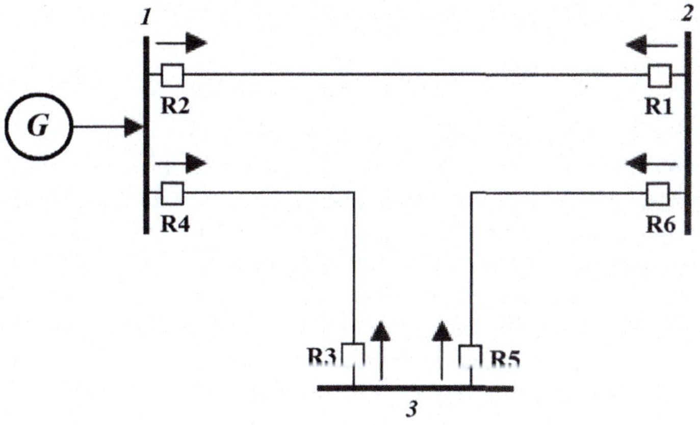


**Fig. 4** Case-1: Schematic of the 3-bus test model

**Table 1** Fault currents observed in the 3-bus test model

| Primary relay | Fault current (A) | Backup relay | Fault current (A) | Primary rating of CT |
|---|---|---|---|---|
| 1 | 9.46 | 5 | 100.63 | 2.06 |
| 2 | 26.91 | 4 | 14.08 | 2.06 |
| 3 | 8.81 | 1 | 136.23 | 2.23 |
| 4 | 37.68 | 6 | 12.07 | 2.23 |
| 5 | 17.93 | 3 | 19.20 | 0.8 |
| 6 | 14.35 | 2 | 25.90 | 0.8 |

**Table 2** Fault currents for P/B relays in the 3-bus test model

| Primary Relay | Fault current (A) | | Backup Relay | Fault current (A) | |
|---|---|---|---|---|---|
| | Close-in | Far-bus | | Close-in | Far-bus |
| 1 | 9.46/2.06 | 14.08/2.06 | 5 | 9.46/0.8 | 14.08/0.8 |
| 3 | 8.81/2.23 | 12.07/2.23 | 6 | 8.81/0.8 | 12.07/0.8 |
| 5 | 17.93/0.8 | 25.9/0.8 | 4 | 17.93/2.23 | 25.9/2.23 |
| 6 | 14.35/0.8 | 19.2/0.8 | 2 | 14.35/2.06 | 19.2/2.06 |

**Table 3** Optimal PS and TMS values for the 3-bus test model

| Relay | N = 40 | | | | N = 60 | | | |
|---|---|---|---|---|---|---|---|---|
| | TMS | PS | $OT_{Close\text{-}in}$ | $OT_{Far\text{-}bus}$ | TMS | PS | $OT_{Close\text{-}in}$ | $OT_{Far\text{-}bus}$ |
| 1 | 0.0500 | 1.2500 | 0.2655 | 0.2025 | 0.0500 | 1.2500 | 0.2655 | 0.2025 |
| 2 | 0.2198 | 1.2500 | 0.6403 | 0.4045 | 0.2194 | 1.2500 | 0.6392 | 0.4038 |
| 3 | 0.0500 | 1.2500 | 0.3007 | 0.2353 | 0.0500 | 1.2500 | 0.3007 | 0.2353 |
| 4 | 0.2391 | 1.2500 | 0.6262 | 0.4138 | 0.2299 | 1.2500 | 0.6021 | 0.3979 |
| 5 | 0.1950 | 1.2500 | 0.4593 | 0.4059 | 0.1950 | 1.2500 | 0.4593 | 0.4059 |
| 6 | 0.1953 | 1.2500 | 0.4997 | 0.4491 | 0.1953 | 1.2500 | 0.4997 | 0.4491 |
| OT in sec | 4.9027 | | | | 4.8613 | | | |

When analyzing the results obtained for two population sizes of the suggested WOA, it is evident that the WOA with a population size of 60 with 1000 iterations produces the optimal settings. In addition, the Run-Time (RT) of the WOA is also recorded in Table 4. As seen in Table 4, the RT value of WOA with a 40-population size is very much lesser than the WOA with a 60-population size. It is obvious that the algorithm with less population size takes lesser computation time. However, the results produced by the 60 population size are superior though the RT is high.

**Table 4** CTI for the 3-bus test model

| Faults | N = 40 | | | | N = 60 | | | |
|---|---|---|---|---|---|---|---|---|
| | Relay | | CTI | RT | Relay | | CTI | RT |
| | Backup | Primary | | | Backup | Primary | | |
| Close-in | 5 | 1 | 0.3000 | 0.4905 | 5 | 1 | 0.3000 | 0.6259 |
| | 6 | 3 | 0.3000 | | 6 | 3 | 0.3000 | |
| | 4 | 5 | 0.3283 | | 4 | 5 | 0.3001 | |
| | 2 | 6 | 0.3805 | | 2 | 6 | 0.3790 | |
| Far-bus | 5 | 1 | 0.3283 | | 5 | 1 | 0.3284 | |
| | 6 | 3 | 0.3141 | | 6 | 3 | 0.3140 | |
| | 2 | 6 | 0.4233 | | 2 | 6 | 0.3893 | |
| | 4 | 5 | 0.3013 | | 4 | 5 | 0.3001 | |

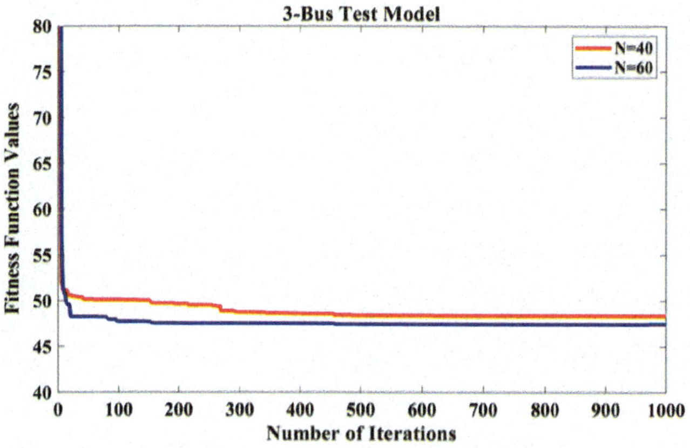


**Fig. 5** Convergence curves for different population sizes (3-bus model)

## 4.2 *Case-2: 9-Bus Test Model*

The schematic of the 9-bus system is shown in Fig. 6. With the normally inverse features, and there seem to be 24 digital DORs in this case study. TMS is recorded in the range of 0.01–1.0, and PS is recorded in the range of 0.5–2.5. In this case, the $CTI_{min}$ is fixed to a duration of 0.2 s. The value of the CT ratio is 500/1.

Short-circuit faults are formed in the centre of each line. The fault locations are labelled with the letters A-L in Fig. 6. Table 5 lists the fault currents observed in each P/B relay. In this optimization problem, there are a total of 48 decision vectors, i.e., TMS1-TMS24 and PS1-PS24 are considered. Due to problem complexity, the maximum iteration is selected as 2000.

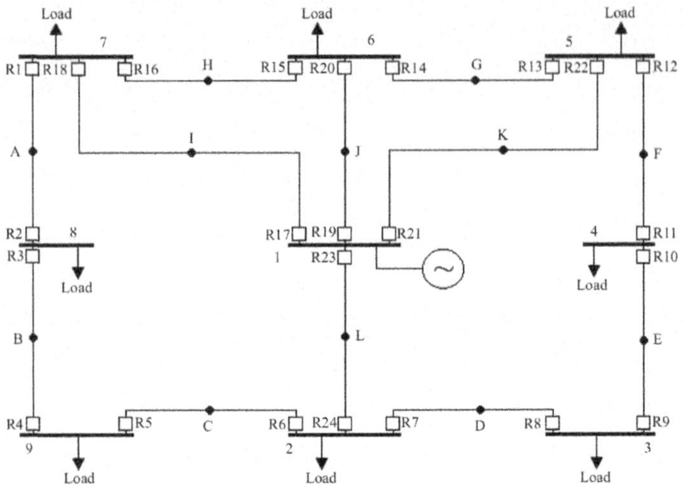


**Fig. 6** Schematic of the 9-bus test model

**Table 5** Fault currents in P/B relays for Case-2

| Location | P-Relay | B-Relay | $I_{f,p}$ (A) | $I_{f,s}$ (A) | Location | P-Relay | B-Relay | $I_{f,p}$ (A) | $I_{f,s}$ (A) |
|---|---|---|---|---|---|---|---|---|---|
| A | 1 | 15 | 24,779 | 9150 | G | 13 | 11 | 16,087 | 3088 |
| | 1 | 17 | 24,779 | 15,632 | | 13 | 21 | 16,087 | 13,000 |
| | 2 | 4 | 8327 | 8327 | | 14 | 16 | 18,213 | 6285 |
| B | 3 | 1 | 16,390 | 16,390 | | 14 | 19 | 18,213 | 11,934 |
| | 4 | 6 | 144,671 | 14,671 | H | 15 | 13 | 18,218 | 6285 |
| C | 5 | 3 | 9454 | 9454 | | 15 | 19 | 18,218 | 11,935 |
| | 6 | 8 | 23,280 | 4777 | | 16 | 2 | 16,087 | 3088 |
| | 6 | 23 | 23,280 | 18,507 | | 16 | 17 | 16,087 | 13,000 |
| D | 7 | 5 | 23,280 | 4777 | I | 18 | 2 | 8161 | 2426 |
| | 7 | 23 | 23,280 | 18,507 | | 18 | 15 | 8161 | 5736 |
| | 8 | 10 | 9454 | 9454 | J | 20 | 13 | 9286 | 4644 |
| E | 9 | 7 | 15,304 | 15,304 | | 20 | 16 | 9286 | 4644 |
| | 10 | 12 | 16,490 | 16,490 | K | 22 | 11 | 8161 | 2426 |
| F | 11 | 9 | 8326 | 8326 | | 22 | 14 | 8161 | 5736 |
| | 12 | 14 | 24,779 | 24,779 | L | 24 | 5 | 6149 | 3075 |
| | 12 | 21 | 24,779 | 24,779 | | 24 | 8 | 6149 | 3075 |

The suggested WOA is used to solve the DORs coordination problem, as discussed earlier. Table 6 shows the optimal TMS and PS values, as well as the optimal objective function value. The OT values are within an acceptable range of 0.1–1.1 s in all primary relays. Comparisons are made between the optimum values of decision variables (TMS and PS) and OT using the results produced by WOA with two different population sizes, such as 150 and 250. Table 7 displays the respective CTI values. When comparing the WOA with two different population sizes, the results produced by WOA with 250 population size demonstrate that the total OT of primary DORs is lowered. As can be seen in Table 7, the optimal outcomes provide no miscoordination. Furthermore, while utilizing the WOA with a 250-population size, the CTI improves since the total CTI values are lowered when compared to WOA with a 150-population size, hence improving the CTI. To examine the convergence behaviour of WOA, the convergence curve is plotted and depicted in Fig. 7. From Fig. 7, it is observed that the convergence rate of the WOA with large population size is higher than lower population size.

### 4.3 *Case-3: 30-Bus Test Model*

The IEEE 30-bus system is taken into consideration to assess the efficacy of the WOA in tackling a larger power system problem. As illustrated in Fig. 8, the system can be thought of as a meshed sub-transmission/distribution system having distributed-generated units interconnected. For the 30-bus test model, a total of 38 DORs with normal inverse characteristics are examined, with one at every end of the lines. Table 8 shows the short-circuit current values for close-in faults. TMS is measured in the range of 0.1–1.1, and PS is measured in the range of 1.5–6. The CT ratio for each relay was supposed to be 1000/5. The $CTI_{min}$ value has been set to 0.3 s. Due to complexity, the maximum iteration is selected as 5000.

Table 9 shows the results achieved by WOA with two different population sizes, such as 300 and 400. Table 10 illustrates the CTI obtained from the optimum TMS and PS values, proving there is no miscoordination. This shows that results produced using WOA with a 400-population size are superior to those acquired using WOA with a 300-population size, as can be seen in the table. This demonstrates that the suggested WOA may be utilized to effectively handle the DORs' coordination problem for large-scale power systems. The RT values are recorded, and it is obvious that WOA with a 400-population size takes more computation time than WOA with a 300-population size.

To observe the convergence behaviour of WOA while handling large-scale problems, the convergence curve is shown in Fig. 9. From Fig. 9, it is noticed that the convergence rate of the WOA with large population size is higher than the lower population size. The suggested WOA with a larger population size outperforms the WOA with a lower population size in terms of producing more stable and better solutions for all three test systems.

**Table 6** Optimal PS and TMS values for the 9-bus test model

| Relay | N = 40 | | | N = 60 | | |
|---|---|---|---|---|---|---|
| | TMS | PS | $OT_{primary}$ | TMS | PS | $OT_{primary}$ |
| 1 | 0.5396 | 1.4796 | 1.0383 | 0.5463 | 1.3679 | 1.0274 |
| 2 | 0.4032 | 0.5653 | 0.8064 | 0.2645 | 1.0616 | 0.6542 |
| 3 | 0.4545 | 1.3600 | 0.9683 | 0.4353 | 1.4819 | 0.9538 |
| 4 | 0.4531 | 1.1242 | 0.9410 | 0.4070 | 0.6595 | 0.7226 |
| 5 | 0.3845 | 1.2198 | 0.9553 | 0.3965 | 1.1589 | 0.9665 |
| 6 | 0.5696 | 1.3054 | 1.0762 | 0.4844 | 1.1781 | 0.8887 |
| 7 | 0.5691 | 1.3172 | 1.0780 | 0.5593 | 1.4422 | 1.0881 |
| 8 | 0.3987 | 1.1234 | 0.9610 | 0.4558 | 0.5537 | 0.8722 |
| 9 | 0.4669 | 1.1836 | 0.9725 | 0.3993 | 1.2013 | 0.8355 |
| 10 | 0.4353 | 1.4644 | 0.9482 | 0.5203 | 0.7805 | 0.9368 |
| 11 | 0.5207 | 0.5000 | 1.0038 | 0.2507 | 1.5719 | 0.7261 |
| 12 | 0.5593 | 1.2179 | 1.0178 | 0.5337 | 1.3684 | 1.0039 |
| 13 | 0.4566 | 1.1768 | 0.9346 | 0.5360 | 1.0687 | 1.0650 |
| 14 | 0.4927 | 1.2430 | 0.9869 | 0.5119 | 1.1360 | 0.9979 |
| 15 | 0.4620 | 1.4353 | 0.9680 | 0.5482 | 1.2777 | 1.1075 |
| 16 | 0.3884 | 1.3391 | 0.8282 | 0.3830 | 1.4219 | 0.8331 |
| 17 | 0.5960 | 1.2268 | 0.0000 | 0.5747 | 1.3607 | – |
| 18 | 0.4099 | 1.2405 | 1.0849 | 0.2316 | 1.3989 | 0.6437 |
| 19 | 0.4872 | 1.4617 | 0.0000 | 0.5814 | 1.1687 | – |
| 20 | 0.4501 | 1.2310 | 1.1297 | 0.4522 | 1.3215 | 1.1663 |
| 21 | 0.5639 | 1.3666 | 0.0000 | 0.5751 | 1.2284 | – |
| 22 | 0.4577 | 1.4813 | 1.3034 | 0.3860 | 1.0707 | 0.9652 |
| 23 | 0.6367 | 1.2716 | 0.0000 | 0.6361 | 1.4368 | – |
| 24 | 0.4843 | 1.1824 | 1.4140 | 0.2192 | 1.4098 | 0.6931 |
| OT in sec | 2041694 | | | 1814768 | | |

## 4.4 *Discussions*

Finding the best settings for these devices to save operation time and improve the durability of power systems is the objective of DOCRs' optimal coordination. This is a challenging, nonlinear optimization problem with many constraints. WOA may be a superior option for this problem compared to other algorithms for the following reasons. (i) WOA, which uses many solutions (whales) in each iteration and changes them in accordance with the best solutions discovered, is a population-based algorithm. Compared to conventional gradient-based optimization approaches, this enables it to explore the solution space more completely and decreases the likelihood of getting stuck in local minima, (ii) Unlike what is frequently the case in relay

**Table 7** CTI for the 9-bus test model

| N = 40 | | | | N = 60 | | | |
|---|---|---|---|---|---|---|---|
| Relay | | CTI | RT | Relay | | CTI | RT |
| Backup | Primary | | | Backup | Primary | | |
| 15 | 1 | 0.2001 | 9.8596 | 15 | 1 | 0.3763 | 13.5062 |
| 17 | 1 | 0.2088 | | 17 | 1 | 0.2162 | |
| 4 | 2 | 0.3389 | | 4 | 2 | 0.2000 | |
| 1 | 3 | 0.2008 | | 1 | 3 | 0.2000 | |
| 6 | 4 | 0.3006 | | 6 | 4 | 0.2985 | |
| 3 | 5 | 0.2220 | | 3 | 5 | 0.1999 | |
| 8 | 6 | 0.2000 | | 8 | 6 | 0.2000 | |
| 23 | 6 | 0.2018 | | 23 | 6 | 0.4378 | |
| 5 | 7 | 0.2029 | | 5 | 7 | 0.2000 | |
| 23 | 7 | 0.2000 | | 23 | 7 | 0.2383 | |
| 10 | 8 | 0.1999 | | 10 | 8 | 0.2344 | |
| 7 | 9 | 0.2544 | | 7 | 9 | 0.4073 | |
| 12 | 10 | 0.2000 | | 12 | 10 | 0.2002 | |
| 9 | 11 | 0.2000 | | 9 | 11 | 0.3092 | |
| 14 | 12 | 0.2304 | | 14 | 12 | 0.2498 | |
| 21 | 12 | 0.2043 | | 21 | 12 | 0.2000 | |
| 11 | 13 | 0.4793 | | 11 | 13 | 0.2000 | |
| 21 | 13 | 0.3663 | | 21 | 13 | 0.2139 | |
| 16 | 14 | 0.2001 | | 16 | 14 | 0.2058 | |
| 19 | 14 | 0.2005 | | 19 | 14 | 0.3110 | |
| 13 | 15 | 0.3499 | | 13 | 15 | 0.3777 | |
| 19 | 15 | 0.2194 | | 19 | 15 | 0.2014 | |
| 2 | 16 | 0.3242 | | 2 | 16 | 0.2000 | |
| 17 | 16 | 0.4967 | | 17 | 16 | 0.4907 | |
| 2 | 18 | 0.2000 | | 2 | 18 | 0.5563 | |
| 15 | 18 | 0.4390 | | 15 | 18 | 1.0667 | |
| 13 | 20 | 0.3859 | | 13 | 20 | 0.5318 | |
| 16 | 20 | 0.2469 | | 16 | 20 | 0.2357 | |
| 11 | 22 | 0.2644 | | 11 | 22 | 0.5744 | |
| 14 | 22 | 0.2142 | | 14 | 22 | 0.5489 | |
| 5 | 24 | 0.2229 | | 5 | 24 | 0.9422 | |
| 8 | 24 | 0.2000 | | 8 | 24 | 0.6005 | |

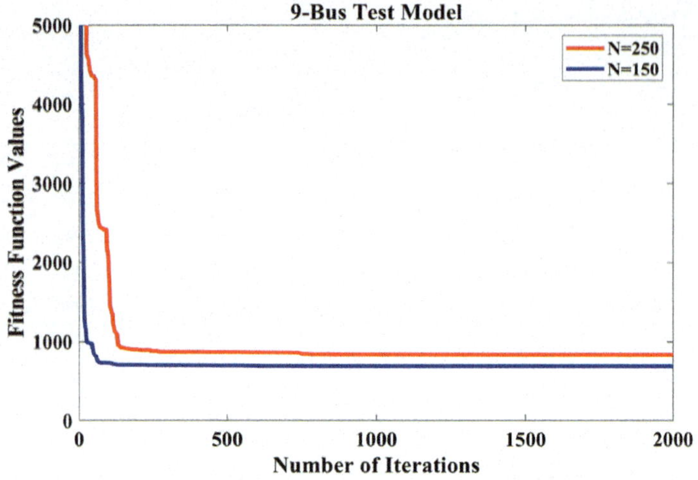


**Fig. 7** Convergence curves for different population sizes (9-bus model)

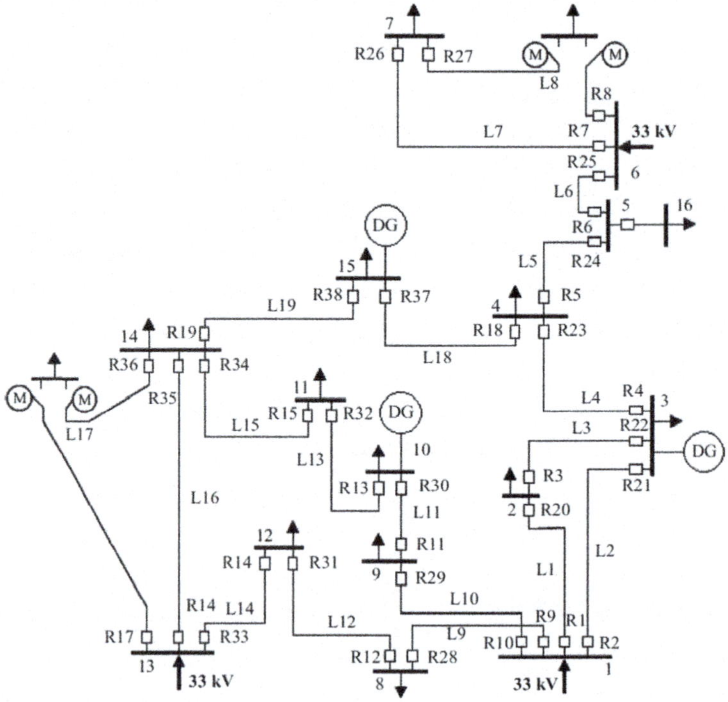


**Fig. 8** Schematic of the 30-bus test model

coordination problems, WOA does not demand that the problem be differentiable or even continuous. This provides it with a considerable edge over techniques like the Newton–Raphson approach, which relies on the gradient of the problem, (iii) Compared to more sophisticated algorithms, WOA is simpler to use and is not as subject to overfitting because it has few parameters that need to be tuned, (iv) WOA has demonstrated strong scalability characteristics, indicating that it can successfully address issues of varied sizes. This is important since relay coordination problems can have a lot of different factors and limitations.

Comparing WOA with different sizes of populations and numbers of iterations can show how well it works and help tune its settings for the DOCR coordination problem. Note that an increased population size with more iterations usually leads to better solutions but at the cost of using more computing resources. Because of this, it is important to find an equilibrium that fits the needs and limits of your application.

## 5 Conclusions

Coordinating DORs in a distributed system is a challenging nonlinear optimization problem addressed in this research. The objective is to reduce the total OTs of all primary relays required to clear faults at their assigned locations. The TMS and the PS of each relay, which serves as the variables in this optimization problem, are two crucial choice criteria. The bus model's complexity causes the dimensionality to rise. Examples include the 3-bus model, which provides a six-dimensional problem, and the 9-bus and 30-bus models, which contain 24 and 38 dimensions. Due to multimodal landscapes' enormous complexity and complicated nature, it is extremely difficult to

**Table 8** Fault currents in P/B relay for Case-3

| P-Relay | B-Relay | $I_{f,p}$ (A) | $I_{f,s}$ (A) | P-Relay | B-Relay | $I_{f,p}$ (A) | $I_{f,s}$ (A) |
|---|---|---|---|---|---|---|---|
| 3 | 1 | 4086.7 | 4086.7 | 9 | 20 | 7212.6 | 1103.5 |
| 4 | 2 | 5411.2 | 2138.8 | 10 | 20 | 7339.3 | 1095.8 |
| 22 | 2 | 4333 | 2147 | 1 | 21 | 7665.3 | 698.8 |
| 4 | 3 | 5411.2 | 3272.5 | 9 | 21 | 7212.6 | 721.2 |
| 21 | 3 | 5411.8 | 3243.6 | 10 | 21 | 7339.3 | 716.1 |
| 5 | 4 | 4960.8 | 3001.3 | 20 | 22 | 3481.5 | 3481.5 |
| 18 | 4 | 4719.4 | 3002.1 | 21 | 23 | 5411.8 | 2193.5 |
| 6 | 5 | 2416 | 2416 | 22 | 23 | 4333 | 2204.6 |
| 7 | 6 | 5669 | 1790.9 | 18 | 24 | 4719.4 | 1717.7 |
| 8 | 6 | 5607 | 1774.8 | 23 | 24 | 3689.7 | 1724.2 |
| 27 | 7 | 1472.3 | 1472.3 | 24 | 25 | 2695 | 2695 |
| 26 | 8 | 1026.8 | 1026.8 | 1 | 28 | 7665.3 | 1552 |

(continued)

**Table 8** (continued)

| P-Relay | B-Relay | $I_{f,p}$ (A) | $I_{f,s}$ (A) | P-Relay | B-Relay | $I_{f,p}$ (A) | $I_{f,s}$ (A) |
|---|---|---|---|---|---|---|---|
| 12 | 9 | 5034.9 | 5034.9 | 2 | 28 | 7985.7 | 1545.8 |
| 11 | 10 | 3457.1 | 3457.1 | 10 | 28 | 7339.3 | 1538 |
| 13 | 11 | 3727.3 | 2875 | 1 | 29 | 7665.3 | 1380.6 |
| 14 | 12 | 2906.5 | 2906.5 | 2 | 29 | 7985.7 | 1375.2 |
| 15 | 13 | 2660.5 | 2660.5 | 9 | 29 | 7212.6 | 1379 |
| 16 | 14 | 6185.6 | 1668.1 | 29 | 30 | 2518.9 | 2518.9 |
| 17 | 14 | 7492.9 | 1641.1 | 28 | 31 | 2036.8 | 2036.8 |
| 19 | 15 | 5445.2 | 1527.3 | 30 | 32 | 2998.8 | 2149 |
| 35 | 15 | 4222 | 1533.2 | 31 | 33 | 3263.6 | 3263.6 |
| 36 | 15 | 6420.2 | 1509.7 | 32 | 34 | 2930.4 | 2930.4 |
| 19 | 16 | 5445.2 | 3128.3 | 17 | 35 | 7492.9 | 1885.4 |
| 34 | 16 | 5796.6 | 3123.9 | 33 | 35 | 6456.2 | 1954.5 |
| 36 | 16 | 6420.2 | 3052.4 | 16 | 36 | 6185.6 | 490.9 |
| 19 | 17 | 5445.2 | 801.3 | 33 | 36 | 6456.2 | 500.6 |
| 34 | 17 | 5796.6 | 800.1 | 5 | 37 | 4960.8 | 1961 |
| 35 | 17 | 4222 | 794 | 23 | 37 | 3689.7 | 1968.5 |
| 38 | 18 | 3133.2 | 2292.2 | 34 | 38 | 5796.6 | 1886.8 |
| 37 | 19 | 3788.9 | 2940.9 | 35 | 38 | 4222 | 1896.7 |
| 2 | 20 | 7985.7 | 1053.9 | 36 | 38 | 6420.2 | 1867.7 |

**Table 9** Optimal PS and TMS values for the 30-bus test model

| Relay | N = 40 | | | N = 60 | | |
|---|---|---|---|---|---|---|
| | TMS | PS | $OT_{primary}$ | TMS | PS | $OT_{primary}$ |
| 1 | 0.4545 | 3.1060 | 1.2345 | 0.4300 | 3.7593 | 1.2665 |
| 2 | 0.3491 | 2.8194 | 0.8976 | 0.3975 | 3.7308 | 1.1462 |
| 3 | 0.3693 | 3.1515 | 1.3573 | 0.3312 | 3.5703 | 1.3059 |
| 4 | 0.4115 | 2.8116 | 1.2437 | 0.4382 | 2.2366 | 1.1999 |
| 5 | 0.4076 | 2.5546 | 1.2268 | 0.3612 | 3.5775 | 1.2806 |
| 6 | 0.3341 | 2.6194 | 1.5068 | 0.3607 | 2.9170 | 1.7516 |
| 7 | 0.4875 | 3.2885 | 1.5505 | 0.1956 | 3.6755 | 0.6566 |
| 8 | 0.3335 | 1.9492 | 0.8526 | 0.2965 | 3.5927 | 0.9897 |
| 9 | 0.5983 | 3.9446 | 1.8509 | 0.4684 | 2.2033 | 1.1404 |
| 10 | 0.4803 | 2.9998 | 1.3092 | 0.3544 | 3.7711 | 1.0658 |
| 11 | 0.4006 | 3.0418 | 1.5861 | 0.3091 | 3.3825 | 1.3049 |

(continued)

**Table 9** (continued)

| Relay | N = 40 | | | N = 60 | | |
|---|---|---|---|---|---|---|
| | TMS | PS | $OT_{primary}$ | TMS | PS | $OT_{primary}$ |
| 12 | 0.4253 | 2.3212 | 1.2194 | 0.2845 | 3.4124 | 0.9766 |
| 13 | 0.3820 | 3.1498 | 1.4777 | 0.3594 | 2.2864 | 1.1739 |
| 14 | 0.2994 | 2.7847 | 1.2475 | 0.1720 | 3.6088 | 0.8522 |
| 15 | 0.3583 | 2.6108 | 1.5154 | 0.2477 | 2.8316 | 1.1035 |
| 16 | 0.4812 | 3.2821 | 1.4683 | 0.3515 | 3.6347 | 1.1248 |
| 17 | 0.3879 | 3.2718 | 1.0869 | 0.4244 | 2.9792 | 1.1440 |
| 18 | 0.3337 | 2.6775 | 1.0501 | 0.2753 | 3.3028 | 0.9609 |
| 19 | 0.4292 | 3.0461 | 1.3417 | 0.3590 | 3.4523 | 1.1918 |
| 20 | 0.3427 | 2.6882 | 1.2602 | 0.2962 | 2.3907 | 1.0239 |
| 21 | 0.4256 | 2.6480 | 1.2523 | 0.3506 | 2.3450 | 0.9790 |
| 22 | 0.4289 | 2.6350 | 1.3952 | 0.3107 | 3.4814 | 1.1681 |
| 23 | 0.3777 | 2.3734 | 1.2629 | 0.3142 | 3.5574 | 1.3145 |
| 24 | 0.3350 | 2.9221 | 1.5107 | 0.2298 | 3.2147 | 1.1063 |
| 25 | 0.4311 | 2.8751 | 0.0000 | 0.3235 | 3.5868 | - |
| 26 | 0.3213 | 1.5000 | 1.8052 | 0.1808 | 2.9973 | 2.3387 |
| 27 | 0.4306 | 3.0602 | 3.4039 | 0.3468 | 1.5000 | 1.5020 |
| 28 | 0.2267 | 3.2851 | 1.3869 | 0.4231 | 1.5000 | 1.5167 |
| 29 | 0.4278 | 2.8702 | 1.9951 | 0.3174 | 3.4029 | 1.6759 |
| 30 | 0.4621 | 3.1394 | 2.0365 | 0.4269 | 3.1829 | 1.8986 |
| 31 | 0.4459 | 2.5311 | 1.6439 | 0.3054 | 3.1830 | 1.2865 |
| 32 | 0.4499 | 2.8409 | 1.8887 | 0.4260 | 2.8187 | 1.7793 |
| 33 | 0.4032 | 3.9091 | 1.3089 | 0.4031 | 2.8416 | 1.1333 |
| 34 | 0.4777 | 3.2525 | 1.4956 | 0.3989 | 3.8942 | 1.3633 |
| 35 | 0.3561 | 2.4247 | 1.1271 | 0.2341 | 3.6121 | 0.9119 |
| 36 | 0.3348 | 1.5000 | 0.7417 | 0.3219 | 1.5000 | 0.7132 |
| 37 | 0.3036 | 2.5729 | 1.0434 | 0.3693 | 3.1141 | 1.4060 |
| 38 | 0.3362 | 2.5866 | 1.2833 | 0.2975 | 2.7390 | 1.1735 |
| OT in sec. | 2041694 | | | 4592653 | | |

reduce associated cost functions. We use the WOA to discover the best DOR coordination to address this. The WOA successfully avoids frequent errors, including premature convergence to inferior solutions and exhibits improved performance in multimodal situations. The WOA was additionally tested across various population sizes in a sensitivity analysis. The results showed that larger population sizes produced the best outcomes while raising the RT value. This highlights the WOA's capacity to deal with the difficult job of DOR coordination among remote networks and offers a viable path for further study and application. When handling these

**Table 10** CTI for the 30-bus test model

| N = 40 | | | | N = 60 | | | |
|---|---|---|---|---|---|---|---|
| Relay | | CTI | RT | Relay | | CTI | RT |
| Backup | Primary | | | Backup | Primary | | |
| 1 | 3 | 0.2999 | 68.1456 | 1 | 3 | 0.4422 | 87.9689 |
| 2 | 4 | 0.5648 | | 2 | 4 | 1.4148 | |
| 2 | 22 | 0.4080 | | 2 | 22 | 1.4370 | |
| 3 | 4 | 0.3001 | | 3 | 4 | 0.3000 | |
| 3 | 21 | 0.2999 | | 3 | 21 | 0.5298 | |
| 4 | 5 | 0.4646 | | 4 | 5 | 0.3002 | |
| 4 | 18 | 0.6411 | | 4 | 18 | 0.6197 | |
| 5 | 6 | 0.3011 | | 5 | 6 | 0.3008 | |
| 6 | 7 | 0.3289 | | 6 | 7 | 1.5693 | |
| 6 | 8 | 1.0409 | | 6 | 8 | 1.2545 | |
| 7 | 27 | 0.7972 | | 7 | 27 | 0.4554 | |
| 8 | 26 | 0.5823 | | 8 | 26 | 3.4554 | |
| 9 | 12 | 0.9985 | | 9 | 12 | 0.3368 | |
| 10 | 11 | 0.3000 | | 10 | 11 | 0.3000 | |
| 11 | 13 | 0.3000 | | 11 | 13 | 0.3000 | |
| 12 | 14 | 0.3459 | | 12 | 14 | 0.5023 | |
| 13 | 15 | 0.3144 | | 13 | 15 | 0.3000 | |
| 14 | 16 | 0.4211 | | 14 | 16 | 0.3001 | |
| 14 | 17 | 0.8314 | | 14 | 17 | 0.3095 | |
| 15 | 19 | 0.9701 | | 15 | 19 | 0.5387 | |
| 15 | 35 | 1.1764 | | 15 | 35 | 0.8118 | |
| 15 | 36 | 1.5956 | | 15 | 36 | 1.0378 | |
| 16 | 19 | 0.7821 | | 16 | 19 | 0.4699 | |
| 16 | 34 | 0.6303 | | 16 | 34 | 0.3000 | |
| 16 | 36 | 1.4166 | | 16 | 36 | 0.9773 | |
| 17 | 19 | 4.9723 | | 17 | 19 | 7.3382 | |
| 17 | 34 | 11.9825 | | 17 | 34 | 8.6861 | |
| 17 | 35 | 12.8853 | | 17 | 35 | 9.4062 | |
| 18 | 38 | 0.2998 | | 18 | 38 | 0.3561 | |
| 19 | 37 | 0.8350 | | 19 | 37 | 0.3030 | |
| 20 | 2 | 2.6423 | | 20 | 2 | 1.4567 | |
| 20 | 9 | 1.4610 | | 20 | 9 | 1.3182 | |
| 20 | 10 | 2.0356 | | 20 | 10 | 1.4138 | |
| 21 | 1 | 9.4819 | | 21 | 1 | 4.8623 | |
| 21 | 9 | 7.7675 | | 21 | 9 | 4.5373 | |

(continued)

**Table 10** (continued)

| N = 40 | | | | N = 60 | | | |
|---|---|---|---|---|---|---|---|
| Relay | | CTI | RT | Relay | | CTI | RT |
| Backup | Primary | | | Backup | Primary | | |
| 21 | 10 | 8.5363 | | 21 | 10 | 4.7075 | |
| 22 | 20 | 0.3001 | | 22 | 20 | 0.3059 | |
| 23 | 21 | 0.4485 | | 23 | 21 | 0.9528 | |
| 23 | 22 | 0.3000 | | 23 | 22 | 0.7550 | |
| 24 | 18 | 1.1014 | | 24 | 18 | 0.6599 | |
| 24 | 23 | 0.8810 | | 24 | 23 | 0.3000 | |
| 25 | 24 | 0.4128 | | 25 | 24 | 0.5818 | |
| 28 | 1 | 0.5959 | | 28 | 1 | 0.5059 | |
| 28 | 2 | 0.9414 | | 28 | 2 | 0.6306 | |
| 28 | 10 | 0.5409 | | 28 | 10 | 0.7167 | |
| 29 | 1 | 2.1479 | | 29 | 1 | 1.8529 | |
| 29 | 2 | 2.5001 | | 29 | 2 | 1.9907 | |
| 29 | 9 | 1.5360 | | 29 | 9 | 1.9841 | |
| 30 | 29 | 0.3011 | | 30 | 29 | 0.4671 | |
| 31 | 28 | 0.8241 | | 31 | 28 | 0.2999 | |
| 32 | 30 | 0.2998 | | 32 | 30 | 0.3000 | |
| 33 | 31 | 0.3032 | | 33 | 31 | 0.2999 | |
| 34 | 32 | 0.2997 | | 34 | 32 | 0.3000 | |
| 35 | 17 | 0.7240 | | 35 | 17 | 0.5478 | |
| 35 | 33 | 0.4546 | | 35 | 33 | 0.4968 | |
| 36 | 16 | 3.2667 | | 36 | 16 | 3.4282 | |
| 36 | 33 | 3.2443 | | 36 | 33 | 3.2449 | |
| 37 | 5 | 0.3406 | | 37 | 5 | 0.9475 | |
| 37 | 23 | 0.3000 | | 37 | 23 | 0.9061 | |
| 38 | 34 | 0.2998 | | 38 | 34 | 0.3000 | |
| 38 | 35 | 0.6610 | | 38 | 35 | 0.7443 | |
| 38 | 36 | 1.0681 | | 38 | 36 | 0.9640 | |

complicated cases, the persuasive results demonstrate that WOA outperforms. So WOA is an intriguing alternative to the standard optimizers that are widely employed in a wide range of real-world complex power system problems.

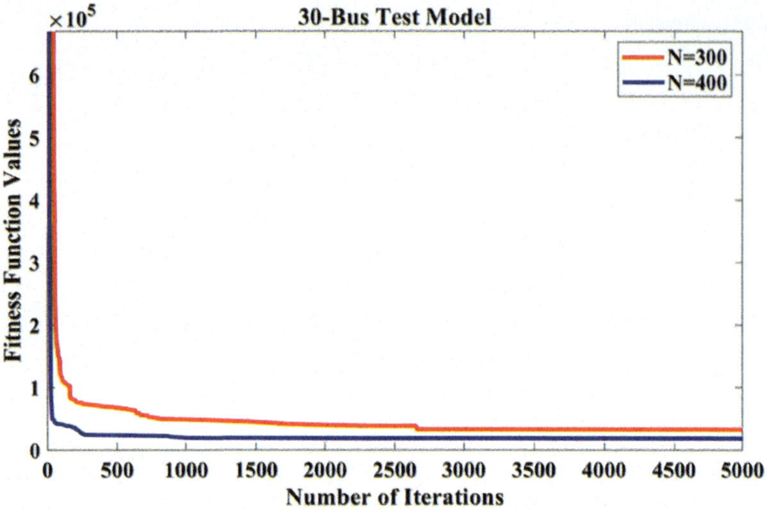


**Fig. 9** Convergence curves for different population sizes (30-bus model)

## References


1. Elsadd MA, Kawady TA, Taalab AMI, Elkalashy NI (2021) Adaptive optimum coordination of overcurrent relays for deregulated distribution system considering parallel feeders. Electr Eng 103:1849–1867. https://doi.org/10.1007/S00202-020-01187-0/TABLES/7
2. Abdelaziz AY, Talaat HEA, Nosseir AI, Hajjar AA (2002) An adaptive protection scheme for optimal coordination of overcurrent relays. Electric Power Syst Res 61:1–9. https://doi.org/10.1016/S0378-7796(01)00176-6
3. Chelliah TR, Thangaraj R, Allamsetty S, Pant M (2014) Coordination of directional overcurrent relays using opposition based chaotic differential evolution algorithm. Int J Electr Power Energy Syst 55:341–350. https://doi.org/10.1016/J.IJEPES.2013.09.032
4. Castillo Salazar CA, Conde Enríquez A, Schaeffer SE (2015) Directional overcurrent relay coordination considering non-standardized time curves. Electric Power Syst Res 122:42–49. https://doi.org/10.1016/J.EPSR.2014.12.018
5. Ibrahim AM, El-Khattam W, ElMesallamy M, Talaat HA (2016) Adaptive protection coordination scheme for distribution network with distributed generation using ABC. J Elect Syst Inform Technol 3:320–332. https://doi.org/10.1016/J.JESIT.2015.11.012
6. Khond SV, Dhomane GA (2019) Optimum coordination of directional overcurrent relays for combined overhead/cable distribution system with linear programming technique. Protect Control Modern Power Syst 4:1–7. https://doi.org/10.1186/S41601-019-0124-6/TABLES/3
7. Sarkar D, Kudkelwar S, Saha D (2019) Optimal coordination of overcurrent relay using crow search algorithm 7:282–297. https://doi.org/10.1080/23080477.2019.1694802
8. Acharya D, Das DK (2021) Optimal coordination of over current relay using opposition learning-based gravitational search algorithm. J Supercomp 77:10721–10741. https://doi.org/10.1007/S11227-021-03705-8/TABLES/13
9. Saldarriaga-Zuluaga SD, López-Lezama JM, Muñoz-Galeano N (2020) Optimal coordination of overcurrent relays in microgrids considering a non-standard characteristic. Energies 13:922. 13, 922. https://doi.org/10.3390/EN13040922
10. Ramaswami R, Venkata SS, Damborg MJ, Postforoosh JM (1984) Computer aided transmission protection system design: Part II: Implementation and Results. IEEE Transactions on Power Apparatus and Systems. PAS-103, 60–65. https://doi.org/10.1109/TPAS.1984.318577

11. Muhammad Y, Raja MAZ, Shah MAA, Awan SE, Ullah F, Chaudhary NI, Cheema KM, Milyani AH, Shu CM (2021) Optimal coordination of directional overcurrent relays using hybrid fractional computing with gravitational search strategy. Energy Rep 7:7504–7519. https://doi.org/10.1016/J.EGYR.2021.10.106
12. Urdaneta AJ, Nadira R, Pérez Jiménez LG (1988) Optimal coordination of directional overcurrent relays in interconnected power systems. IEEE Trans Power Del 3:903–911. https://doi.org/10.1109/61.193867
13. Gao F, Han L (2012) Implementing the nelder-mead simplex algorithm with adaptive parameters. Comput Optim Appl 51:259–277. https://doi.org/10.1007/s10589-010-9329-3
14. Singaravelan AMK, Ram JPBG, Kim YJ (2021) Application of two-phase simplex method (TPSM) for an efficient home energy management system to reduce peak demand and consumer consumption cost. IEEE Access 9:63591–63601. https://doi.org/10.1109/ACCESS.2021.3072683
15. Momoh JA, El-Hawary ME, Adapa R (1999) A review of selected optimal power flow literature to 1993 part i: nonlinear and quadratic programming approaches. IEEE Trans Power Syst 14:96–103. https://doi.org/10.1109/59.744492
16. Borzabadi AH, Alemy H (2016) Dual simplex method for solving fully fuzzy linear programming problems. 4th Iranian Joint Congress on Fuzzy and Intelligent Systems, CFIS 2015 (2016). https://doi.org/10.1109/CFIS.2015.7391653
17. Luenberger DG, Ye Y (2015) Linear and nonlinear programming. Springer Publishing Company, Incorporated
18. Birla D, Maheshwari RP, Gupta HO (2006) A new nonlinear directional overcurrent relay coordination technique, and banes and boons of near-end faults based approach. IEEE Trans Power Del 21:1176–1182. https://doi.org/10.1109/TPWRD.2005.861325
19. Ravikumar Pandi V, Zeineldin HH, Xiao W (2013) Determining optimal location and size of distributed generation resources considering harmonic and protection coordination limits. IEEE Trans Power Syst 28:1245–1254. https://doi.org/10.1109/TPWRS.2012.2209687
20. Premkumar M, Srikanth Babu V, Somwya R (2019) Scheduling task to heterogeneous processors by modified ACO algorithm. Adv Intell Syst Comput 758:565–576. https://doi.org/10.1007/978-981-13-0514-6_55
21. Premkumar M, Jangir P, Sowmya R, Elavarasan RM (2021) Many-objective gradient-based optimizer to solve optimal power flow problems: analysis and validations. Eng Appl Artif Intell 106:104479. https://doi.org/10.1016/J.ENGAPPAI.2021.104479
22. Premkumar M, Jangir P, Santhosh Kumar B, Sowmya R, Haes Alhelou H, Abualigah L, Riza Yildiz A, Mirjalili S, Premkumar M (2021) A new arithmetic optimization algorithm for solving real-world multiobjective CEC-2021 constrained optimization problems: diversity analysis and validations. IEEE Access. 9:84263–84295. https://doi.org/10.1109/ACCESS.2021.3085529
23. Devi RM, Premkumar M, Jangir P, Elkotb MA, Elavarasan RM, Nisar KS (2022) IRKO: an improved Runge-Kutta optimization algorithm for global optimization problems. Comp Mater Continua 70:4803–4827. https://doi.org/10.32604/CMC.2022.020847
24. Hussain K, Mohd Salleh MN, Cheng S, Shi Y (2018) Metaheuristic research: a comprehensive survey. Artif Intell Rev 52:4. 52, 2191–2233. https://doi.org/10.1007/S10462-017-9605-Z
25. Mirjalili S, Lewis A (2016) The whale optimization algorithm. Adv Eng Softw 95:51–67. https://doi.org/10.1016/j.advengsoft.2016.01.008
26. Jordan R (2018) Optimal coordination of directional overcurrent relays. In: Jordan R (ed) Metaheuristic optimization in power engineering, pp 449–473. Institution of Engineering and Technology. https://doi.org/10.1049/PBPO131E_CH13
27. Sarwagya K, Nayak PK, Ranjan S (2020) Optimal coordination of directional overcurrent relays in complex distribution networks using sine cosine algorithm. Electric Power Syst Res 187:106435. https://doi.org/10.1016/J.EPSR.2020.106435
28. Amraee T (2012) Coordination of directional overcurrent relays using seeker algorithm. IEEE Trans Power Del 27:1415–1422. https://doi.org/10.1109/TPWRD.2012.2190107

29. Moravej Z, Adelnia F, Abbasi F (2015) Optimal coordination of directional overcurrent relays using NSGA-II. Electric Power Syst Res 119:228–236. https://doi.org/10.1016/J.EPSR.2014.09.010
30. Premkumar M, Sumithira R (2018) Humpback whale assisted hybrid maximum power point tracking algorithm for partially shaded solar photovoltaic systems. J Power Elect 18:1805–1818. https://doi.org/10.6113/JPE.2018.18.6.1805

# Optimization and Comparative Analysis of Hybrid Renewable Energy Generation (Solar-Wind-Biomass) Using HOMER

**Chandan Singh and Shelly Vadhera**

**Abstract** Energy scarcity is one of the biggest challenges for the developing countries due to rapid industrialization and urbanization. Considering current electricity demand growth patterns, the world will undoubtedly face resource and carbon constraints. Renewable Resources are an attractive option to cater these issue and have potential to supply this growing demand with clean and safe energy in affordable price. This research describes the comparative techno-economic analysis of grid connected and stand-alone hybrid (Solar-Wind-Biomass) renewable energy resources (HRES) system using Hybrid Optimization of Multiple Energy Resources (HOMER) Pro software for electrification of a rural village in northern India which has limited access of the main grid. Solar, wind and biomass resources are used in the hybrid system. The main purpose of this study is to minimize the net present cost (NPC) and cost of energy (COE) of the hybrid system. The necessary data is obtained and inspected to perform the modelling and sizing the HRES system.

**Keywords** HOMER pro · HRES · Hybrid microgrid · Renewable energy · Wind energy · Biomass

## 1 Introduction

Energy is an essential component to analyze the development of any nation in the world. It has an important role in economical as well as social development [1]. Also, the energy demand of any developing country like India is increasing rapidly day by day due to urbanization and industrialization. In order to fulfill these demands using conventional sources has a vast impact on the environment and has possibility to exhaust all the coal and petroleum sources in upcoming few decades. RES microgrid

C. Singh (✉) · S. Vadhera
National Institute of Technology Kurukshetra, Haryana, India
e-mail: chandancngh@gmail.com

S. Vadhera
e-mail: shelly_vadhera@nitkkr.ac.in



O. H. Gupta et al. (eds.), *Soft Computing Applications in Modern Power and Energy Systems*, Lecture Notes in Electrical Engineering 1107,
https://doi.org/10.1007/978-981-99-8007-9_18

seems to be an attractive option to fulfill the demand of remote areas as well as support the main grid during peak demand without having any harmful impact on the environment and health. The conventional power systems are now being replaced with hybrid systems that also uses Renewable Energy Sources (RES) to promote sustainable development [2].

Solar PV, biomass, wind, small hydropower, and other Renewable Energy Sources (RES) have recently been used to minimize environmental pollution and reduce the strain of fossil fuel generation [3]. Off-grid HRES systems have emerged as a desirable alternative for rural and isolated locations with limited or no connection to the main grid. Additionally, it is highly essential to take into account RES because of both the environmental problems caused by greenhouse gas emissions and the rising prices of conventional fuel [4, 5]. Due to the seasonal, inconsistent, and time-dependent character of RES, dependability and power quality have emerged as an additional issue. Therefore, due to the intermittent nature of each RES, it is not practical to supply a demand continuously alone by any one system [6].

Many studies on stand-alone hybrid systems and grid connected hybrid systems have been conducted globally for the electrification of rural regions, institutions, and islands. According to Hajer Mannai et al. [7], a grid-connected solar PV/wind hybrid system is more economical and eco-friendlier for a Tunisian administrative building. For Pratas island in Taiwan, Chin-Ta Tsai et al. [3] analyzed several microgrid systems and recommended that the PV-diesel hybrid system has the lowest COE with a low excess electricity fraction. Sayedus Salehin et al. [8] studied the PV-biogas-diesel hybrid system with battery backup on an unelectrified island in Bangladesh. The Nomadic People Optimization (NPO) method was used by Abbas Q. Mohammed et al. [9] to analyze a residential complex in Thi-Qar and suggested that the PV-Wind-Large Diesel Generator-Battery system scenario is considerably more appealing than the others. In [10], a hybrid solar PV-biomass system for generating power in a remote location of Sarawak, Malaysia, is explored. In the North-West of Afghanistan, in the village of Awbeh, Mohammad Moien Sedeegi et al. [2] it was shown that ON-grid hybrid (solar wind battery) is more cost-effective and generally accepted than the off-grid hybrid system. The research of several hybrid RES systems in different parts of India has been documented in [11–15].

Despite several literature survey on grid connected hybrid microgrid, it is observed that the impact of frequent grid outages causing limited access of the grid and compromising reliability of power supply in rural areas has not been adequately explored. In this study, two different hybrid renewable power systems have been modelled and techno-economic analysis is performed to find the best optimal solution for the rural area having limited access of the grid.

This paper is structured in three sections. Section 1 includes introduction and methodology in which motivation of research and resources used is explained. Section 2 describes the proposed system models with optimal economic result. Section 3 covers the comparative analysis of both systems and at the end the paper is concluded.

# 2 Methodology

HOMER Pro software which is used to model and optimize the different renewable energy systems is industrialized by National Renewable Energy Laboratory (NREL). This software is very helpful in analysis of electric as well as economic aspect of any microgrid to find optimal microgrid. Furthermore, user can also define different input parameters and constraints for guiding the modelling of economic microgrid. Figure 1 shows the simulation process architecture diagram of HOMER Pro software.

## *2.1 Study Area*

Even though India has reached its goal of 100% electrification [16], still many rural and remote areas are not getting reliable supply because of frequent outages and connection issue. A similar rural area for which the case study is carried out is Bayal village in Mahendragarh District of Haryana. The latitude and longitude for this location is 12°58.5'N, 77°35.2'E. The village has a total area of 1550 hectares and has a population of roughly 5000 people divided into 750 families.

## *2.2 Load and Resource Assessment*

A real-time load data for a village having nearly same population with 24-h power supply is provided by State Utility. The daily average energy consumption is 2345.10

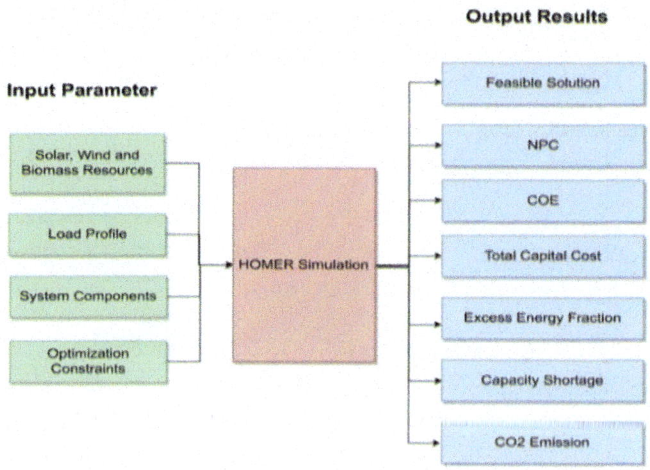


**Fig. 1** HOMER software simulation process architecture diagram

kWh with peak and average demand of 282.13 kW and 97.71 kW respectively. 0.35 is the load factor. Figures 2 and 3 show the daily average profile of the load and seasonal load profile. Time series chart of a year load is shown in Fig. 4.

The solar irradiance, temperature and wind speed related data is obtained online from NASA server by HOMER itself. Monthly average solar radiation data available at the study area with clearness index is shown in Fig. 5. It is observed that the available solar insolation has a lowest radiation and extreme radiation of 3.6 kWh/m2/day and 6.3 kWh/m2/day for the months of December and May respectively with an annual average of 5.04 kWh/m2/day and has a clearness index of 0.5713.

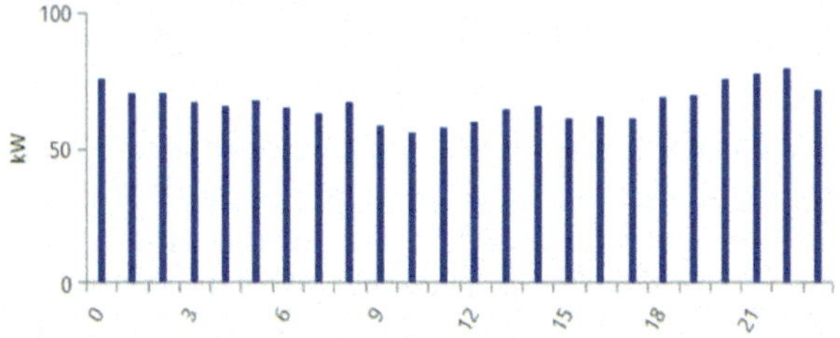


**Fig. 2** Daily average profile of load

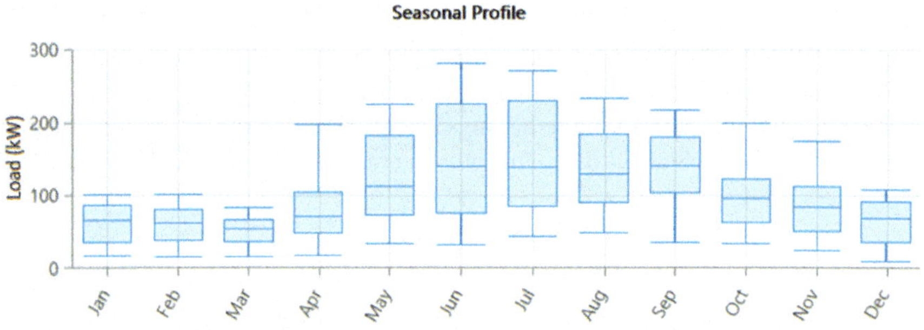


**Fig. 3** Seasonal profile of load

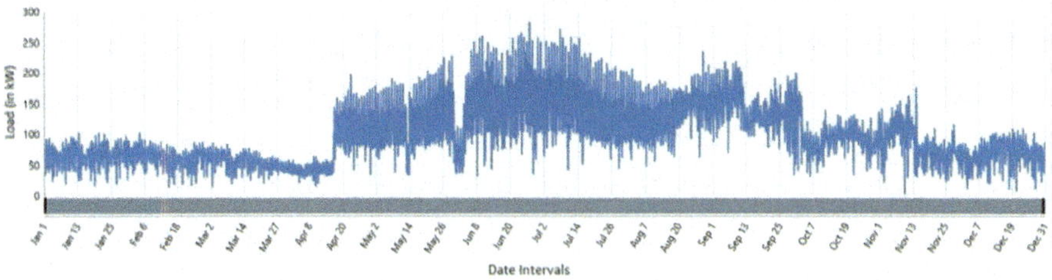


**Fig. 4** Time series chart of load data

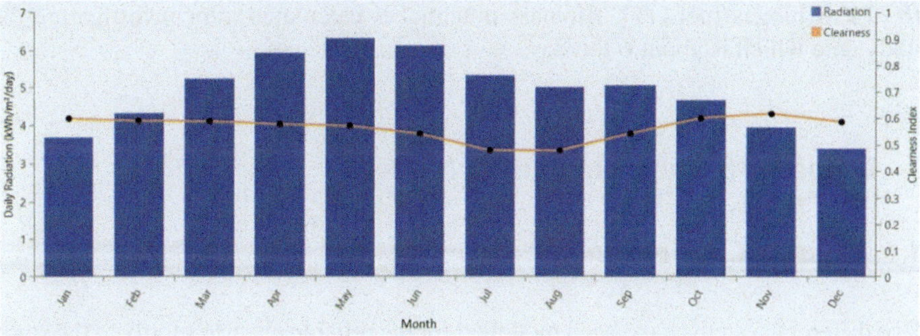


**Fig. 5** Monthly average solar global irradiance (GHI) plot

Figures 6 and 7 depict the monthly average wind speed and temperature of the study area. The average wind speed is coming out to be 4.62 m/s.

Biomass, which is defined as organic waste from plants and animals, is a renewable energy source that is used to produce electric power. These biomass resources in India have a bio-energy potential of 7.79 × 102 PJ as cellulosic ethanol fuel and 1.29 ×

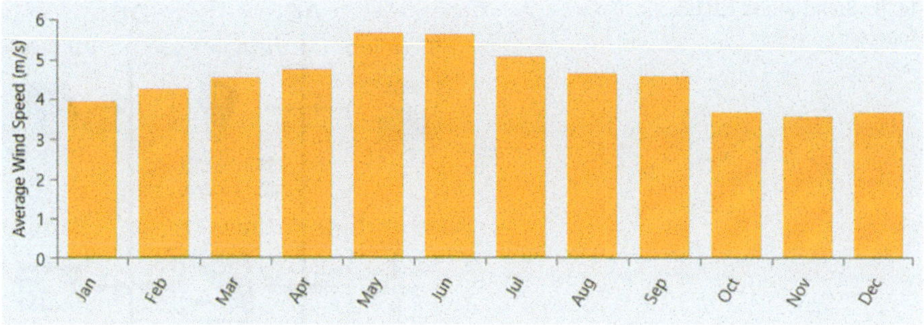


**Fig. 6** Monthly average wind speed data

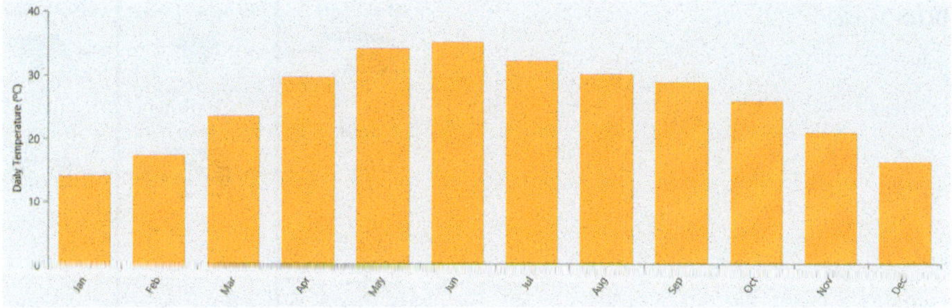


**Fig. 7** Monthly average temperature data

103 PJ as biogas fuel [17]. Biomass potential is calculated after investigating the study area which is about 6 ton/day.

## 3 Proposed System Simulation Model

### 3.1 System Configuration

Two different systems have been modelled for the reliable electric supply to the study area by local resources utilization. Figure 8 shows a stand-alone hybrid system which includes solar panel, wind turbine and biogas generator to meet the load demand of village. Battery is added to store the energy when the generated power is in surplus and supply during the peak hours to maintain the supply continuity. The converter acts as a bridge between AC and DC bus and ensures the conversion. Figure 9 shows a grid connected hybrid system considering the grid outages. Both system models are shown in Figs. 8 and 9. The lifetime of the proposed hybrid systems is considered to be 25 years with the discount rate of 5% and inflation rate of 2% [5].

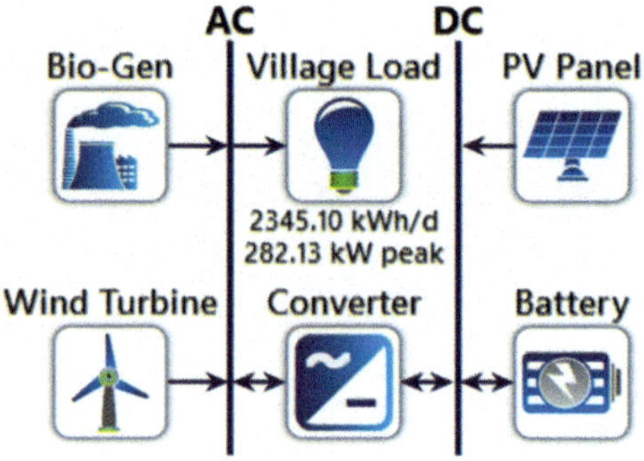


**Fig. 8** Stand-alone HRES Model

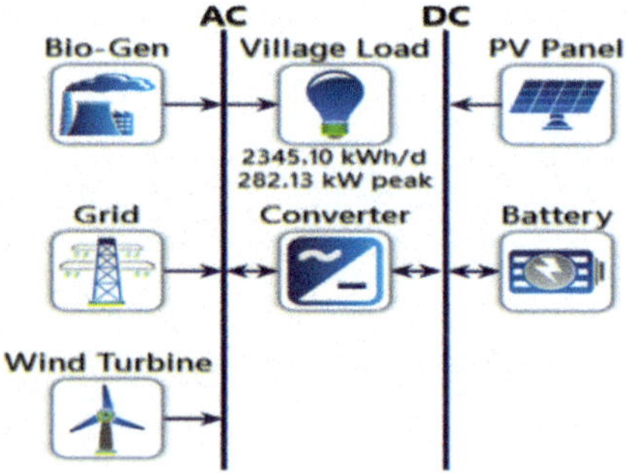


**Fig. 9** Grid connected HRES Model

**Table 1** Solar PV Specifications

| Solar PV panel | Description |
|---|---|
| Type | Mono Perc Half Cut |
| Model | Shark 445 |
| Maximum power ($P_{max}$) | 445 W |
| Short circuit current ($I_{SC}$) | 11.29 A |
| Open circuit voltage ($V_{OC}$) | 49.7 V |
| Maximum power current ($I_{MP}$) | 10.80 A |
| Efficiency of module (%) | 20.14 |
| Temperature coefficient of $P_{max}$ | –0.36%/°C |
| Capital cost | ₹20,000 |
| Replacement cost through life span project | ₹0 |
| Life cycle | 25 years |
| Average reduction in efficiency through 25 years (RF) | 86% |

### 3.2 *PV Module*

The solar PV performance is highly influenced by atmospheric conditions like solar radiation and temperature. Solar PV panel output power can be calculated using Eq. 1 [9].

$$P_{PV} = P_R \times RF \times (G / G_{STC}) \times (1 + \beta (T_C - T_{STC}) \tag{1}$$

where $P_{PV}$ is the power output of the panel, PR is the rated power of the panel, RF is the module degrading factor, G is solar radiation, GSTC is solar radiation at standard test condition, β is the temperature coefficient of Pmax, TC is the cell temperature and TSTC is the cell temperature at standard test condition. Many PV panel modules of different power rating are available in the market. Table 1 displays the description of the panel that is employed in this research [18].

### 3.3 *Wind Turbine*

The wind turbine used in this research is Eocycle EO20. Figure 10 depicts the power curve of this turbine. Permanent magnet synchronous alternator is used in this turbine with cut-in speed, cut-out speed and rated speed of 2.8 m/s, 20 m/s and 7.5 m/s respectively. The hub height is 26 m and rated power is 20 kW. A capital cost of ₹13,00,000 and O&M cost of ₹3,000 has been considered with a life cycle 25 years.

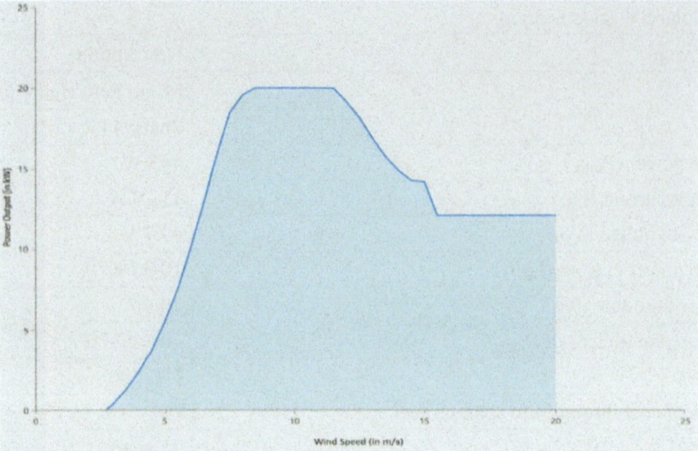


**Fig. 10** Power curve of wind turbine

### 3.4 *Bio-Generator*

Biogas obtained from the biogas digestion of the biomass (cow dung) is used by the bio-generator to generate the electricity. A generic generator with a biogas digester is considered with a capital cost of ₹10,000 and replacement cost of ₹4,000. The total operating time of the generator is 20,000 h with operating cost of ₹2/hour/kW.

### 3.5 *Battery*

Battery act as a storage device which is used to maintain the supply continuity when the generation through different RESs is not enough to meet the load demand and during the grid outage. A 1kWh generic Li-Ion battery is taken with 6 V of nominal voltage having roundtrip efficiency of 90%. It has 3000 kWh life duration with 10-year lifetime. The capital cost and replacement cost are assumed to be ₹20,000 and ₹14,000 with negligible operation and maintenance cost.

### 3.6 *Converter*

A 15kVA UTL Alfa + Solar PCU converter is considered in both models [19]. The converter specification is shown in Table 2.

**Table 2** Converter specification

| Converter | Description |
|---|---|
| Rated power | 15 kVA |
| Efficiency | 95% |
| Purchasing cost | ₹1,80,000 |
| Maintenance cost | ₹0 |
| Replacement cost | ₹1,50,000 |
| Life cycle | 20 ears |

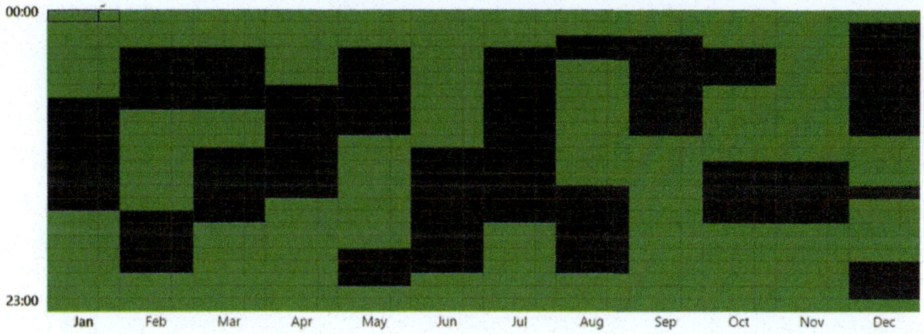


**Fig. 11** Grid outages

### *3.7 Grid*

The grid used in grid connected HRES acts as a source to fulfill the peak demand when RES alone is not able to fulfill the load and also the surplus energy can be fed back to the grid to generate revenue. However, accessibility of grid is an issue as there is frequent electricity outage in the study area. Figure 11 shows the reliability of the grid considered in this work. Black area shows the grid outage. The price of electricity from the grid is taken ₹5.5 and sellback price is ₹2 per kWh.

## 4 Optimization and Economic Analysis

Different optimized models of different sizings are provided by HOMER Pro in a sorted manner of least to high NPC and displayed in the Fig. 12.

### *4.1 Stand-Alone HRES*

A stand-alone system has been developed to meet the village's load demand using the local resources available and uses batteries as backup source as shown in Fig. 8. It

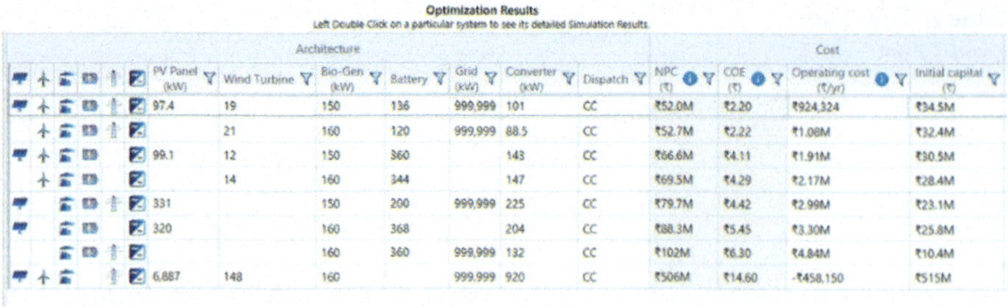
**Optimization Results**
Left Double Click on a particular system to see its detailed Simulation Results.

| PV Panel (kW) | Wind Turbine | Bio-Gen (kW) | Battery | Grid (kW) | Converter (kW) | Dispatch | NPC (₹) | COE (₹) | Operating cost (₹/yr) | Initial capital (₹) |
|---|---|---|---|---|---|---|---|---|---|---|
| 97.4 | 19 | 150 | 136 | 999,999 | 101 | CC | ₹52.0M | ₹2.20 | ₹924,324 | ₹34.5M |
| | 21 | 160 | 120 | 999,999 | 88.5 | CC | ₹52.7M | ₹2.22 | ₹1.08M | ₹32.4M |
| 99.1 | 12 | 150 | 360 | | 143 | CC | ₹66.6M | ₹4.11 | ₹1.91M | ₹30.5M |
| | 14 | 160 | 344 | | 147 | CC | ₹69.5M | ₹4.29 | ₹2.17M | ₹28.4M |
| 331 | | 150 | 200 | 999,999 | 225 | CC | ₹79.7M | ₹4.42 | ₹2.99M | ₹23.1M |
| 320 | | 160 | 368 | | 204 | CC | ₹88.3M | ₹5.45 | ₹3.30M | ₹25.8M |
| | | 160 | 360 | 999,999 | 132 | CC | ₹102M | ₹6.30 | ₹4.84M | ₹10.4M |
| 6,887 | 148 | 160 | | 999,999 | 920 | CC | ₹506M | ₹14.60 | -₹458,150 | ₹515M |

**Fig. 12** Hybrid system optimization result

**Table 3** Optimal stand-alone system result composition

| Component | Size/Numbers |
|---|---|
| Solar PV panel | 99.1 kW |
| Wind Turbine | 12 units |
| Bio-generator | 150 kW |
| Battery | 360 units |
| Converter | 143 kW |

is observed from the simulation result shown in Fig. 12 that the most optimal stand-alone HRES structure has an NPC of ₹6,66,42,230 with COE of ₹4.11 per kWh. The initial investment and operation cost for this system is ₹3,04,72,360 and ₹94,78,830/year with system component sizing being described in Table 3. Figure 13 shows the graph of different costs associated to the system.

The electricity generated by the proposed stand-alone system is 1,256,675 kWh/year whereas village consumption is 855,804 kWh/year. The difference of 377,004

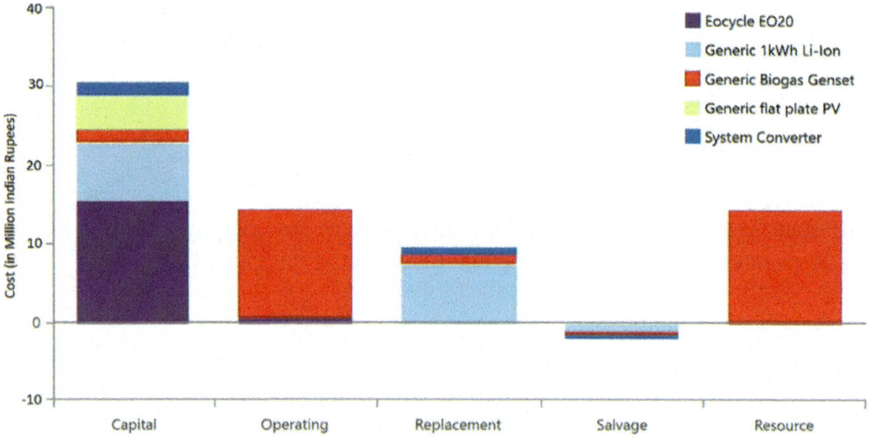


**Fig. 13** Cost summary of optimal stand-alone microgrid

**Table 4** Electrical details of optimal stand-alone system

| Production | kWh/year | Percentage share |
|---|---|---|
| Solar PV | 159,984 | 12.7 |
| Bio-generator | 315,796 | 25.1 |
| Wind Turbine | 780,896 | 62.1 |
| Total | 1,256,675 | **100** |
| Consumption (Load) | 855,804 | 68.1 |
| Excess electricity | 377,004 | 30.0 |
| Unmet electric load | 157 | 0.0184 |
| Capacity shortage | 724 | 0.0846 |

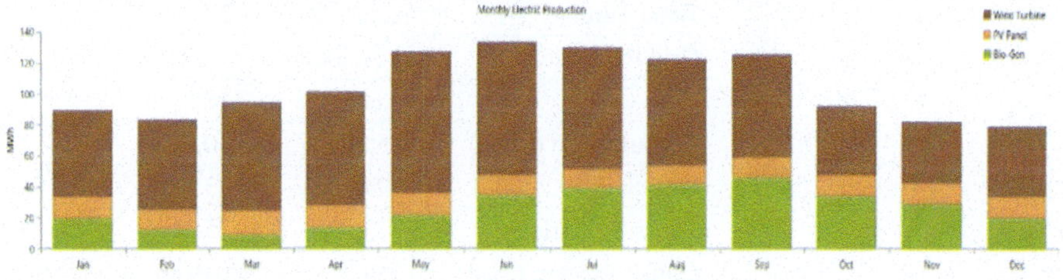


**Fig. 14** Monthly electric production by sources in stand-alone system

kWh/year which is 30.0% of total is excess electricity which is not utilized and considered as waste in off-grid systems and can be minimized by connecting this system to the grid.

The generated electricity from various renewable resources by stand-alone HRES system, its consumption and other parameters are shown in Table 4. Figure 14 shows the monthly electric production by different sources.

## 4.2 *Grid Connected HRES*

The same stand-alone system with the same parameter is connected to the grid to fulfill the demand using both the RES and grid as shown in Fig. 9. The excess energy generated by RES are sellback to the grid and during peak energy is taken from the grid. From Fig. 12, it is observed that NPC of the grid connected HRES is ₹5,20,29,000 with COE of ₹2.20 per kWh. The initial investment and operation cost for this system is ₹3,45,11,557 and ₹72,09,930/year with system component sizing as described in Table 5.

The generated electricity by the grid connected HRES system, its consumption and other parameters are shown in Table 6. Figure 15 shows the monthly electric production by different sources.

**Table 5** Optimal grid connected system composition

| Component | Size/Numbers |
|---|---|
| Solar PV panel | 97.4 kW |
| Wind Turbine | 19 units |
| Bio-generator | 150 kW |
| Battery | 136 units |
| Converter | 101 kW |

**Table 6** Electrical details of optimal grid connected system

| Production | kWh/year | Percentage Share |
|---|---|---|
| Solar PV | 157,204 | 9.35 |
| Bio-generator | 163,355 | 9.72 |
| Wind Turbine | 1,236,418 | 73.50 |
| Grid Purchase | 124,400 | 7.40 |
| Total | 1,681,378 | 100 |
| Consumption (Load) | 855,802 | 50.90 |
| Grid Sales | 394,176 | 23.44 |
| Excess Electricity | 423,998 | 25.21 |
| Unmet Electric Load | 159 | 0.0186 |
| Capacity Shortage | 854 | 0.0997 |

**Table 7** Comparative analysis of stand-alone and grid connected HRES

| Parameter | Stand-alone HRES | Grid connected HRES |
|---|---|---|
| Solar panel required | 99.1 kW | 97.4 kW |
| No of wind turbine required | 12 units | 19 units |
| Bio-generator required | 150 kW | 150 kW |
| Battery | 360 units | 136 units |
| Converter | 143 kW | 101 kW |
| Net present cost (NPC) of system | ₹6,66,42,230 | ₹5,20,29,000 |
| Cost of energy (COE) | ₹4.11 per kWh | ₹2.20 per kWh |

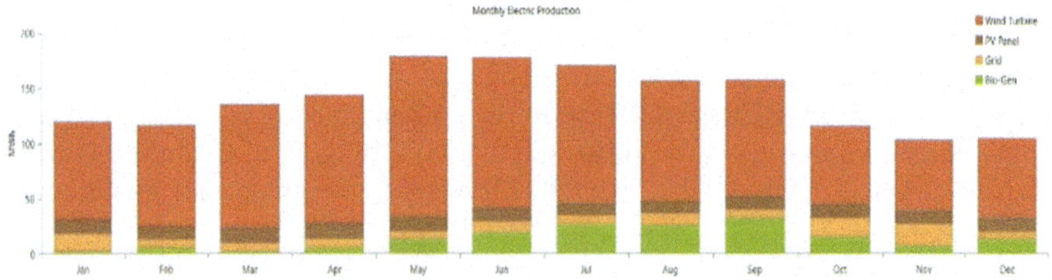


**Fig. 15** The monthly electric production by different sources in grid connected system

## 5 Comparative Analysis

Performing the techno-economic analysis of both the OFF grid and grid connected system, it is observed that the grid connected HRES system costs less and is feasible for the study area. The NPC of the grid connected HRES system is nearly ₹1,46,00,000 lower than the stand-alone system which is about 22% of NPC of the stand-alone system. Furthermore, COE of this system is ₹2.20 which is 46.5% less.

## 6 Conclusion

This study successfully demonstrated the modelling, analysis and functioning of stand-alone as well as grid connected HRES for a village using HOMER Pro and shows the different possibilities of utilization of the grid and other local renewable resources available. Implementation cost of the HRES (solar-wind-biomass) system is identified by calculating all cost related to each system component. Different hybrid systems havebeen developed and simulated to minimize the total NPC and COE to get the best outcome. It is found that the grid connected HRES (solar-wind-biomass) has minimum NPC of Rs. 5,20,29,000 and COE of Rs. 2.20 which is less than the utility price taken. The proposed grid connected HRES model will give the feasible solution for the Bayal village.

## References

1. Mary VB, Christopher IW, Themozhi G (2019) RHES-economic analysis and power management for a technical institution using homer. In *2019 Fifth International Conference on Electrical Energy Systems (ICEES)* (pp. 1–6). IEEE
2. Sedeeqi MM, Vaisakh K, Sahoo NC (2022) Homer pro based techno-economic analysis for remote area electrification of Awbeh Village in Afghanistan. In *2022 IEEE 2nd International Conference on Sustainable Energy and Future Electric Transportation (SeFeT)* (pp. 1–6). IEEE
3. Tsai CT, Beza TM, Molla EM, Kuo CC (2020) Analysis and sizing of mini-grid hybrid renewable energy system for islands. IEEE Access 8:70013–70029
4. Rohani G, Nour M (2014) Techno-economical analysis of stand-alone hybrid renewable power system for Ras Musherib in United Arab Emirates. Energy 64:828–841
5. Marais S, Kusakana K, Koko SP (2019) Techno-economic feasibility analysis of a grid-interactive solar PV system for South African residential load. In *2019 International Conference on the Domestic Use of Energy (DUE)* (pp. 163–168). IEEE
6. Kellogg WD, Nehrir MH, Venkataramanan G, Gerez V (1998) Generation unit sizing and cost analysis for stand-alone wind, photovoltaic, and hybrid wind/PV systems. IEEE Trans Energy Convers 13(1):70–75
7. Mannai H, Oueslati H, Mabrouk SB (2020) Homer based optimization of pv wind grid connected hybrid system in administrative building. In *2020 6th IEEE International Energy Conference (ENERGYCon)* (pp. 830–835). IEEE
8. Salehin S, Islam AS, Hoque R, Rahman M, Hoque A, Manna E (2014) Optimized model of a solar PV-biogas-diesel hybrid energy system for Adorsho Char Island, Bangladesh.

In *2014 3rd International Conference on the Developments in Renewable Energy Technology (ICDRET)* (pp. 1–6). IEEE
9. Mohammed AQ, Al-Anbarri KA, Hannun RM (2020) Optimal combination and sizing of a stand–alone hybrid energy system using a nomadic people optimizer. IEEE Access 8:200518–200540
10. Huda N, Nabipour-Afrouzi H, Kieh TS, Mehranzamir K, Ahmed J, Wooi CL (2019) Optimization analysis of hybrid renewable energy system using homer software for rural electrification in Sarawak. In *2019 International UNIMAS STEM 12th Engineering Conference (EnCon)* (pp. 77–82). IEEE
11. Makwana V, Kotwal C, Arora K (2019) Optimum sharing of solar power and grid for an academic institute using HOMER. In *2019 IEEE 5th International Conference for Convergence in Technology (I2CT)* (pp. 1–5). IEEE
12. Shilpa N, Sridevi HR (2019) Optimum design of rooftop PV system for an education campus using HOMER. In *2019 Global Conference for Advancement in Technology (GCAT)* (pp. 1–4). IEEE
13. Nayanatara C, Baskaran J, Dharani S, Sri VK, Kanmani E (2019) Optimization of hybrid energy resources using homer software. In *2019 International Conference on Computation of Power, Energy, Information and Communication (ICCPEIC)* (pp. 227–231). IEEE
14. Kumar S, Upadhyaya P, Kumar A (2019) Performance analysis of solar energy harnessing system using homer energy software and PV syst software. In *2019 2nd International Conference on Power Energy, Environment and Intelligent Control (PEEIC)* (pp. 156–159). IEEE
15. Vendoti S, Muralidhar M, Kiranmayi R (2018) HOMER based optimization of solar-wind-diesel hybrid system for electrification in a rural village. In *2018 International Conference on Computer Communication and Informatics (ICCCI)* (pp. 1–6). IEEE
16. https://cea.nic.in/dashboard/?lang=en
17. Singh AD, Gajera B, Sarma AK (2022) Appraising the availability of biomass residues in India and their bioenergy potential. Waste Manage 152:38–47
18. https://www.loomsolar.com/collections/loom-solar-panel/products/loom-solar-panel-shark-super-high-efficiency-module-pack-of-2
19. https://www.upsinverter.com/utl/solar-inverter/alfa-plus-solar-pcu/

# A Review on Battery Management System

**Jatin Sheoran, Javed Dhillon, and Sachin Mishra**

**Abstract** In the modern era, Electric Vehicles (EVs) are widely considered as a superior alternative to traditional automobiles that rely on internal combustion engines (ICEs). The rapid advancement of EV technologies has made battery systems a crucial aspect of their development. In recent times, the upgradation of battery technology along with the increase in demand for high-performance and safe battery system has driven various developments in the battery management system (BMS). The development of a BMS system is also required for the integration of smart technologies such as IoT and machine learning. A BMS is a control process used for managing and monitoring rechargeable batteries to ensure their effective and safe operation. It is responsible for monitoring variables like voltage level, temperature, and charge status and taking necessary actions to maintain them within limits. In this paper SOC, SOH, thermal management, and cell balancing techniques are discussed, which protect against excessive charging and discharging, and overcurrent conditions. This study presents a literature survey on a BMS, for dependable and secure operation of battery systems in the EV.

**Keywords** State of health · Cell balancing · State of charge · Electric vehicle · Battery management system

## 1 Introduction

Electrical vehicles are being moved out in place of conventional vehicles since EVs are more environmentally friendly because it is composed of gases, which are completely efficient vehicles due to the depletion in terms of non-renewable energy

J. Sheoran · J. Dhillon (✉) · S. Mishra
School of Electronics and Electrical Engineering, Lovely Professional University Phagwara, Phagwara, India
e-mail: javeddah@gmail.com



O. H. Gupta et al. (eds.), *Soft Computing Applications in Modern Power and Energy Systems*, Lecture Notes in Electrical Engineering 1107,
https://doi.org/10.1007/978-981-99-8007-9_19

sources. Battery Management System (BMS) is an essential component of an electric vehicle since it consists of numerous circuits, both electric and electronic that maintain and achieve a battery system's effective output.

BMS is a critical component in modern rechargeable battery systems, designed to assure effective and safe operation. The initial purpose of a BMS is to monitor the performance of the battery's performance, ensuring its longevity and protecting it from damage. The BMS continuously monitors parameters like cell voltage, temperature, and state of charge (SOC), and adjusts the charging and discharging processes accordingly to maintain these parameters within safe limits.

A BMS can also prevent overcharging, over-discharging, and overcurrent conditions, which can cause permanent damage and reduce the lifespan of the battery. Additionally, by balancing cell voltage and optimizing the charging and discharging processes, a BMS can enhance performance and efficiency regarding the battery system. The specific design of a BMS depends on the requirements of the battery system, including its size, capacity, and intended application.

In the case of regenerative braking, the BMS monitors the battery's SOC and limits the amount of energy that can be stored in the battery to prevent overcharging. The BMS achieves this by controlling the charging rate during regenerative braking, ensuring that the battery is charged at a safe rate that does not exceed its maximum capacity. The BMS also monitors the temperature of the battery and adjusts the charging rate accordingly to prevent overheating.

BMS can be implemented in various forms, including hardware, software, or a combination of both. The design of a BMS depends on the specific requirements of the battery system and can range from simple standalone units to complex multi-layered systems that integrate with other subsystems.

Some of the key trends in the field of BMS include the integration of smart technologies, such as the Internet of Things and machine learning, for improved monitoring and control of battery systems. Overall, the BMS is an essential part of modern battery systems, providing essential protection and performance optimization to ensure functioning that is secure, efficient, and effective. Figure 1 shows the diagram of the battery management system [1].

The BMS configuration comprises of:

- battery monitoring
- battery state
- battery management
- Battery monitoring is used to check and show the data such as voltage level, energy consumption, current consumption, and battery temperature are all useful. The data acquisition observing layer of the battery module includes various types of sensors.
- The battery state controls the charging time, discharging technique, and cell balancing, while the user interface will receive the state determination. It provided the user with an estimate of how long a battery will remain before needs to be recharged when it is full.

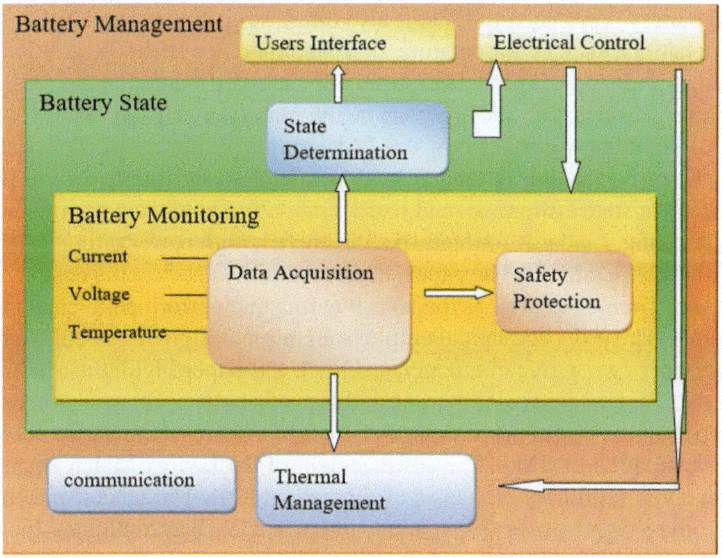


**Fig. 1** Illustration of the battery management system [1]

- Battery management is a system for monitoring and managing the battery. It controls the I/P power of the cells, to meet modern battery requirements.

The paper covers the following topics: Section II consists of a literature review. Section III explains the battery management system. Section IV consists of the conclusion of the paper.

## 2 Literature Review

The [1] research examined the primary issues related to BMSs, which encompass battery state assessment, modeling, and cell balancing. Among these concerns, the paper emphasized the significance of developing effective methodologies for evaluating the status of batteries. In [2], the research gives the idea about BMS which assures the security and dependable operation of the batteries in EVs. The utilization of Lithium-ion (Li-ion) batteries for energy storage in EVs is prevalent. This study examined BMS in EV and contrasted various monitoring methods used with BMS. In [3], the author introduces the idea of a "New BMS," in which a battery is split into two halves, with one half being used to charge using renewable energy and the other half being used to discharge at the same time. The concept of BMS [4] consists of several electronic and electrical circuits to track and maximize a battery system's output; it is a crucial component of an electric vehicle. The BMS in an EV describes the battery algorithm block implementation, Li-ion battery model, and common battery types used in EVs. It also discusses problems and difficulties with the Li-ion battery, such as temperature, cost, and memory effect. In addition to many other things, the study [5] has been used for balanced battery states, cell voltage, proper battery discharging and charging health management, and many other things.

In [6], the author discusses how the BMS manages and controls the battery and how its primary goals are to ensure its dependability and safety. In the paper, the methodologies, state estimation, and battery modeling prediction are reviewed. This includes meeting standards for high specific energy, high power rating, low cost, and low self-discharge rate [7]. The paper [8] discusses the need for the design of battery management circuits, as well as the tests that must be performed to characterize Li-ion batteries. The proposed modal employs a unique design of cell balancing which comprises only two active components. In [9], the paper highlights the maximum capacity, health, longevity, and charge condition of a battery The study focuses on the optimization of the power performance of electric vehicles, with particular emphasis on the investigation of BMS. The advancement of the Energy Management System (EMS) in EV is a primary emphasis of this article [10]. The objective of this study [11] is to offer a concise overview of many important BMS technologies, including an estimation of the battery's state and charging. The research [12] examines the battery properties and reduces manufacturing inhomogeneities and process uncertainties; the research should help EVs to operate better by producing more dependable and stable BMS. The study [13] addressed how changes in SOC, dynamic currents, voltages, and temperatures affect battery performance. In [1], the main BMS concerns were discussed. It includes cell balancing, modeling, and battery condition evaluation. As a result, comparable studies on the battery SOC, state of health (SOH), and SOL were examined. The paper [14] discussed the BMS, a BMS is used to monitor the battery voltage, current, and temperature and also estimate the SOC of a Li-ion battery to safeguard the battery from overcharged and over-discharged situations. The study work [15] has covered the BMS, which can be modified to protect batteries of various types and can offer all the safety characteristics.

The battery algorithm component in the Li-ion battery model is described in the context of BMS for an EV in [16] and its main purpose is to observe open-circuit voltage with the coulomb counting method to determine SOC. It also discusses the common battery types used in EVs, as well as the problems and difficulties of Li-ion batteries. The electrical energy storage capacity in EV of Li–Sulfur, molten salt, nickel–metal hydride, and Li-ion are the four different types of batteries which are primarily employed in BMS [17]. In [18], the research discusses the vulnerability of BMS for Li-ion and Li-polymer (LiPo) batteries employed in upcoming electric and hybrid vehicles to mitigate electromagnetic interference (EMI). Based on the problem statement the battery operating conditions are analyzed in [19], and an optimization technique is created using adaptive dynamic programming. The importance of the battery in an EV must be discussed in detail while detailing the EMS. To deliver essential power for transportation, the battery is crucial in electric vehicles. In [20], the research discusses the problem statement and an analysis of the battery operation modes; an optimization solution is created using adaptive dynamic programming. The ADP uses an actor module and a critic network to solve the BMS optimal control problem in an EV. In [21], the review provides a thorough overview of Li-ion batteries. The paper describes battery monitoring methods, input voltage, and current control strategies, both charging and discharging strategies and battery safety precautions.

In the paper [22], the characteristics of management systems for batteries are discussed, along with the technology's benefits and drawbacks for use in electric vehicle applications. The review covers battery cell tracking, state estimates, and charging/discharging controls. The author discusses [23] the highly exact SOC determination algorithms that were created with the use of models with adaptable batteries and serve as a solid foundation for real-world application. According to the research [24], a BMS module is suggested that enables individual cell charging while the battery is in operation and allows cell-based opening and closing of the battery. A charging unit with a limited area on the circuit is designed. The EKF and MATLAB programs are used to monitor and control how the BMS is operating. The modified AEKF approach is used to increase the estimation's accuracy when the specific properties of static noise in unknown or evolving Li-ion battery packs over time [25]. The purpose of the study [26] is to create a Li-ion cell model using current information and to make it simpler, for obtaining a simpler model that may be used to design the full battery model and charging mechanism. In [27], a discussion of balancing techniques for various battery operation processes and putting balancing algorithms into practice at a necessary time, irrespective of the balancing circuit for optimizing the effects of balancing, has been covered. When developing a dependable and stable automobile system for EVs, there are usually both intrinsic and extrinsic problems with batteries linked to correctly maintaining temperature, charging and discharge, and satisfying acceptable operating conditions [28]. Zinc–bromine flow battery, Li-ion, nickel–cadmium, sodium–nickel chloride, sodium–sulfur, lead–acid. and vanadium-redox flow batteries are just a few of the widely used battery types discussed in the work [29].

The research [30] discussed Li-ion battery which has a long lifespan, high efficiency, low discharge rate, high power density, and high dependability. Battery Management System. In [31], the author proposes a system design for BMS that is responsible for managing and controlling the operation of the battery system factors such as during charging/discharging: Measuring the temperature, current, and voltage of the battery is a crucial aspect of monitoring and controlling its operation. In [32], the author addressed the power-type batteries, which have a lower capacity but can supply more power instantly and are used in light HEV. SOC gives information about the battery's current level of stored energy. The study [33] describes the Kalman filtering state observation approach, which is used to assess a battery's SOC. This method provides an accurate SOC measurement for Li-ion batteries. Power-type batteries, which are utilized in light HEV, have a lower capacity but can supply high power in an instant [34]. The four types of the SOC estimate approach include model-based estimation, ampere-hour integral estimation, data-driven estimation, and characteristic parameters estimation. The methods used to estimate a battery level of charging [35]. The estimating techniques were divided into three groups based on their theoretical and experimental traits. The findings imply that the focus of study in this area is on algorithms that are based on theory, particularly intelligent algorithms. For a precise SOC estimation, the author in [36] presents a combination of fuzzy logic and the coulomb counting method. In [37], it provides a concise introduction of different types of rechargeable batteries and presents a comparison of the many types of rechargeable batteries.

The study [38] covered a variety of intrinsic and extrinsic battery problems linked to optimal charging/discharging and temperature maintenance in EVs. In [39], dual extended Kalman filter is presented and demonstrated by the use of simulation using actual battery voltage or current data. The proposed method viability was examined through simulation using actual battery voltage and current data. The four primary categories for SOP estimate methods are the voltage-based technique, the multi-constrained dynamic method, the SOC-based method, and the hybrid pulse power characterization methods are discussed in the research [40]. According to the research [41], the SOC has a significant impact on EV energy management optimization, improving battery size and energy consumption. It proposes real-time estimation of the SOC as a performed technique for the Li-ion battery that combines coulomb counting during the equilibrium state and direct electromotive force (EMF) detection [42]. The research [43] focuses on how BMS monitors battery current, temperature, and voltage to protect the battery from overcharged and over-discharged circumstances. It also explains how it is used to calculate the SOC of Li-ion batteries. To observe the parameter of the modified rundles circuit battery model, an expanded Kalman filter-based real-time identification method was presented [44].

It is used to extend Kalman filters to predict the bulk capacitance range, a measurement of the battery capacity to hold an over-time charging capacity, to detect cell capacity degradation and, as a result, provide a way to monitor SOH [45]. In [46], the primary battery SOH estimation techniques are reviewed in this paper, together with their key benefits and drawbacks for real-time automobile compatibility. The research [47] provided an analysis of battery health management for EVs. Initially, battery terminologies introduce health management techniques for batteries. The study [48] suggested a dual-sliding-mode observer to estimate the SOC, terminal voltage, and polarization effects in real-time, for the calculation of SOH in terms of capacity fading and resistance degradation. According to the research [49], batteries are safe when utilized with a power source that has safety safeguards and automatic termination. Thus, BMS has been explored in the research on EV. In order to be determined through correlations between cell performance and predictor variables, to predict SOC, SOH, and RUL to provide the uncertainty bounds., it proposed a Bayesian framework [50].

The study in the paper [51] proposes that the SOH can be evaluated by adaptively acquiring the battery's high-frequency resistance. The study [52] discusses how SOH is typically used to determine the health of a battery system and how well it is performing in comparison to its nominal or original state. In [53], is proposed the thermal control of a lithium battery based on secondary loop cooling be accomplished by modifying compressor speed in response to battery temperature intermission. In [54], the paper describes how the Li-ion cells are balanced by the BMS in an EV using cell balancing procedures. The Li-ion cell within the battery must be charged and discharged simultaneously. The author [55] discusses the many cell balancing methods that have been developed for battery packs. By dissipating energy from higher SOC cells, the passive cell balancing method equalizes the cell's SOC which makes all cells' SOC equal. The different battery cell balancing techniques are reviewed in the study [56].

The study [57] has talked about the idea of cell balancing. Each cell's voltage ranges differ when the battery is charging and discharging. As a result, cell balancing is essential and can be accomplished through active or passive balancing. The study [58] describes the comparisons of the various cell balancing techniques. In the study, active cell balancing techniques covered DC–DC converters, inductor/transformers, and capacitor-based cell balancers. In [59], the author discusses the technique of passive cell balancing. The charging at a fixed voltage technique and the charging at a constant current approach are analyzed. Capacitor charging and discharging are depicted in the first section of the paper. The second section of the paper describes the cell balancing procedure, and the method of charging with constant voltage and current is represented in the third section. In [60], the paper discusses energy storage components that can accommodate additional energy storage and be used in the active cell balancing technique. In the research, an improved active cell voltage balancing approach is put forward. This will result in improved voltage equalization, speed, and overall efficiency of the battery. The paper [61] gives a practical active cell balancing strategy, to minimize the inconsistent behavior of Li-ion batteries interconnected in parallel and series, a forward converter-based cell balancing solution or switch-matrix employing active clamp driver topology is provided. The research [62] investigates the concept of an inductor balancing-based cell equalization technique. Compared to alternative equalization techniques like switched capacitor balance, the inductor-based equalizer offers the advantage of a shorter equalization time period. In [63], the author discusses the different topologies of BMS. Decentralized control has significant controllability restrictions, but centralized control has challenged extensibility and the single fact of failure difficulties.

## 3 Battery Management System

### *3.1 Types of Batteries in EVs*

BMSs can use different types of cell chemistries or battery packs depending on the requirements of the application.

Li-ion batteries, which are favored for their high energy density, low self-discharge rate, and long cycle life.

Nickel–Metal Hydride (NiMH) batteries have a higher energy density than Ni–Cd batteries and are less toxic, but are more expensive and have a shorter cycle life than Li-ion batteries.

Lead–acid batteries are another commonly used battery chemistry in automotive applications due to their low cost and ability to provide high current output, but they have a low energy density, short cycle life, and are relatively heavy and bulky.

Batteries are divided into two categories based on their ability to be recharged: primary and secondary batteries. When the primary battery is fully discharged, it can only be used once; however, the secondary battery can be recharged after already been discharged. Secondary batteries with a high power density, a long cycle lifespan, low energy loss, and sufficient safety standards are required for EV and HEV applications. Li-ion, lead–acid, and NiMH batteries are among the frequently used battery types in electric vehicles. Table 1 provides a summary of some of the popular battery types [23].

Every battery has an operational temperature range, and the temperature of the battery rises when it is discharged or charged. A lot of energy can be saved by lowering the temperature of the batteries. When designing the packaging for a rechargeable energy storage system (REESS), it is important to consider certain conditions to ensure safety and optimal performance. These conditions include:

1. Thermal management: It is crucial to regulate the temperature of the battery cells and prevent overheating by using proper thermal insulation and cooling systems.
2. Electrical isolation: The battery pack should be designed to provide electrical isolation between the battery cells and the external circuitry to prevent electrical shorts, fire, or explosion.
3. Cell balancing: Incorporating a cell balancing system helps to ensure that each cell is charged and discharged evenly, thus preventing overcharging or over-discharging of individual cells that can cause safety hazards and reduce battery life.
4. Monitoring and control: The battery pack should be equipped with a monitoring and control system to track the battery's SOC, temperature, and other important parameters. This information can be used to optimize the battery's performance and prevent safety hazards.

**Table 1** Types of batteries

| Types | Energy efficiency (%) | Energy density | Power density | Depth of discharge (%) | Cycle life |
|---|---|---|---|---|---|
| Lead–acid | 82 | 50–80 | 100–400 | 50 | 1500 |
| NI–Cd (Nickel–Cadmium) | 83 | 60–150 | 80–600 | 85 | 2500 |
| Li-ion (Lithium-ion) | 96 | 200–400 | 1500–10,000 | 95 | 10,000 |

## 3.2 *Function of BMS*

A BMS is responsible for overseeing and regulating the operation of rechargeable batteries to ensure reliable and safe performance. This includes tasks such as monitoring cell voltage, temperature, and current, balancing cells to ensure equal charge levels, and providing protection against overcharge, over-discharge, and overcurrent conditions. The BMS also communicates with the battery pack's external management system to provide battery status and other relevant information.

- **Data Acquisition:** Data acquisition in BMS refers to the process of collecting and storing data from various sensors and measurement devices within the battery pack. This data can include cell voltage, temperature, current, and SOC, among others. The BMS uses this data to monitor the health and performance of the battery and to make decisions related to charging and discharging. The process of data acquisition is crucial to ensure the safe, efficient, and reliable operation of a battery, and for providing valuable information for battery maintenance and management.
- **SOC estimation:** BMS involves determining the quantity of energy that a battery can hold at a given time. The SOC of a battery represents the proportion of the current available charge in the battery compared to its maximum capacity. SOC is calculated to ensure that the batteries are never under or overcharged. This is a crucial parameter to monitor, as it provides insight into the condition and functionality of a battery. Several factors affect the accuracy of SOC estimation in a battery, including temperature, battery condition, and discharge/charge current. The BMS constantly monitors the SOC and adjusts the charging/discharging process to maintain optimal performance of battery and longevity. The extended Kalman filter (EKF) is a highly accurate method for estimating the SOC in batteries. The EKF uses a mathematical model of the battery, along with measurements, to recursively estimate the SOC. This is achieved through two main steps: prediction and update.

  During the prediction step, the battery model is used to predict the expected voltage and SOC based on current and voltage measurements. This involves two calculations: state prediction and covariance prediction. The state prediction calculates the expected battery state based on the model and the previous state estimate, while the covariance prediction calculates the predicted error covariance matrix, which represents the expected uncertainty in the predicted state.

  During the update step, the measured voltage and current are compared to the predicted values, and the difference is used to update the SOC estimate. This step also involves two calculations: measurement residual and Kalman gain. The measurement residual calculates the difference between the measured and predicted values, while the Kalman gain determines how much weight to give to the measurement residual in updating the SOC estimate.
- **SOH estimation:** It is the key function of a BMS. The SOH is the amount of the health or battery degradation and is a percentage of its total original capacity. The BMS uses various techniques, such as voltage analysis, data processing, and

coulomb counting to calculate the battery SOH over time. This information is important for determining when a battery needs to be replaced, and for predicting its remaining useful life. By monitoring the SOH, the BMS can also provide an early warning of potential issues, such as an impending failure, which can help prevent unexpected downtime and extend the life of the battery.

- **Thermal Management:** Thermal management is an important aspect of BMS design and operation. The temperature of batteries can greatly affect their performance, capacity, and safety. High temperatures can lead to thermal runaway and even combustion, while low temperatures can decrease the capacity of the battery and raise its internal resistance. To ensure safe and optimal battery performance, a BMS typically includes temperature sensors that monitor the battery temperature and control the thermal environment. BMS commonly use temperature sensors that are based on thermistors, which are resistors that are sensitive to temperature and have a resistance that varies with temperature. RTDs (Resistance Temperature Detectors) are temperature sensors made of metal and their resistance changes with temperature, while thermocouples are devices that generate a voltage in response to a temperature difference between two different metals.

  The BMS can use this information to regulate the battery temperature by controlling the charging and discharging currents, and by controlling the cooling system, if present. In some cases, the BMS may also turn off the battery to prevent damage if the temperature has risen above a particular limit. Thermal management is especially important for large batteries, such as the ones found in EVs, which can generate significant heat during charging and discharging. Proper temperature management is crucial for the optimal performance and longevity of a battery, a BMS can help extend its life and improve its overall performance. To maintain battery temperature within a specific range in a BMS, different methods are employed, including:

  1. Thermal Management System (TMS): This cooling and/or heating system helps regulate battery cell temperature. It can be active or passive and may utilize air or liquid cooling/heating.
  2. Temperature Sensors: These sensors measure the battery cell temperature and provide feedback to the BMS, which can then adjust the temperature using the TMS.
  3. Charging/Discharging Current Control: The BMS can control the charging and discharging current to prevent battery cells from overheating or becoming too cold.
  4. Operating Mode Control: The BMS can regulate the operating mode of the battery system, such as switching between charging and discharging, to maintain the desired temperature.

The recommended temperature range for Li-ion batteries is generally 0–45 °C, with an optimal temperature range of 20–25 °C. The BMS utilizes a set-point temperature value and tolerance range to maintain the temperature within the desired range. If the temperature exceeds the tolerance range, the BMS will trigger an alarm or take corrective action to prevent battery cell damage.

- **Fault Diagnosis:** It performs a crucial role in the BMS. BMS uses various methods, including voltage analysis, current analysis, and temperature analysis, to help recognize and handle battery issues and their associated components. The BMS can detect faults such as cell imbalance, overcharging, over-discharging, short circuits, and more. When a fault is detected, the BMS can take appropriate actions to protect the batteries, including shutting down the battery, reducing the charging or discharging current, or triggering an alarm. This helps prevent further damage to the battery and minimizes the possibility of safety risks, such as thermal runaway or fire. By detecting faults early, the BMS can also help prolong the life of the battery and improve its overall reliability. The BMS can also log fault information, including the type of fault, time of occurrence, and other relevant data, to help with fault analysis and troubleshooting. This information can be used to improve the design of future BMS systems and to maximize the performance and lifespan of a battery.
- **Cell Balancing:** It is a key component of a BMS. In a battery system, each individual cell can experience different amounts of charge and discharge, which can cause an imbalance in the voltage of the cells. This cell voltage imbalance can cause reduced performance, reduced battery life, and even safety hazards, such as thermal runaway.

The BMS uses cell balancing techniques to maintain a uniform voltage level across all cells in a battery. Cell balancing can be achieved through several methods, such as passive balancing, active balancing, or a combination of both. Passive cell balancing is a technique used by the BMS to equalize the voltage of cells in a battery pack. This technique uses resistors or other passive components to discharge the higher voltage cells in a battery pack., Active balancing is a cell balancing technique used by the BMS to maintain the voltage balance of cells, active balancing employs electronic circuits to transfer charge between cells. For the industry purpose mainly passive cell, active cell hybrid cell balancing are used.

Cell balancing is typically carried out when the battery is charged/discharged. The BMS collects data on the voltage of every cell and adjusts the balancing circuits as needed to maintain a balanced voltage. By ensuring cell balancing, the BMS can enhance efficiency, extend battery life, and minimize the risk of safety hazards.

**Passive Cell Balancing:-** It is a method used to equalize the voltage of the individual cells. In passive cell balancing, every cell inside the battery system has a resistor connected in parallel. During charging, the cells with higher voltage will charge the resistors connected to the cells with lower voltage, bringing the voltage of all cells closer to each other.

This process continues until the voltage of all cells is equal. It is a slow process and does not provide fast and precise voltage control, so it may not be suitable for high-power battery packs. Additionally, passive cell balancing generates heat, which can increase the temperature of the battery and may not be suitable for high-temperature applications. Overall, passive cell balancing is a good solution for small, low-power battery packs, but, for larger and higher power battery packs, active cell balancing

**Table 2** Passive cell balancing techniques

| Type | Balancing principle | Advantages | Disadvantages |
|---|---|---|---|
| Fixed-shunt resistor | Balancing continuous discharge of high-voltage cells through a fixed resistor | Simple and inexpensive, does not require additional hardware | Limited balancing performance, slow balancing speed |
| Switched-shunt resistors | Intermittent discharge of high-voltage cells through a switchable resistor | | Effective balancing, fast balancing speed |

technique could be a more suitable option. Passive cell balancing is proposed in Table 2 [55].

**Active Cell Balancing:-**This method equalizes the individual cell voltage in a battery pack. Unlike passive cell balancing, which relies on resistors to discharge cells with higher voltage, active cell balancing uses electronic circuits to transfer charge between cells to maintain voltage balance.

The basic concept of active cell balancing is to move charge from cells with higher voltage to cells with low voltage. The electronic circuits, typically based on power electronics, sense the voltage of each cell and transfer charge between cells as needed to balance the voltage. Active cell balancing can be performed in real time, so it is much faster and more precise than passive cell balancing.

Active cell balancing is commonly used in high-power battery packs, such as those used in EVs, to ensure optimal performance and extend the life of the battery. It also provides better safety features, such as the ability to detect and protect the battery, it is important to avoid overcharging or over-discharging it, as these actions can cause damage to the battery.

Overall, the benefits of active cell balancing, such as faster, more precise voltage control, longer battery life, and improved safety, make it a good investment for many high-power battery applications. Active cell balancing is proposed in Table 3 [55].

## *3.3 BMS Topologies*

There are several topologies commonly used in BMS, including:

- **Centralized BMS:** A centralized BMS is a type of BMS architecture in which single central unit is responsible for performing all supervision and control functions in a system. In this type of BMS, the central unit communicates with all the individual cells and makes decisions about how to manage the battery.
- **Distributed BMS:** A distributed BMS is a type of BMS architecture in which monitoring and control activities are distributed among multiple units. In this type

**Table 3** Active cell balancing techniques

| Type | Balancing principle | Merits | Demerits |
|---|---|---|---|
| Single capacitors | Transfer of charge between the cells through a single capacitor | Simple and inexpensive, low cost | Limited balancing performance, slow balancing speed, requires additional hardware |
| Switched capacitors | Transfer of charge between the cells through a switchable capacitor | Effective balancing, fast balancing speed | More complex requires additional hardware, increases battery pack size |
| Double-tiered switched capacitors | Transfer of charge between the cells through a multiple switchable capacitor | Effective balancing, fast balancing speed | More complex and expensive, requires additional hardware, increases battery pack size |
| Single inductors | Transfer of charge between cells through a single capacitor | Simple and less expensive compared to multiple inductors balancing | Limited balancing performance, slow balancing speed, requires additional hardware |
| Several Inductors | Transfer of charge between cells through multiple inductors | Effective balancing, fast balancing speed | More complex and expensive, requires additional hardware, increases battery pack size |
| Single transformers | Uses electrical balance to distribute the current equally among cells | Lower cost, simple design, low power loss | Inefficient for high current applications, limited balancing accuracy |
| Single magnetic core transformers | Uses magnetic balance to distribute the current equally among cells | Improved accuracy, more stability, improved reliability | Higher cost, more complex design, higher power loss |
| Single capacitor | Transfer of charge between the cells through a single capacitor | Simple and inexpensive, low cost | Limited balancing performance, slow balancing speed, requires additional hardware |
| Switched capacitors | Transfer of charge between the cells through a switchable capacitor | Effective balancing, fast balancing speed | More complex requires additional hardware, increases battery pack size |
| Double tiered switched capacitors | Transfer of charge between the cells through a multiple switchable capacitor | Effective balancing, fast balancing speed | More complex and expensive, requires additional hardware, increases battery packs size |

(continued)

**Table 3** (continued)

| Type | Balancing principle | Merits | Demerits |
|---|---|---|---|
| Single inductor | Transfer of charge between cells through a single capacitor | Simple and less expensive compared to multiple inductors balancing | Limited balancing performance, slow balancing speed, requires additional hardware |
| Several inductors | Transfer of charge between cells through multiple inductors | Effective balancing, fast balancing speed | More complex and expensive, requires additional hardware, increases battery pack size |
| Single transformer | Uses electrical balance to distribute the current equally among cells | Lower cost, simple design, low power loss | Inefficient for high current applications, limited balancing accuracy |
| Single magnetic core transformer | Uses magnetic balance to distribute the current equally among cells | Improved accuracy, more stability, improved reliability | Higher cost, more complex design, higher power loss |

of BMS, each unit monitors a group of cells, and the units communicate with each other to exchange data and make decisions about how to manage the battery.

- **Hybrid BMS:** In a hybrid BMS, some functions are performed centrally, while others are performed by distributed units. This topology combines the benefits of centralized and distributed BMS, providing both precise control and scalability.
- **Modular BMS:** A modular BMS is a type of BMS architecture in which monitoring and control activities are divided into separate, modular components. In this type of BMS, each module monitors a specific battery feature, such as voltage, temperature, or SOC.

The broad comparison of BMS topologies is proposed in Table 4 [63].

**Table 4** Comparison of different BMS topologies

| Topology | Centralized | Distributed | Hybrid | Modular |
|---|---|---|---|---|
| Cost | Low | Moderate | Moderate | High |
| Complexity | Low | Moderate | Moderate | High |
| Monitoring | Centralized | Distributed | Both | Modular |
| Management | Centralized | Distributed | Both | Modular |
| Scalability | Limited | Good | Good | Excellent |
| Performance | Good | Excellent | Good | Good |
| Topology | Centralized | Distributed | Hybrid | Modular |

### 3.4 *BMS Architecture*

The architecture of a BMS can change according to the specific requirements of the battery system. The architecture of BMS is developed to ensure the proper functioning and health of the battery operates in safe and optimal conditions and to provide reliable, long-lasting power for the system. However, most BMSs have the following basic components:

- **Sensors:** The BMS uses sensors to gather data about the battery, including voltage, current, temperature, and more. This data is used to make decisions about how to manage the battery.
- **Control Unit:** It is the heart of the BMS. It processes the data from the sensors and makes decisions about how to manage the battery. The control unit can include a microcontroller, microprocessor, or other processing units.
- **Communication Interface:** The BMS typically communicates with other components in the system, such as the battery charger or the host system, to exchange data and receive commands. The communication interface can use various communication protocols, such as CAN bus, I2C, or other serial communication methods.
- **Power Electronics:** Power electronics are used in the battery to transmit charge between cells as needed, to maintain voltage balance. This is part of the active cell balancing function.
- **Protection Circuits:** The BMS includes protection circuits to verify that the battery operates safely. These protection circuits can detect overcharging, over-discharging, over-temperature, and other conditions that can damage the battery.

## 4 Conclusion

This paper conducted a survey focusing on battery management systems. The assessment of SOC, SOH, thermal control, and cell balancing is discussed, which provides essential protection and performance optimization to ensure safe, reliable, and efficient operation. Different BMS topologies are also discussed and each BMS topology has its unique advantages and disadvantages. The paper examines and contrasts various battery methods in terms of energy efficiency, energy density, power density, and cycle life. Based on this analysis, the Li-ion battery is more suitable option. The paper discusses different parameters but mainly focuses on cell balancing and its techniques and compares different approaches for cell balancing, taking into account factors such as the principle of balancing, the capability for charging and discharging, cost-effectiveness, and their suitability for specific applications and the analysis shows that an active cell balancing technique has superior energy transmission efficiency and a shorter balancing period in comparison to passive cell balancing. Consequently, active cell balancing is identified as a more favorable alternative for cell balancing.

## References


1. Xing Y, Ma EWM, Tsui KL, Pecht M (2011) Battery management systems in electric and hybrid vehicles. Energies 4(11):1840–1857
2. Spoorthi SB, Pradeepa P (2022) Review on battery management system in EV, International Conference on Intelligent Controller and Computing for Smart Power (ICICCSP), pp 1–4, Hyderabad, India
3. Gade AR (2021) the new battery management system in electric vehicle. Intern J Eng Res Technol (IJERT) 10(7)
4. Ananthraj CR, Ghosh A (2021) Battery management system in electric vehicle, 2021 4th Biennial International Conference on Nascent Technologies in Engineering (ICNTE), pp 1–6, NaviMumbai, India
5. Hossain Lipu MS, Hannan MA, Karim Tahia, Hussain Aini, Md Saad, Mohamad Hanif, Ayob Afida, Miah Md. Sazal, Mahlia T M Indra (2021) Intelligent algorithms and control strategies for battery management system in electric vehicles: Progress, challenges and future outlook. J Clean Prod 292
6. Mishra S, Swain SC, Samantaray RK (2021) A review on battery management system and its application in electric vehicle, 2021 International Conference on Advances in Computing and Communications (ICACC), pp 1–6, Kochi, Kakkanad, India
7. Lu X, Wang H (2020) Optimal sizing and energy management for cost-effective PEV hybrid energy storage systems. IEEE Trans Ind Inform 16(5):3407–3416
8. Ravikumar R, Ghatge S, Soni SR, Nadar J (2020) Design of battery management system. IEEE Pune Section International Conference (PuneCon), pp 48–52, Pune, India
9. Sandeep R, Hariprasad A (2020) Battery management system in electric vehicles. Intern J Eng Res Technol (IJERT) 9(5)
10. Aruna P, Vasan PV (2019) Review on Energy management system of electric vehicles, 2nd International Conference on Power and Embedded Drive Control (ICPEDC), Chennai, India, pp 371–374
11. Liu K, Li K, Peng Q et al (2019) A brief review on key technologies in the battery management system of electric vehicles. Front Mech Eng 14:47–64
12. Lelie M, Braun T, Knips M, Nordmann H, Ringbeck F, Zappen H, Sauer DU (2018) Battery management system hardware concepts: an overview. Appl Sci 8:534
13. Suntanto D, Sarrafan K, Muttaqi KM, Graham T (2017) Accurate range estimation for an electric vehicle including changing environmental conditions and traction system efficiency. IET Elect Syst Transp 7:117–124
14. Cheng KWE, Divakar BP, Wu H, Ding K, Ho HF (2011) Battery-management system (BMS) and SOC development for electrical vehicles. IEEE Trans Veh Technol 60(1):76–88
15. Duryea S, Islam S, Lawrance W (2001) A battery management system for stand-alone photovoltaic energy systems. IEEE Ind Appl Mag 7(3):67–72
16. Kuang X et al. (2020) Research on control strategy for a battery thermal management system for electric vehicles based on secondary loop cooling. IEEE Access 8:73475–73493
17. Chen C, Man KL, Ting TO, Lei Chi-Un, Krilavicius T, Jeong TT, Seon JK, Guan Sheng-Uei, Wong Prudence WH (2012) Design and realization of a smart battery management system. Intern Multiconf Eng Comput Sci II(IMECS 2012):1–4, Hong Kong
18. Aiello O (2020) Electromagnetic susceptibility of battery management systems' ICs for electric vehicles: experimental study. Electronics 9(3):510
19. Xue J, Yan C, Wang D, Wang J, Wu J, Liao Z (2020) Adaptive dynamic programming method for optimal battery management of battery electric vehicle, IEEE 9th Data Driven Control and Learning Systems Conference (DDCLS), pp 65–58, Liuzhou, China
20. Korkas CD, Baldi S, Yuan S, Kosmatopoulos EB (2018) An adaptive learning-based approach for nearly optimal dynamic charging of electric vehicle fleets. IEEE Trans Intell Transp Syst 19:2066–2075

21. Vaideeswaran V, Bhuvanesh S, Devasena M (2019) Battery management systems for electric vehicles using lithium ion batteries, innovations in power and advanced computing technologies (i-PACT). Vellore, India, pp 1–9
22. Du J, Ouyang M (2013) Review of electric vehicle technologies progress and development prospect in china. Elect Vehicle Symp Exhib (EVS27), pp 1–8, China
23. Shete S, Jog P, Kumawat RK, Palwalia DK (2021) Battery management system for SOC estimation of lithium-ion battery in electric vehicle: a review. 6th IEEE International Conference on Recent Advances and Innovations in Engineering (ICRAIE), pp 1–4, Kedah, Malaysia
24. Goksu OF, Acar Vural R (2018) Battery management module with active balancing and cell switching, 6th International Conference on Control Engineering & Information Technology (CEIT), pp 1–6, Istanbul, Turkey
25. Iclodean C, Varga B, Burnete N, Cimerdean D, Jurchiay B (2017) Comparison of different battery types for electric vehicles. IOP Conf Series: Mater Sci Eng 252
26. Parikshith S, Shailesh KR (2016) Reduction of parameters in a Lithium ion cell model by experimental validation of relationship between OCV and SOC. Online International Conference on Green Engineering and Technologies (IC-GET), pp 1–5
27. Zheng L, Zhu J, Wang G (2016) A comparative study of battery balancing strategies for different battery operation processes, IEEE Transportation Electrification Conference and Expo (ITEC), Dearborn. MI, USA, pp 1–5
28. Ekici Y, Tan N (2019) Charge and discharge characteristics of different types of batteries on a hybrid electric vehicle model and selection of suitable battery type for electric vehicles. Intern J Autom Sci Techn 3(4):62–70
29. Manzetti S, Mariasiu F (2015) Electric vehicle battery technologies: from present state to future systems. Renew Sustain Energy Rev 51:1004–1012
30. Lu L, Han X, Li, Jianqiu, Hua J, Ouyang M (2012) A review on the key issues for lithium-ion battery management in electric vehicles. J Power Sourc 226:272–288
31. Wei Y, Ling L (2021) State of charge estimation for lithium-ion battery based on artificial neural network. IEEE 5th Advanced Information Technology, Electronic and Automation Control Conference (IAEAC), pp 2454–2458, Chongqing, China
32. Zhang M, Fan X (2020) Review on the state of charge estimation methods for electric vehicle battery. World Electr Veh J 11(1):23
33. Hariprasad A, Priyanka I, Sandeep R, Ravi V, Shekar O (2020) Battery management system in electric vehicles. Intern J Eng Res Techn (IJERT) 9(5)
34. Liu X, Zheng C, Wu J, Meng J, Stroe D-I, Chen J (2020) an improved state of charge and state of power estimation method based on genetic particle filter for lithium-ion batteries. Energies 13(2):478
35. Zhang Mingyue, Fan Xiaobin (2020) Review on the state of charge estimation methods for electric vehicle battery. World Elect Vehicle J 11(1):23
36. Saji D, Babu PS, Ilango K (2019) SoC estimation of lithium ion battery using combined coulomb counting and fuzzy logic method, 4th International Conference on Recent Trends on Electronics, Information, Communication & Technology (RTEICT), pp 948–952, Bangalore, India
37. Umair Ali M, Hussain Nengroo S, Adil Khan M, Zeb K, Ahmad Kamran M, Kim H-J (2018) A real-time simulink interfaced fast-charging methodology of lithium-ion batteries under temperature feedback with fuzzy logic control. Energies 11(5):1122
38. Hannan MA, Hossain Lipu MS, Hussain Aini, Mohamed Azah (2017) A review of lithium-ion battery state of charge estimation and management system in electric vehicle applications: Challenges and recommendations. Renew Sust Energy Rev 78:834–854
39. Kim Y, Yun S, Lee J (2017) SOC estimation and BMS design of Li-ion battery pack for driving, 14th International Conference on Ubiquitous Robots and Ambient Intelligence (URAI), pp 216–218, Jeju, Korea (South)
40. Malysz P, Ye J, Gu R, Yang H, Emadi A (2016) Battery state-of-power peak current calculation and verification using an asymmetric parameter equivalent circuit model. IEEE Trans Veh Technol 65(6):4512–4522

41. Taborelli C, Onori S, Maes S (2016) Advanced battery management system design for SOC/SOH estimation for e-bikes applications. Intern J Powert 5:325
42. Pop V, Bergveld HJ, Notten PHL, Op het Veld JHG, Regtien PPL (2009) Accuracy analysis of the State-of-Charge and remaining run-time determination for lithium-ion batteries. Measurement 42(8):1131–1138
43. Xing Y, Ma EWM, Tsui KL, Pecht M (2011) Battery management systems in electric and hybrid vehicles. Energies 4:1840–1857
44. Do DV, Forgez C, Benkara KEK, Friedrich G (2009) Impedance observer for a Li-ion battery using Kalman filter. IEEE Trans Veh Technol 58(8):3930–3937
45. Bhangu BS, Bentley P, Stone DA, Bingham CM (2005) Nonlinear observers for predicting state-of-charge and state-of-health of lead-acid batteries for hybrid-electric vehicles. IEEE Trans Veh Technol 54(3):783–794
46. Noura N, Boulon Loic, Jemei S (2020) A review of battery state of health estimation methods: hybrid electric vehicle challenges. World Elect Veh J 11(4):66
47. Omariba ZB, Zhang L, Sun D (2018) Review on health management system for lithium-ion batteries of electric vehicles. Electronics 7(5):72
48. Kim ILS (2010) A technique for estimating the state of health of lithium batteries through a dual-sliding-mode observer. IEEE Trans Power Electron 25(4):1013–1022
49. Gould CR, Bingham CM, Stone DA, Bentley P (2009) New battery model and state-of-health determination through subspace parameter estimation and state-observer techniques. IEEE Trans Veh Technol 58(8):3905–3916
50. Saha B, Goebel K, Poll S, Christophersen J (2009) Prognostics methods for battery health monitoring using a bayesian framework. IEEE Trans Instrum Meas 58(2):291–296
51. Verbrugge Mark W et al. (2005) Adaptive energy management of electric and hybrid electric vehicles. J Electrochem Soc 152
52. Verbrugge M, Tate E (2004) Adaptive state of charge algorithm for nickel metal hydride batteries including hysteresis phenomena. J Power Sources 126(1–2):236–249
53. Kuang X et al (2020) Research on control strategy for a battery thermal management system for electric vehicles based on secondary loop cooling. IEEE Access 8:73475–73493
54. Nath A, Rajpathak B (2022) Analysis of cell balancing techniques in bms for electric vehicle. International Conference on Intelligent Controller and Computing for Smart Power (ICICCSP), pp. 1–6, Hyderabad, India
55. Hemavathi S (2021) Overview of cell balancing methods for li-ion battery technology. Energy Storage 3(2)
56. Omariba ZB, Zhang L, Sun D (2019) Review of battery cell balancing methodologies for optimizing battery pack performance in electric vehicles. IEEE Access 7:129335–129352
57. Omer Faruk Goksu, Revna Acar Vural (2018) Battery management module with active balancing and cell switching. 6th International Conference on Control Engineering & Information Technology (CEIT), pp 1–6, Istanbul, Turkey
58. Lee Y, Jeon S, Lee H, Bae S (2016) Comparison on cell balancing methods for energy storage applications. Indian J Sci Technol 9(17):1–7
59. Shukla AP, Patel RA (2022) Battery management system by passive cell balancing for electric vehicle, 2nd International Conference on Power Electronics & IoT Applications in Renewable Energy and its Control (PARC), pp. 1–6, Mathura, India
60. Singirikonda S, Obulesu YP (2021) Active cell voltage balancing of electric vehicle batteries by using an optimized switched capacitor strategy. J Energy Stor 38:102521
61. Shah S, Murali M, Gandhi P (2018) A practical approach of active cell balancing in a battery management system. IEEE International Conference on Power Electronics, Drives and Energy Systems (PEDES), pp 1–6, Chennai, India
62. Moghaddam AF, Van Den Bossche A (2018) An active cell equalization technique for lithium ion batteries based on inductor balancing, 2018 9th International Conference on Mechanical and Aerospace Engineering (ICMAE), pp 274–278, Budapest, Hungary
63. Ramkumar MS, Reddy CSR, Ramakrishnan A, Raja K, Pushpa S, Jose S, Jayakumar M (2022) Review on Li-Ion battery with battery management system in electrical vehicle. Adv Mater Sci Eng 2022:1–8

# Photovoltaic Faults Prediction by Neural Networks

**M. Santhosh, P. Bhargav, and Arun Balodi**

**Abstract** Solar energy is one of the most important forms of renewable energy. The most significant method of converting solar energy is photovoltaic (PV) systems. Fault diagnosis is crucial for the dependable and efficient operation of PV systems. Early fault detection and diagnosis can save maintenance costs, avoid or minimise system downtime, and generally improve system performance. Artificial neural networks, a key artificial intelligence methodology, have been developed and applied in a variety of fields, including the fault diagnosis of PV systems, because of their robust self-learning capability, outstanding generalisation performance, and high fault tolerance. This study shows how artificial neural networks (ANN) can be used to predict solar panel problems.

**Keywords** Artificial intelligence · Fault diagnosis · Artificial neural network · Photovoltaic · Solar energy

## 1 Introduction

An energy revolution is currently taking place in modern society in an effort to switch from fossil fuels to renewable energy sources in order to stop catastrophic climate change [1]. The development of urban areas and technological advancements nearby should strongly encourage the use of renewable energy sources. As a result, the demand for renewable energy is rising daily [2]. Solar photovoltaic (PV) energy has quickly replaced conventional energy sources in recent decades due to its global availability, modularity, lack of pollution, ease of installation, and low cost. Particularly in terms of efficiency, affordability, and optimising the power that can be collected from PV cells, the study of PV systems has evolved tremendously. Never-

M. Santhosh · P. Bhargav
Atria Institute of Technology, Bengaluru, India

A. Balodi (✉)
Dayananda Sagar University, Bengaluru, India
e-mail: drbalodi@gmail.com



O. H. Gupta et al. (eds.), *Soft Computing Applications in Modern Power and Energy Systems*, Lecture Notes in Electrical Engineering 1107,
https://doi.org/10.1007/978-981-99-8007-9_20

(a) The solar panel

(b) Concentrating solar thermal power

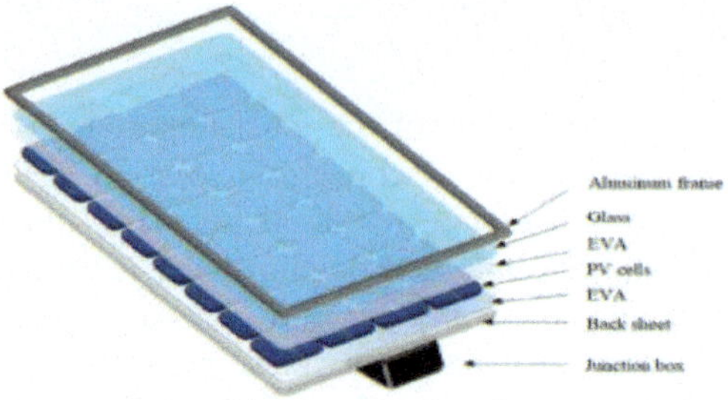


(c) Solar Panel

**Fig. 1** Solar plant and solar panel

theless, a variety of faults frequently affect PV systems, which can negatively impact their conversion efficiency and safe operation [3].

The solar panels in large areas produce high energy but when these solar panels are getting aged faults will be occurring in every solar panel. It is a difficult problem to identify in which solar panel faults are occurring. The faults in the solar panel make a solar panel give less energy. The faults in a solar panel damage the entire solar panel. After the damage to a solar panel, the full solar panel has to change [9]. To increase the efficiency of solar panels many methods came like MPPT (maximum power point tracking) [6], using SMD (smart monitoring devices) [7], and using many algorithm methods, etc. (Fig. 1).

There are three different categories of failures in PV systems: electrical, chemical, and physical faults [10]. Therefore, experts have proposed a range of strategies to quickly and accurately identify and diagnose a variety of faults in PV systems. The two primary categories of defect diagnosis procedures are the electrical approach and the thermal and optical methods [11]. Open circuit faults, line-to-line faults (short circuit), arc faults (series and parallel), ground faults, bridging faults, dust (soil formation), and cell-level degradation (quality control) are common defects that affect all solar panels [8]. The open circuit fault is the disconnection of the current from all the solar panels associated with the same string zero. The detection methods used for the open circuit faults are analysed based on measured ambient conditions, current, voltage, etc. For small-scale plants, optical and thermal methods

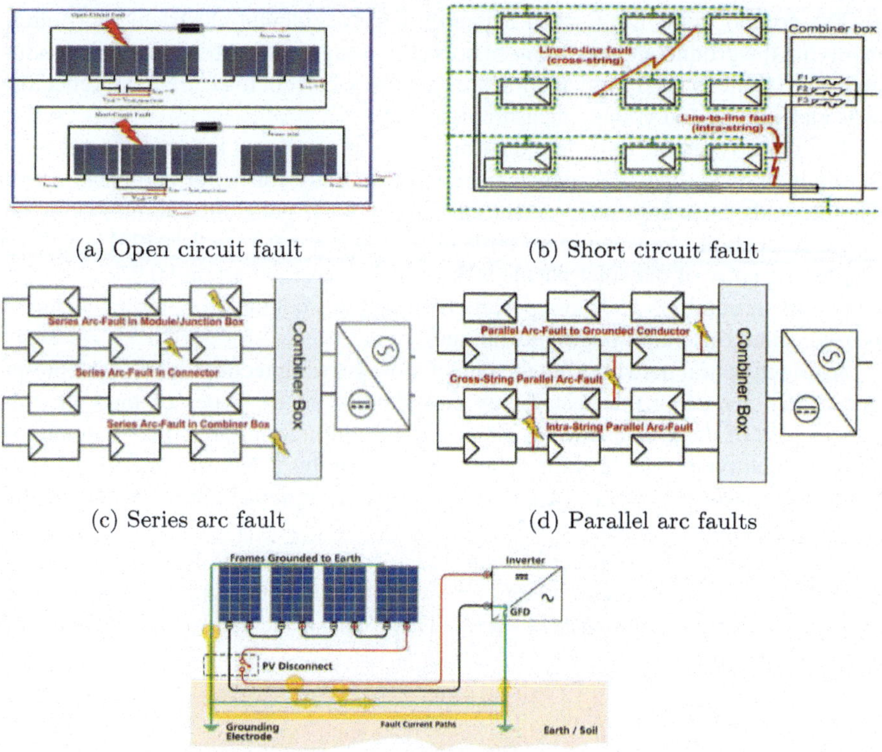


(a) Open circuit fault (b) Short circuit fault

(c) Series arc fault (d) Parallel arc faults

(e) Ground faults

**Fig. 2** Different faults in solar panel

are sufficient for fault monitoring and diagnosis, whereas electric methods are better suited for PV system monitoring and diagnosis [12].

Line-to-line faults develop when connections made between two sites of difference in an electric network or system have unintentionally low resistance. The method used for the line-to-line faults is an analysis based on the ambient conditions, current, voltage, etc. The line-to-line faults in a solar panel are also identified as short circuit faults (Fig. 2).

The arc faults are divided into two categories they are series arc faults and parallel arc faults. Series arc faults are formed due to a loss of the continuity of a conductor, connection, module, or other PV system components. The detection methods are the current and voltage measurement and frequency domain analysis of a PV array current. But the parallel arc faults are between the two conductors or between the conductor and the ground. Frequency domain analysis of a current and the detection of a sharp reduction in the PV array's voltage or current are the two methods used to find parallel arc defects.

The ground faults are formed because of the accidental electrical short circuit involving the ground and one or more normally designated current-carrying conductors. The methods used for detection are frequency current measurement, reflectometry, and current and voltage measurement [10].

The other types of faults like dust on the solar panel can reduce 30% of output in one month. So, the solar panels have to clean daily for better and clean output. The methods are using for to detect dust on the solar panel are thermography and visual inspection. And one more fault in the solar panel is cell-level degradation. The high quality of the solar panels is degraded at a rate of around 0.5% every year generating around 12–15% less power. To detect this degradation, fault a technique is used called different imaging techniques.

Artificial neural networks are the one of most growing technologies in the present generation. By using this technology can identify all the types of faults in solar panels. One of the greatest things about the ANN it not only identifies faults it also shows which type of fault in the solar panel. Using this ANN can predict the faults in the solar before the faults are getting bigger. The implementation of ANN will be shown in the methodology [13–15].

The remainder of this paper is organised as follows: The problem statements are covered in Sect. 2. The methodology for PV system defect diagnostics is introduced in Sect. 3. Results are discussed in Sect. 4. This work's conclusion and its possibilities are presented in Sect. 5.

## 2 Problem Statement

This research focuses on the use of ANN for solar PV system fault detection. With the help of ANN technology, machines, and computers can become intelligent on their own. ANN has nodes in it called Neurons. An ANN is composed of three layers: the input layer, the output layer, and the hidden layer [4, 5]. Through the use of an activation function, the hidden layer generates output from a set of weighted inputs. Multiple input layers, multiple output layers, and numerous hidden levels are also possible with ANN.

## 3 Methodology

The solar panel of a PV cell forms nonlinear I- V characteristics and it can be obtained by a simple model which consists of a constant current source, diode, and a resistor associated in both the series and in the parallel.

Typically, the electrical behaviour of a solar cell is described using the PV module of one diode and two diodes models. Figure 3b roughly depicts the equivalent circuit of a photovoltaic cell (PV). Several factors affect how sunlight is converted into electricity. Table 1 lists these factors.

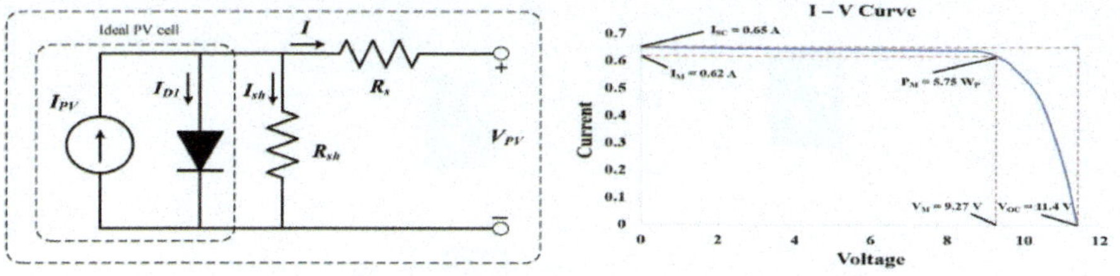


(a) PV module (b) I-V characteristic

**Fig. 3** PV module and I-V characteristic

**Table 1** PV parameters

| Amps | Volts | Watts |
|---|---|---|
| 0 | Voc = 11.4 | 0 |
| 0.2 | 11.06 | 2.21 |
| 0.4 | 10.59 | 4.24 |
| 0.5 | 10.24 | 5.12 |
| 0.6 | 9.54 | 5.72 |
| 0.61 | 9.39 | 5.73 |
| $I_M = 0.62$ | $V_M = 9.27$ | $P_M = 5.75$ |
| 0.63 | 9.08 | 5.72 |
| 0.64 | 8.72 | 5.58 |
| $I_{SC} = 0.65$ | 0 | 0 |

In Table 1 and graph, the major parameters are identified as the current, voltage irradiance, and temperature. If there is a change in any parameter there will be a change in the output also. The STC (standard test conditions) manufacturers provide the cell parameters. The comparable solar radiation under the STC is 1000 W/m$^2$. The working temperature of the cell is 25 °C. Maximum power point (MPP), is the working peak of each solar panel. The fault detection using I-V data can be accomplished by measuring MPPs and observing the variation of a measured MPP from an actual MPP. The unsupervised algorithm was ineligible to be categorised as a defect. Therefore, in order to categorise unlabeled data, we need a method that is also employed for partially labelled data. Using artificial neural networks (ANN), we are able to accurately identify the type of fault as well as detect it. Use MATLAB software for the code and SIMULINK as described in Fig. 4 to implement and understand how ANNs in solar panels work.

To implement ANN model the following steps are involved.

1. Selecting of inputs and targets.
2. Validation of the data set.
3. Deciding the number of hidden Neurons.
4. Training the network by using any training algorithm.
5. Network testing.

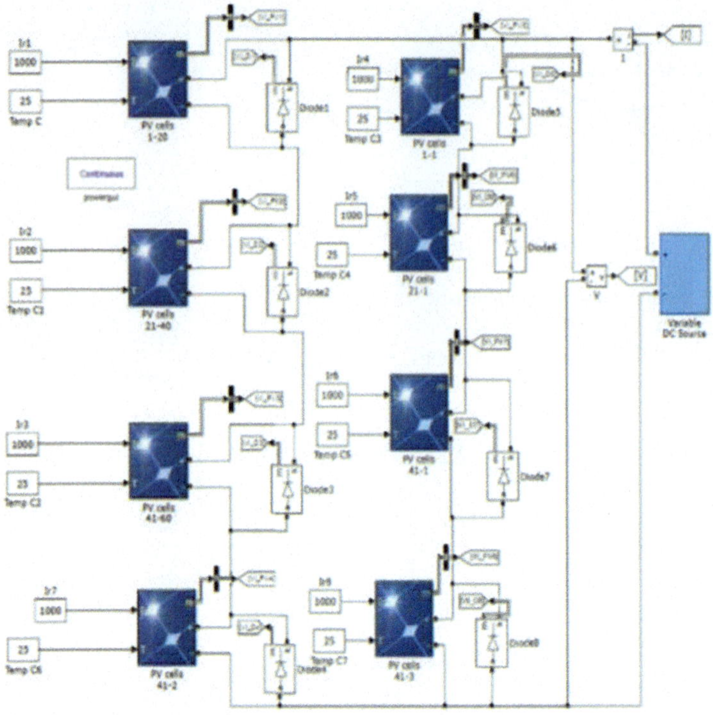


**Fig. 4** I-V characteristic

To implement the ANN we have to collect the data from the solar panel. And we know the important parameters of the solar panel. To build the ANN the data has to be trained for the detection and for identifying the faults. The collection of data is a major thing for ANN. The collected data of the different parameters will be collected as the maximum data and minimum data as shown in Table 2.

## 3.1 Building Model

Data from the solar panels should be gathered in order to create the ANN Model. To forecast the errors, the data should be trained using the obtained data. The dataset is divided into 7 layers ranging in value from $x_1$ to $x_7$ for an input layer of an ANN. The current (A) in branch 1 of the PV system is $x_1$, while branch 2's current (A) is $x_2$. In the PV system, branch 1 has a voltage of $x_3$ while branch 2 has a value of $x_4$. The irradiation level ($k$ lux) is $x_5$. The average temperature of the PV system is $x_6$. The weather is represented by the integer $x_7$ (snow, sunny, overcast, rainy). Two hidden layers, $h_1$ and $h_2$, combine to form the hidden layer. We select the ReLU

**Table 2** Sensor data

| State | | SI1 | SI2 | Sv1 | Sv2 | Irradiation klux | Temperature | Weather |
|---|---|---|---|---|---|---|---|---|
| | | Amps | | Volts | | | | |
| Normal | Max | 7.6 | | 110 | | 110 | 41 | Sunny/summer |
| | | 5.4 | | 102 | | 97 | 23 | Cloudy/summer |
| | Min | 5.5 | | 105 | | 108 | 15 | Sunny/winter |
| | | 0.4 | | 80 | | 9 | −3 | Cloudy/winter |
| Open | Max | 7.6 | | 110 | | 108 | 15 | Sunny/winter |
| | Min | 0 | | 0 | | 9 | −3 | Cloudy/winter |
| Line-line | Max | 6.1 | | 90 | | 110 | 41 | Sunny/summer |
| | | 1.8 | | 72 | | 97 | 23 | Cloudy/summer |
| | Min | 5.5 | | 75 | | 108 | 15 | Sunny/winter |
| | | 0.4 | | 62 | | 9 | −3 | Cloudy/winter |
| Variance | | 0.52 | | 7.23 | | 35.56 | 8.7 | Summer |
| | | 4.82 | | 9.54 | | 47.61 | 15.24 | Winter |

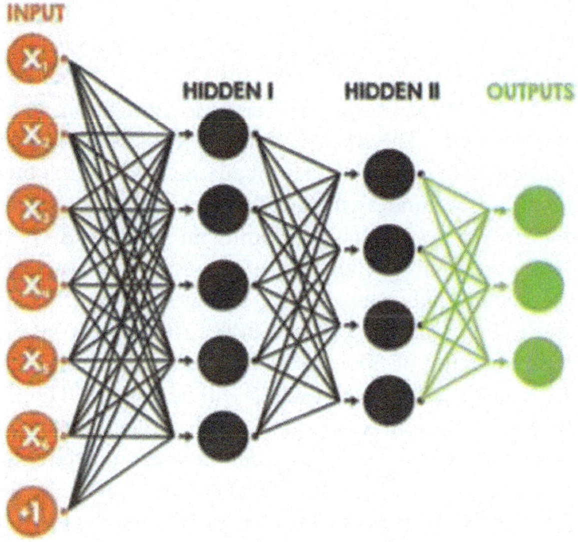


**Fig. 5** Layers of ANN

(rectified linear unit) as the activation function since it offers several benefits in multidimensional nonlinear datasets. According to the ReLU, $y = max(0, x)$. The three layers $y_1$, $y_2$, and $y_3$ make up the output layer. To create the ANN model for defect detection, the various parameter data sets that have been collected will go through the following processes. The multilayer perceptron (MLP), which has a backward network and a feedforward network, is depicted in Fig. 5. The ANN module learns a lot from the trained data, and the output is compared to the tested data.

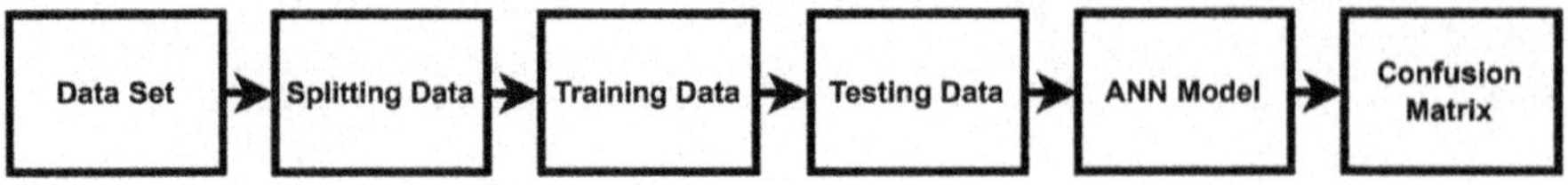


**Fig. 6** Flow chart of the fault detection

**Table 3** I-V characteristic

| | Time | Ipv | Vpv | Vdc | ia | ib | ic | va |
|---|---|---|---|---|---|---|---|---|
| 0 | 0.000041 | 2.369843 | 90.429688 | 147.949219 | 0.616820 | −0.530396 | −0.160283 | −146.003418 |
| 1 | 0.010040 | 2.378357 | 90.368652 | 148.242188 | −0.678956 | 0.570679 | 0.047847 | 148.546906 |
| 2 | 0.020039 | 2.373627 | 90.460205 | 148.535156 | 0.603393 | −0.537109 | −0.187139 | −145.027008 |
| 3 | 0.030038 | 2.368896 | 90.423584 | 148.242188 | −0.685670 | 0.570679 | 0.041133 | 147.341461 |
| 4 | 0.040037 | 2.362274 | 90.435791 | 148.242188 | 0.630248 | −0.523682 | −0.166997 | −143.279114 |

| | vb | vc | Iabc | If | Vabc | Vf | Label |
|---|---|---|---|---|---|---|---|
| 0 | 120.255127 | 25.595601 | 1.000000 | 50.000000 | 1.000000 | 50.000000 | FOM |
| 1 | −119.881439 | −24.723663 | 1.000000 | 50.000000 | 1.000000 | 50.000000 | FOM |
| 2 | 120.315399 | 26.109924 | 1.000000 | 50.000000 | 1.000000 | 50.00000 | FOM |
| 3 | −119.736789 | −26.105906 | 0.440380 | 50.078491 | 118.275197 | 49.921509 | FOM |
| 4 | 120.942230 | 26.680501 | 0.618996 | 50.248673 | 90.022162 | 49.606494 | FOM |

Machine Learning (ML) error identification and diagnosis methods have been used recently. The training data determines how effective the ML technique is. The training data and PV data typically have very high prediction accuracy rates that can reach 100%. Finding the key features of the input dataset is of the utmost importance for the ML system while building an ML model.

Figure 6 shows the steps of the code for fault detection. Collect the data of the parameters like temperature, voltage, current, and irradiance. Import the data by spitting the data into 70 and 30%. 70% of the data is used to train the system and the remaining of 30% of the data is used for the test data. After training the ANN with the data the ANN module will be builded. Then the predictions will of the form of confusion matrix. The data of all the parameters are taken in the two forms like minimum dataset and maximum dataset in different conditions of the weather. For to predict and to build the ANN module with the knowledge the sample data is given which is shown in Fig. 7. This sample data helps the ANN module to predict and identify the faults by comparing the values. And the ANN module is trained with large datasets containing different data values as shown in Table 3.

Each data column is described as the following: Time: The time of real measurement in seconds. The avg sampling is $T_s = 9.9989\,\mathrm{Ms}$. Ipv: PV array current measurement. $V_pv$: PV array voltage measurement. $V_dc$: DC voltage measurement. $i_a$: Phase-A current measurement. $i_b$: Phase-B current measurement. $i_c$: Phase-C current measurement. $v_a$: Phase-A voltage measurement. $v_b$: Phase-B voltage measurement. $v_c$: Phase-C voltage measurement. $I_abc$: Positive- sequence estimated current magnitude. $I_f$: Positive- sequence estimated current frequency. $V_abc$: Positive- sequence estimated voltage magnitude. $V_f$: Positive- sequence estimated current frequency.

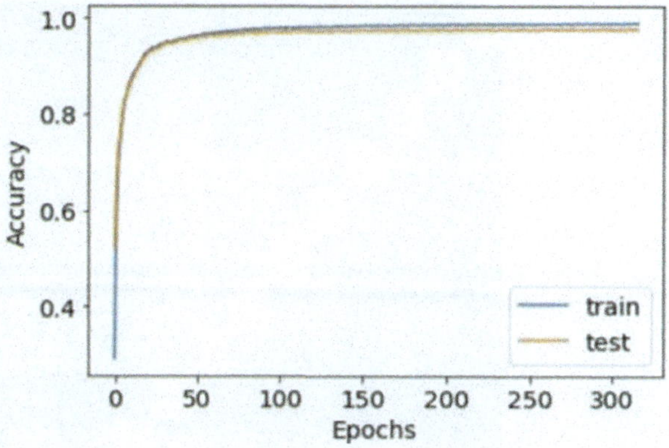


**Fig. 7** Accuracy versus Epochs curve

There are other parameters also like the temperature and the irradiance which are given in the solar panel.

## 4 Results

In an ANN, data travels in two directions: forward propagation and the MLP, which forecasts results for the input dataset. Back propagation also takes the mistake in the projected data into account while adjusting its settings. The graph of the trained and tested data is shown in Fig. 7. A complete iteration of the algorithm over the training data set is referred to as an epoch. Each epoch consists of one or more batches where we train the neural networks using a portion of the dataset. The epoch aids in identifying the model from the data.

Which model is most effective at recognising the connections and patterns in the trained data depends on its accuracy. The ratio between the correct prediction data sets and the overall number of predictions in the trained datasets is predicted by the epoch and accuracy graphs. The precision of a PV cell is not increased by increasing the epochs. The confusion matrix, which is displayed in Fig. 8, compares the trained data's predictions to the tested data. The confusion matrix displays all datasets with insufficient data as well as the most datasets possible. Checking the data's accuracy is necessary.

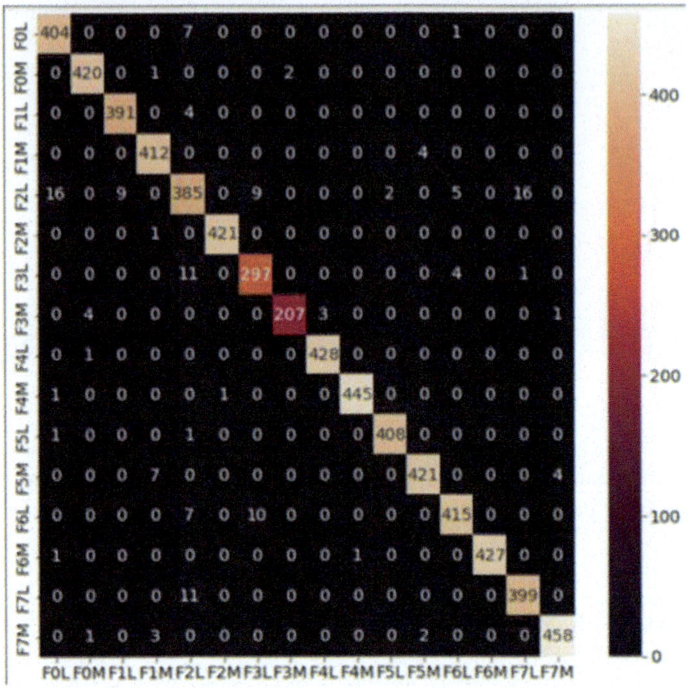


**Fig. 8** Confusion matrix

## 5 Conclusion

The basics of various PV panel failures were explored in this study. The current ANN-based fault detection and diagnostic techniques were tested in trials. The simulation results using neural networks are successfully displayed by recognising and detecting frequently recurring defects. It shows a significant improvement in detection precision. It is quite efficient to utilise ANN to predict and locate problems in or between PV modules. The existing fault diagnosis ANNs face a serious problem with the cost of training time. These ANNs can be employed in conjunction with an embedded system of digital signals to provide real-time diagnostics, maximising the effectiveness of the fault diagnosis system. It is intended to employ this problem detection in massive PV systems in the future.

## References

1. Abdalla AN, Nazir MS, Tao H, Cao S, Ji R, Jiang M, Yao L (2021) Integration of energy storage system and renewable energy sources based on artificial intelligence: an overview. J Energy Storage 40:102811

2. Olabi AG, Abdelkareem MA (2022) Renewable energy and climate change. Renew Sustain Energy Rev 158:112111
3. Gawre SK (2022) Advanced fault diagnosis and condition monitoring schemes for solar pv systems. Planning of hybrid renewable energy systems, electric vehicles and microgrid: modeling, control and optimization. Springer Nature, Singapore, pp 27–59
4. Frizzo Stefenon S, Waldrigues Branco N, Nied A, Wildgrube Bertol D, Cristian Finardi E, Sartori A, Bartnik Grebogi R (2020) Analysis of training techniques of ANN for classification of insulators in electrical power systems. IET Gen Trans Distrib 14(8):1591–1597
5. Mishra K, Balodi A (2022) Comparative analysis of neural network algorithms for severity analysis of valvular regurgitation. In: 2022 IEEE Delhi section conference (DELCON). IEEE, pp 1–5
6. Subudhi B, Pradhan R (2012) A comparative study on maximum power point tracking techniques for photovoltaic power systems. IEEE Trans Sustain Energy 4(1):89–98
7. Shapsough S, Takrouri M, Dhaouadi R, Zualkernan IA (2021) Using IoT and smart monitoring devices to optimize the efficiency of large-scale distributed solar farms. Wirel Netw 27:4313–4329
8. Hlaili M, Mechergui H (2016) Comparison of different MPPT algorithms with a proposed one using a power estimator for grid connected PV systems. Int J Photoenergy 2016
9. Abdul Mawjood K, Refaat SS, Morsi WG (2018) Detection and prediction of faults in photovoltaic arrays: a review. In: 2018 IEEE 12th international conference on compatibility, power electronics and power engineering (CPE-POWERENG 2018). IEEE, pp 1–8
10. Yuan Z, Xiong G, Fu X (2022) Artificial neural network for fault diagnosis of solar photovoltaic systems: a survey. Energies 15(22):8693
11. Tina GM, Cosentino F, Ventura C (2016) Monitoring and diagnostics of photovoltaic power plants. In: Renewable energy in the service of mankind. Vol II: Selected topics from the world renewable energy congress, WREC 2014. Springer International Publishing, pp 505–516
12. Tsanakas JA, Ha LD, Al Shakarchi F (2017) Advanced inspection of photovoltaic installations by aerial triangulation and terrestrial georeferencing of thermal/visual imagery. Renew Energy 102:224–233
13. Katoch S, Muniraju G, Rao S, Spanias A, Turaga P, Tepedelenlioglu C, ... Srinivasan D (2018) Shading prediction, fault detection, and consensus estimation for solar array control. In: 2018 IEEE industrial cyber-physical systems (ICPS). IEEE, pp 217–222
14. Pillai DS, Rajasekar N (2018) A comprehensive review on protection challenges and fault diagnosis in PV systems. Renew Sustaine Energy Rev 91:18–40
15. Pillai DS, Blaabjerg F, Rajasekar N (2019) A comparative evaluation of advanced fault detection approaches for PV systems. IEEE J Photovolt 9(2):513–527

# Neural Network-Based Approach for Islanding Detection in a PV Grid-Connected System

**Kumaresh Pal, A. K. Akella, K. Namrata, and Anshuman Bhuyan**

**Abstract** Modern power networks are more dependent on smaller dispersed generating units as a result of the increased focus on renewable energy. The identification of islanding occurrences in these grids is more difficult than in conventional power grids, which rely on bigger centralised units. In grid-tied, PV systems, the difficulty in grid connection caused by unintentional islanding still presents a barrier. However, keeping the system connected to the power grid is crucial to ensuring high system stability and when an islanding condition occurs, the grid is de-energised to safeguard any personnel who may be on the grid trying to determine what caused the islanding. The three types of methods for islanding detection are communication-based, active, and passive. Active approaches use noise to estimate the size of the grid and measure the impact of that noise to identify islanding, which lowers the grid's quality of electricity. The implementation of communication-based techniques is constrained by their high cost and dependency on the communications infrastructure. Unlike existing techniques, the suggested method of Artificial Neural Network (ANN) depends on the modified negative sequence impedance values and phase angle jump values to properly identify every islanding situation with a 98.9% accuracy. Without using any feature extraction, these data are provided directly into the ANN, enabling quicker and more accurate decision-making. In order to accomplish faster and more accurate detection, the suggested technique is anticipated to combine the high accuracy of ANNs with the rapid change in instantaneous values. A 100 kW of PV integrated grid system is used for the study using MATLAB/Simulink (version 2018a).

K. Pal (✉) · A. K. Akella · K. Namrata
Department of Electrical Engineering, NIT, Jamshedpur, India
e-mail: kumaresh.pal@rediffmail.com

K. Namrata
e-mail: namrata.ee@nitjsr.ac.in

K. Pal
Department of Electrical and Electronics Engineering, Arka Jain University, Jamshedpur, India

A. Bhuyan
Department of EE, Gandhi Institute for Education and Technology, Bhubaneswar, India

 
O. H. Gupta et al. (eds.), *Soft Computing Applications in Modern Power and Energy Systems*, Lecture Notes in Electrical Engineering 1107,
https://doi.org/10.1007/978-981-99-8007-9_21

**Keywords** Photovoltaic (PV) system · Artificial neural network (ANN) · Negative sequence impedance (NSI) · Islanding mode

## 1 Introduction

Sustainable energy sources have exploded in popularity during the last decades. It produces less emissions and releases fewer greenhouse gases than fossil fuels and nuclear power [1]. Examples of distributed generation such as landfill gases, photovoltaic (PV) energy, wind energy, micro-hydro turbines, and biomass categorised as renewable energy [2]. Because of the world's fast expansion, distributed generation (DG) technology is advancing at a rapid pace, and as scientific, technological, and economic progress advances, people are placing greater demands on the environment. Including these DGs in the distribution network provides several advantages, including improved reliability, increased energy efficiency, lower line losses, and improved power quality. Distributed generating technology is not only safe, clean, environmentally friendly, cost-effective, and efficient, but it can also assess the whole power system's dependability and efficiency [3–5].

A PV system becomes "islanded" when a section of the utility system which includes load as well as distributed resources, such as a PV inverter, stays operational even though the utility system is disconnected [6]. Islanding mode can result in a number of challenges, including safety concerns, load damage, reclosing issues, and more [7]. Grid voltage and frequency may fluctuate because a grid-connected inverter cannot regulate the grid's voltage and frequency when it is in islanding mode. Loads may sustain irreparable damage as a result of this. Grid synchronisation is also impossible in islanding mode due to the lack of a standard grid voltage. Additionally, it is challenging to synchronise an isolated grid with the utility grid during recloser operation due to the isolated grid's voltage being out of phase with the electrical utility grid, and the grid might trip again due to an overcurrent fault.

For grid-connected PV inverters, Anti-Islanding Detection (AID) is a necessary function since islanding might pose a hazard to the operation of the grid. When an island is detected, the PV inverter must stop energising the grid within the allotted period. A number of AID algorithms have been commercialised and have been developed to prevent islanding. These algorithms could be classified into passive techniques and active techniques [10]. Any significant system disturbance, a grid failure, or intentional islanding for a planned or unexpected event might cause the islanding to happen. An islanding situation can be an unintentional or intentional. Unintentional islanding situation is a threat to the electrical grid's protection, posing a risk to utilities, equipment, and maintenance personnel. As a result, following island formation, the islanding system must be disconnected in 2 s in accordance with different islanding standards such as IEEE Std. 929-2000, IEEE Std. 1547-2003, IEC 62,116 and distributed generating operations must be stopped [11, 12].

Islanding detection can be categorised as communication-based, passive, active, and hybrid, depending on the methodology. Monitoring of grid parameters is the foundation of passive approaches. By using a relay of frequency and voltage Islanding can be easily detected, relays such as UOV (Under or Over Voltage)/UOF (Under or Over Frequency). Measuring voltage as well as frequency that are not within the suitable range can avoid islanding. The detection of an islanding operation can also be done using a phase jump or a change in harmonics [13]. Active approaches disrupt the grid using a PV inverter and monitor how the grid's parameters respond. Two categories can be made for this. The first active approach uses positive feedback, which causes the grid to become unstable when islanding situation occurs. When an islanding condition arises, a positive feedback loop to power references or current phase angle starts and voltage or frequency instability would be tracked [14–16]. The frequency drift techniques (FDTs) are representative techniques, like active frequency drift (AFD), slip-mode frequency shift (SMS), and Sandia-frequency shift (SFS) [17, 18]. The second active method type injects a signal into a grid and monitors how system parameters respond [19]. A signal used for the injection could be either the current or the voltage. After islanding, the voltage created by the injected components may be felt as the current is injected. Grid impedance could be determined and may be utilised as an indication for detecting islanding as high-frequency voltage is injected. This paper looks into a dilemma that arises at the point where a distributed generation facility meets the rest of the power system. The dilemma could be described by the power systems' recognition of an islanding condition. In recent years, a great number of contemporary methods for determining islanding scenarios have been described.

## 2 Artificial Neural Network (ANN) Model

Artificial neural networks (ANNs) mimic the intricate functioning of organic neural networks in the human brain. The basic building blocks of any artificial neural network are known as neurons, which are formally referred to be identical processing units that are coupled and infinite in number. An input layer, a hidden layer, and finally an output layer make up a fundamental neural network. In addition to this, they also have a neuron, transfer function, and weight. Figure 1 shows a basic architecture of ANN model. An ANN is made up of several layers and nodes. Via hidden layers, data are transferred from the input layer to the output layer. The nodes in succeeding levels are connected via links. The hidden layer and input layers are then supplied the error signals from the output layer [20]. The output of any node in the hidden and output layer is connected to the input node using activation function. The procedures of weight updating and reduction are carried out using a learning algorithm. Many training methods are available for the learning phase of ANN [21].

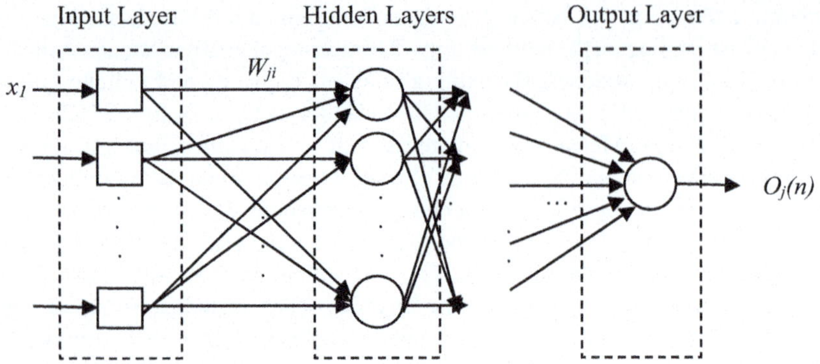


**Fig. 1** Basic architecture of ANN model [21]

## 3 Voltage Phase-Jump Detection

By keeping an eye out for a sudden phase jump in the terminal voltage of inverter ($V_{PCC}$) and output current ($I_{PV\_inv}$), the voltage phase-jump detection (PJD) approach is used as shown in Fig. 2 [22]. When the power system transitions from normal operation to islanding mode, the phase angle of the voltage phase-locked loop (VPCC) will alter and synchronise with the phase angle of the local load in the event of islanding. As a result, an abrupt phase shift occurs at PCC. To identify islanding, the PJD approach will look for this quick shift in phase angle. Moreover, loads frequently have a non-unity power factor, which indicates that the utility grid's voltage is partially hindered rather than fully absorbed [8].

The inverter in PV systems often uses a phase-locked loop (PLL) to track the phase of the grid signal [23]. During islanding operation, the power factor is dependent on

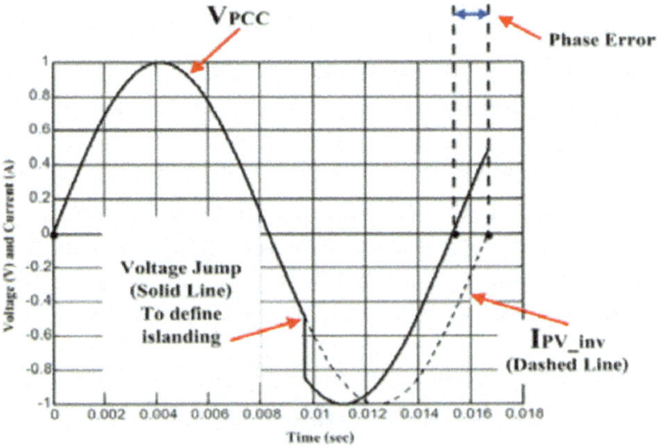


**Fig. 2** The operation of voltage phase-jump detection [22]

the local load; therefore, it is crucial for grid-connected inverters to have a unity power factor as the non-detection zone (NDZ) of the PJD technique is solely dependent on the power factor. If the inverter does not work with unity power factor, the NDZ could change. Hence, the PJD method is often referred to as the transient phase detection or the power factor detection [23].

## 4 Sequence Components of Voltage and Current at PCC

Out of several detection methods, the essential requirement for the existence of every disturbance in the voltage signal and the current signal detected at the PCC is the negative-sequence component approach [20]. Therefore, to be able to effectively identify islanding and notch perturbance caused by islanding occurrences, this research has analysed the negative-sequence component of voltage signals and the current signals discovered at PCC. The symmetrical component of analysis can be used to describe the negative, positive, and zero sequence components of the voltage signal at the PCC as [21].

$$V_n = \frac{1}{3}(V_a + \lambda^2 V_b + \lambda V_c) \tag{1}$$

$$V_p = \frac{1}{3}(V_a + \lambda V_b + \lambda^2 V_c) \tag{2}$$

$$V_z = \frac{1}{3}(V_a + V_b + V_c) \tag{3}$$

where $V_a$, $V_b$, and $V_c$ are the three-phase voltages obtained at the PCC and $V_n$, $V_p$, and $V_z$ are the negative, positive, and zero sequence voltages, respectively, and $\lambda = 1 < 20°$ is the intricate operator.

The symmetrical component of analysis can be used to indicate the zero, negative, and positive sequence components of the current signal at PCC as

$$I_n = \frac{1}{3}(I_a + \lambda^2 I_b + \lambda I_c) \tag{4}$$

$$I_p = \frac{1}{3}(I_a + \lambda I_b + \lambda^2 I_c) \tag{5}$$

$$I_z = \frac{1}{3}(I_a + I_b + I_c) \tag{6}$$

where $I_a$, $I_b$, and $I_c$ are the three-phase currents obtained at the PCC and $I_n$, $I_p$, and $I_z$ are the negative, positive, and zero sequence currents, respectively, and $\lambda = 1 < 20°$ is the intricate operator.

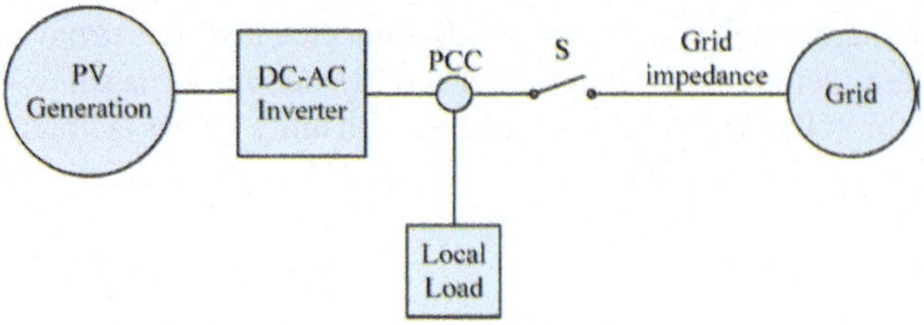


**Fig. 3** Block diagram of islanding in the system [3]

The negative-sequence impedance may be expressed by the symmetrical component of analysis given as

$$Z_n = \frac{V_z}{I_z} \tag{7}$$

The reciprocal of square root of negative-sequence impedance can be expressed by the symmetrical component of analysis given as

$$Z_x = \frac{1}{\sqrt{Z_n}} \tag{8}$$

The three phase voltage signals and the current signals acquired at PCC are given as inputs to the three-phase sequence analyser in MATLAB block. The negative-sequence component is chosen to take into account from among the three sequential components. In this paper negative sequence impedance-based passive technique is used because it emulates the info following perturbance condition. Evaluation of the reciprocal of square root of negative-sequence impedance at the point of common coupling provides excellent noise shielding for islanding detection, therefore it performs well. Variations in magnitude of reciprocal of the square root of NSI can be used to identify the islanding mode. The magnitude of NSI may be calculated using the negative-sequence voltage to negative-sequence current ratio as observed at PCC.

The suggested solution for islanding detection given in this study leverages machine learning to improve protection performance and minimise the detection time. The ANN-based protection offered a dependable quick reaction, high degree of precision, and quick detection (Fig. 3).

## 5 Proposed Islanding Detection Technique

An ANN is used in the presented islanding protection to identify patterns in the input signals and categorise occurrences as being as islanding or not. The inputs for this classifier are the modified negative sequence impedance values computed from negative sequence components of voltage and current found at PCC and voltage phase angle jump values derived at PCC.

This section discusses suggested methods and ANN architectural designs. The suggested IDT is primarily based on two stages. The projected IDT comprises two stages. In the first phase, at PCC, the negative sequence current and voltage data is initially gathered and utilised to calculate the inverse of square root of negative sequence impedance and also the phase angle shift values are collected in various islanding circumstances and non-islanding circumstances. In the second phase, the recorded data for modified negative sequence impedance and phase angle shift are fed into the ANN for inherent learning, training, and testing. And this trained ANN model is utilised to classify an islanding and non-islanding event. When non-islanding interruptions are present, a trustworthy islanding detection technique shouldn't fail. Data quality and quantity have a direct impact on intelligent Methods. For the ANN to generalise learnt features well and perform well on testing data, we needed a lot of training data. A number of short circuit failures, including few of the most frequent disturbances like load switching, load disconnection, and capacitor bank switching at PCC are investigated in order to show how the suggested IDT operates in non-islanding circumstances (Table 1).

The suggested technique is examined on a test system that includes a distributed generation unit powered by photovoltaic system and connected to a distribution network system through 100 KVA, 0.260/25 kV transformers. The total generation capacity of PV system is 100 KW. The distribution network system is operated at 25 kV. The total capacity of PV connected grid system is 200 kW. Loads are rated at 100 KW, 30 KVAR and 50 KW, 40 KVAR.

**Table 1** Analysis of data for the proposed IDT

| | | |
|---|---|---|
| *Islanding condition* | | |
| 1 | Islanding by changing up to 50% active power mismatch | 20,001 |
| *Non-islanding condition* | | |
| 2 | Constant load | 25,000 |
| 3 | Load switching | 16,000 |
| 4 | Short circuit fault (Symmetrical and Unsymmetrical) | 16,000 |
| 5 | Capacitor bank switching | 16,000 |
| 6 | Single pole open | 16,000 |
| 7 | Double pole open | 16,000 |
| *Total cases* | | 125,001 |

## 6 Result and Discussion

In this work, ANN analyses the islanding detection method. The PCC node's output voltage is read by ANN. It is continually worked through each cycle. When ANN senses an islanding situation, ANN will send tripping signal to Circuit Breaker to cut off the PV inverter's connection to the load. It is proposed to avoid the failure of equipments. The overall simulation time is 13 s, and the sampling frequency is held at 10 kHz. The proposed method was tested by analysing 125,001 unique cases, including 20,001 cases of islanding and 105,000 cases of non-islanding scenarios at the point of common coupling (PCC). The islanding data were generated by introducing an active power mismatch of 50% and a reactive power mismatch of 25%. ANN learning is designed arbitrarily with a composition of 50% for the learning process, 25% for testing, and 25% for validation, according to the sample result. Three layers make up the applied ANN: an input layer, a hidden layer, and an output layer. At hidden layer, 20 neurons were taken for training the network. A total of 1000 epochs were taken for iterations. For generating ANN model, Feed-forward back propagation network type is selected, with Levenberg–Marquardt as training function containing 20 validation checks.

Figure 4 shows the training state of the trained ANN network. The plot shows the training and validation accuracy and loss curves over 546 epochs of training. At the beginning of the training, the network's accuracy on the training and validation sets is low, and the loss is high. As the training progresses, the accuracy on both sets gradually increases, while the loss decreases. After around 20 epochs, the training and validation accuracy curves start to plateau, while the training and validation loss curves continue to decrease but at a slower rate.

We notice that the training and validation curves are relatively close together, indicating that the network is not overfitting or underfitting the data.

In Fig. 5, we can see the performance plot of the trained ANN network, and at 546th iteration, best validation performance was achieved with a mean square error of 0.01263.

Figure 6 shows the regression plot of the trained ANN network. The plot shows the predicted values compared to the actual values for the test set. The model achieved a correlation coefficient ($R$) of 0.95, indicating high accuracy in detecting Islanding.

Figure 7 shows the Error Histogram plot of the trained ANN network. The plot shows the distribution of errors (difference between target and output values) for the test set, with a standard deviation of 0.626. The majority of errors are concentrated around zero, indicating that the model is performing well in classifying the islanding condition.

Figure 8 shows the Confusion matrix plot of the trained ANN network. The plot shows the number of true positives, true negatives, false positives, and false negatives for each class of datasets. The model achieved high accuracy in classifying the islanding situation, with a precision of 0.999 and recall of 0.988, indicating that it has low false positive and false negative rates.

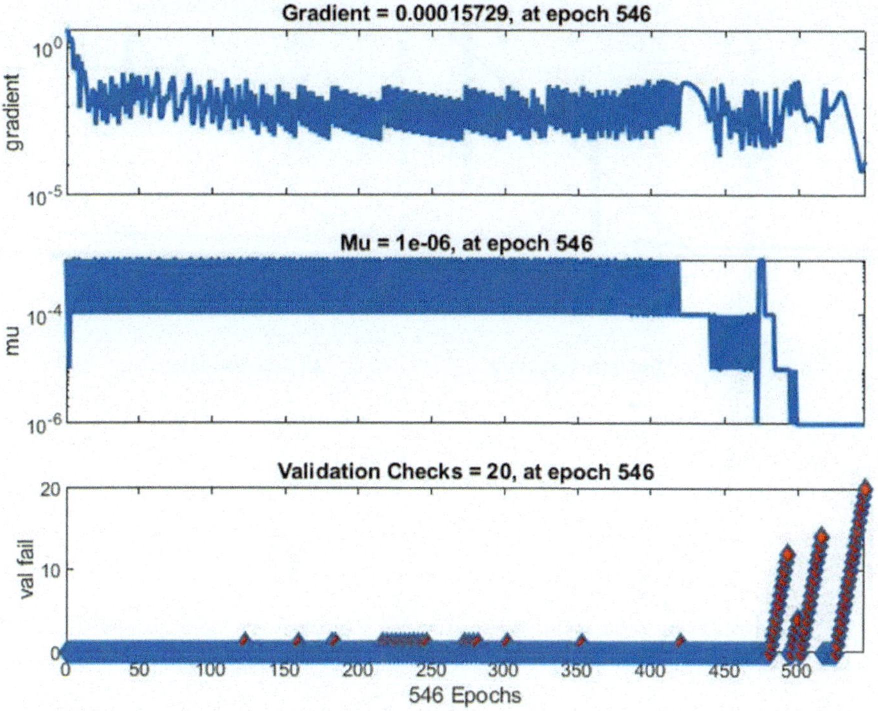


**Fig. 4** Training state plot of ANN network

**Fig. 5** Performance plot of ANN network

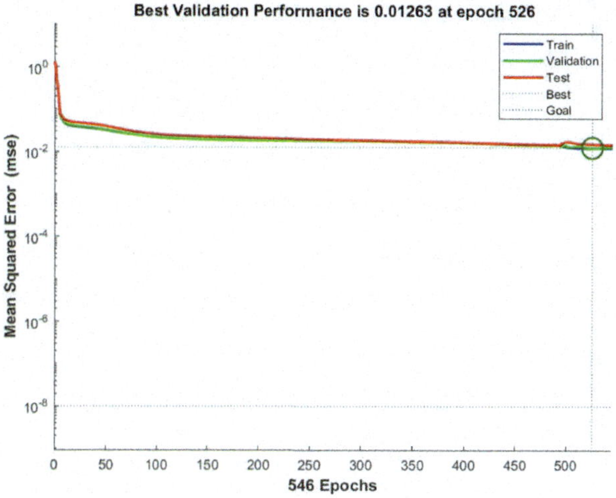

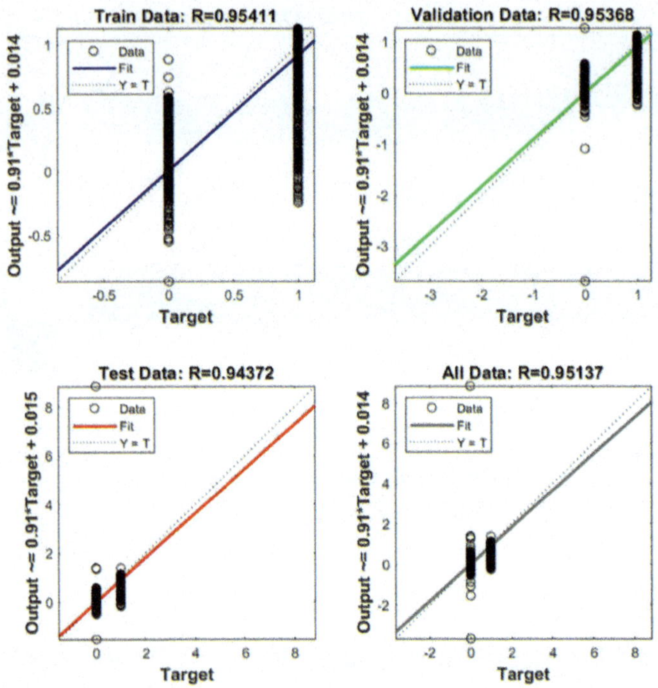


**Fig. 6** Regression plot of ANN network

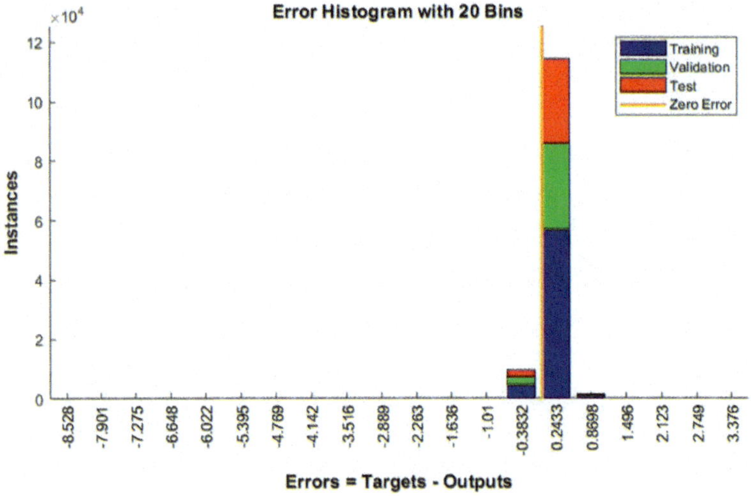


**Fig. 7** Error histogram plot of ANN network

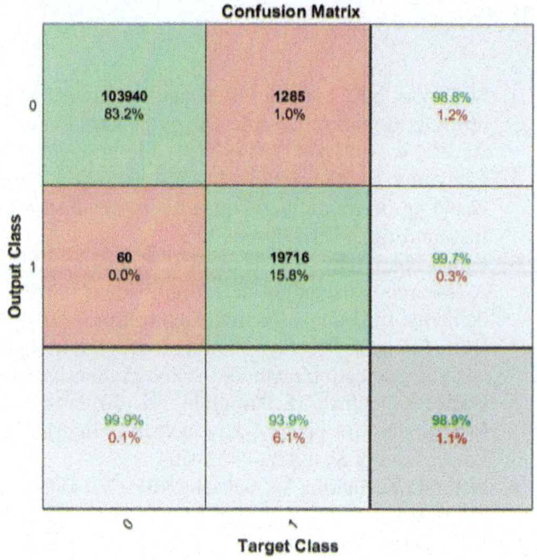


**Fig. 8** Confusion matrix plot of ANN network

The proposed method of Islanding detection using ANN was also tested for a 250 KW & 400 KW of grid-connected PV array systems. The results were found similar and satisfactory.

## 7 Conclusion

The focus of this study is unintentional islanding detection for a 100 kW PV grid-connected system. The results suggest that the ANN performed better than the traditional approaches. The modified negative sequence impedance values and phase angle jump values are provided as the inputs for the ANN. The proposed method of Artificial Neural Network (ANN) detects well an islanding condition with an accuracy of 98.9%. In this proposed technique, no feature extraction nothing is required, so, faster detection for unintentional islanding is done. A quicker and more accurate detection can be accomplished using the suggested technique since it is will leverage the speed of instantaneous data changes and the high accuracy of artificial neural networks (ANNs).

## References

1. Swarnkar, NK, Mahela OP, Khan B, Lalwani M (2021) Identification of islanding events in utility grid with renewable energy penetration using current based passive method. In: IEEE Access, vol. 9, pp. 93781–93794, doi: https://doi.org/10.1109/ACCESS.2021.3092971
2. Panigrahy N, Ilamparithi T, Kashinath MV, Prakash R (2016) Comparison and review of islanding detection techniques for power distribution studies. Int J Adv Res Electr Electron Instrum Eng 5(7):6485–6492
3. Fadzil NA, Hairi MH, Hanaffi F, Kamarudin MN, Isira AS, Packeer Mohamed MF (2019) A research of islanding detection method for distributed generation: mechanism, merits and demerits. Int J Innov Technol Explor Eng 8(12S2) ISSN: 2278–3075
4. Dutta S, Sadhu PK, Jaya Bharata Reddy M, Mohanta DK (2018) Shifting of research trends in islanding detection method—A comprehensive survey. Prot Control Mod Power Syst
5. Dash PK, Padhee M, Panigrahi TK (2012) A hybrid time–frequency approach based fuzzy logic system for power island detection in grid connected distributed generation. Int J Electr Power Energy Syst 42(1):453–464
6. Namrata K, Sharma SP, Seksena SBL (2014) Determining regression constants for calculating global solar radiation at Jharkhand (India) region. In: 2014 International conference on renewable energy research and application (ICRERA), Milwaukee, WI, USA, 2014, pp 795–797, doi: https://doi.org/10.1109/ICRERA.2014.7016494
7. Kumar N, Namrata K, Samadhiya A (2023) Bi-level decision making in techno-economic planning and probabilistic analysis of community-based sector-coupled energy system. Appl Intell 53:6604–6628. https://doi.org/10.1007/s10489-022-03794-9
8. Pal K, Kumar Panigrahi B, Mohapatra S, Mohapatra A (2017) Impact of STATCOM on voltage profile in a DG penetrated grid connected system. In: 2017 International conference on circuit, power and computing technologies (ICCPCT), Kollam, India, pp. 1–5, doi: https://doi.org/10.1109/ICCPCT.2017.8074213
9. Pal K, Akella AK, Namrata K, Pati S (2022) Face detection using artificial neural network and wavelet neural network. In: International conference on intelligent controller and computing for smart power (ICICCSP), Hyderabad, India, 2022, pp. 1–6, doi: https://doi.org/10.1109/ICICCSP53532.2022.9862349
10. Reigosa DD, Briz F, BlancoCharro C, Guerrero JM (2017) Passive islanding detection using inverter nonlinear effects. IEEE Trans Power Electron 32:8434–8445
11. Panigrahi BK, Bhuyan A, Shukla J, Ray PK, Pati S (2021) A comprehensive review on intelligent islanding detection techniques for renewable energy integrated power system. Int J Energy Res 45(10):14085–14116
12. Ahmadipour M, Hizam H, Othman ML, Radzi MAM, Chireh N (2019) A novel islanding detection technique using modified Slantlet transform in multi-distributed generation. Int J Electr Power Energy Syst 112:460–475
13. Bhuyan A, Panigrahi BK, Pal K, Pati S (2022) Convolutional neural network based fault detection for transmission line. In: 2022 international conference on intelligent controller and computing for smart power (ICICCSP), Hyderabad, India, pp. 1–4, doi: https://doi.org/10.1109/ICICCSP53532.2022.9862446
14. Gupta D, Kumari N, Samadhiya A (2020) Photovoltaic modeling using single diode model in MATLAB. In: 2020 IEEE international conference on computing, power and communication technologies (GUCON), Greater Noida, India, 2020, pp. 734–739, doi: https://doi.org/10.1109/GUCON48875.2020.9231165
15. Khamis A, Shareef H, Bizkevelci E, Khatib T (2013) A review of islanding detection techniques for renewable distributed generation systems. Renew Sustain Energy Rev 28:483–493
16. Pal K, Akella AK, Namrata K, Pati S (2022) Classification of fault using artificial neural network and power quality improvement using DVR in a PV integrated hybrid power system. In: 2022 international conference on intelligent controller and computing for smart power (ICICCSP), Hyderabad, India, pp. 1–4, doi: https://doi.org/10.1109/ICICCSP53532.2022.9862499

17. Abokhalil A, Awan A, Al-Qawasmi A-R (2018) Comparative study of passive and active islanding detection methods for PV grid-connected systems. Sustainability 10:1798
18. Pal K, Kumar Panigrahi B, Mohapatra S, Mohapatra A (2017) Impact of STATCOM on voltage profile in a DG penetrated grid connected system. In: 2017 International conference on circuit, power and computing technologies (ICCPCT), Kollam, India, 2017, pp. 1–5, doi: https://doi.org/10.1109/ICCPCT.2017.8074213
19. Prasanna SL, Pal K, Mandal D (2022) Predicting the monthly average incident shortwave solar energy for Hubli, India by using training functions in ANN. In: 2022 international conference on intelligent controller and computing for smart power (ICICCSP), Hyderabad, India, 2022, pp. 1–4, doi: https://doi.org/10.1109/ICICCSP53532.2022.9862510
20. Soumya AV, Belwin Edward J (2021) An intelligent method of anti Islanding detection using ANN. In: 2021 innovations in power and advanced computing technologies (i-PACT), Kuala Lumpur, Malaysia, 2021, pp. 01-06, https://doi.org/10.1109/i-PACT52855.2021.9696825
21. Laghari JA, Mokhlis H, Karimi M, Bakar AHA, Shahriari A (2014) Artificial neural network based islanding detection technique for mini hydro type distributed generation. In: 3rd IET international conference on clean energy and technology (CEAT) 2014, pp 1–6
22. Aljankawey AS, et al. (2010) Passive method-based islanding detection of renewable-based distributed generation: the issues. In: Electric Power Energy Conf. (EPEC), 2010 IEEE 2010, pp 1–8
23. Best R (1993) Phase-locked loops: theory, design, and applications. 2nd ed. McGraw-Hill Inc

# An Investigation of Fault Detection in Electrical Distribution Systems Using Deep Neural Networks

**P. Balamurali Krishna, A. V. Pavan Kumar, and Akkenaguntla Karthik**

**Abstract** The primary goal of the research is to detect and classify defects in electrical distribution networks using deep learning techniques. At a fault situation, fault voltage, fundamental frequency, and current components are considered for fault identification and categorization. An IEEE 33 bus system is used to model distribution network, and fault conditions are created in simulation to obtain fault components. When a fault occurs current and voltage waveforms contain significant high frequency transient signals. Discrete Wavelet Transform (DWT) and Deep learning (DL) approaches are used to detect and classify the fault in distribution system. DWT is applied for decomposition of high frequency transient signals to extract information in both time and frequency domains. Results show that the proposed Deep Neural Network (DNN) model has high accuracy in recognizing and classifying the fault accordingly. The simulation is done through MATLAB software using deep learning.

**Keywords** Electrical distribution system · IEEE 33 bus systems · Discrete wavelet transform and deep learning

## 1 Introduction

The process of identifying and tracking faults or unusual conditions that happen inside the electrical power system is referred to as fault detection in power systems. For electrical systems to remain safe and reliable, fault detection is crucial. Power system operators can more easily locate and segregate faults, as well as take the

P. Balamurali Krishna (✉)
School of Electrical Engineering, KIIT University, Patia, Bhubaneshwar, Odisha, India
e-mail: pbmk20031993@gmail.com

A. V. Pavan Kumar
Department of EEE, Madanapalle Institute of Technology & Science, Madanapalle, AP, India

A. Karthik
PGDM (Energy Management), NTPC School of Business, Noida, UP, India



O. H. Gupta et al. (eds.), *Soft Computing Applications in Modern Power and Energy Systems*, Lecture Notes in Electrical Engineering 1107,
https://doi.org/10.1007/978-981-99-8007-9_22

necessary steps to lessen the effects of the defects on the system. The use of safety relays, circuit breakers, and other monitoring tools, as well as sophisticated software algorithms that examine information collected by sensors and additional sources to detect potential faults, are examples of fault detection approaches. In distribution network, faults can be found in a variety of methods. Few typical approaches are, by using; protective relays, circuit breakers, visual inspection, fault indicators (alarms) and advanced monitoring systems developed using ANN, ML, and Wavelet transformation techniques.

The distribution system in electrical power system is a vital area. Faults occurring in this area should be addressed quickly to have a reliable power supply to end consumers. Currently addressing distribution lines issues and solving the problems using advanced intelligent methods is the key area of research. DL is a start-of-art Machine Learning (ML) technique that provides greater accuracy and performance when compared to traditional ML techniques [1]. The electrical signals are analyzed using neural networks in this study. A methodology for training a neural network model to detect abnormalities in distribution lines with high accuracy and flexibility is required. To understand this issue, it is necessary to understand that distribution line failures are mainly generated by physical occurrences [2]. The effectiveness of a DL model may be improved, by strengthening the model and experimenting with more effective learning algorithms [8]. DNN has been employed for power system fault diagnosis, utilizing Convolutional Sparse Auto Encoders (CSAE) [9, 10] and Stacked Auto Encoders (SAE) [11]. However, they are unsupervised learning techniques, although this study offers a supervised learning-oriented DNN design. DL networks offer a lot of possibilities and are used in a variety of sectors. It is feasible to construct non-obvious interdependence across input & output information by using alternative network configurations. The DL network's, Long Short-Term Memory (LSTM) model, which is utilized for memorizing and processing long sequences of inputs, may therefore be applied to temporal defect diagnosis problem [7].

The design and number of layers, quantity of epoch, and size of a bunch are the fundamental settings of a DL network, as explained in Ref. [8, 9]. The proposed LSTM network design includes a sequence input, hidden units, a completely connected, and a classification layer. In Ref. [9], the author describes how to use the LSTM approach to time-series categorization. The initial stage of the fault detection system's LSTM network is trained on MATLAB-generated fault simulation data. Current and voltage from distribution lines are used as inputs. The main controller collects just the data relating to the fault line after detecting a problem in one of the lines. The fault classification step uses these current and voltage data as inputs. This stage, which includes the LSTM network, provides the defect type. Training a neural network [10] necessitates a vast number of instances. Through such a training step, neural networks start recognizing a certain behavior among a group of instances. Whenever a new sample is introduced, the network output tries to tune more, getting closer to the true output. This necessitates a significant number of samples which will enable the network generalization to the desired pattern.

The identification and categorization of faults in power grids [11] and their localization are regarded as critical prerequisites for implementing the smart grid paradigm

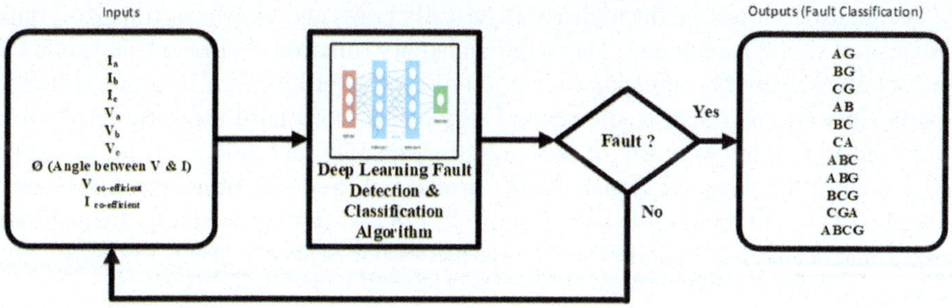


**Fig. 1** Proposed deep learning methodology

in power grids. However, adding self-healing systems to reduce the consequences of interruptions is a plausible approach [12]. In any event, the employment of that kind of sort of technique indicates that the systems are tightly meshed and capable of isolating a problematic portion till the issue has been solved without requiring the entire infrastructure to be down.

We uncovered several relevant materials that seem to have unique importance for the content accomplished in this paper by concentrating on neural networks. Various investigations successfully employed neural networks and machine learning models to detect and classify faults in power systems [13–18]. Whereas Ref. [19, 20] utilizes the DWT, S-Transform (ST), pattern recognition, Fast Fourier Transform (FFT), image processing, and statistical approaches for fault identification and categorization in distribution lines and multi-microgrids. Some research has also been done on defect detection, with the goal of extrapolating voltage measurements from many modeled power systems. These parameters were employed to train a neural network to identify problems in a practical distribution network [21]. Niche Binary Particle Swarm Optimization (NBPSO) technique is employed in [22] to improve power quality by identifying harmonic sources.

The preferred technique in this research is depicted in Fig. 1. Inputs are three-phase voltages, three-phase currents, voltage and current coefficients, and phase angle between voltage and current. These inputs are loaded into the built deep learning model for the identification and classification of faults, which continually checks the input state.

## 2 Discrete Wavelet Transform

A discrete signal is converted to a discrete wavelet form via DWT, which uses time-scale representation, which allows for efficient multi-resolution. Because of WT's success in transient analysis, it has been integrated with other approaches to improve the reliability of protective mechanisms.

A signal $x(n)$ is passed through a high pass filter $a(n)$ and a low pass filter $d(n)$ until the desired signal is obtained. The detail and approximation coefficients at levels D1 and A1 are acquired by splitting the output of each filter by two. The approximation coefficients are later sent to the second stage, and so on until the sixth stage. It is proposed to employ six tiers in this investigation (D6 and A6). F is the sampling frequency of the sampled signal. As illustrated in Fig. 2, the frequency signal data seized by D1 is in the middle of F/4 and F/2, F/8 and F/4 are seized by D2, and D6 seizes data between F/64 and F/32. The frequency bands at each DWT decomposition level are shown in Table 1.

Using a trial-and-error procedure, the researchers discovered that the db6 wavelet provides the greatest classification accuracies for this purpose. As a consequence, the db6 wavelet was selected since it provides the best fault recognition and categorization results. The DWT decomposes voltage and current signals into various frequency bands, allowing for accurate measurement of pre- and post-fault timings.

Figure 3 depicts the fault voltage signal of a multi-resolution DWT. The 6-stage decomposition employs the Db6 wavelet. At every frequency group, the reconstructed

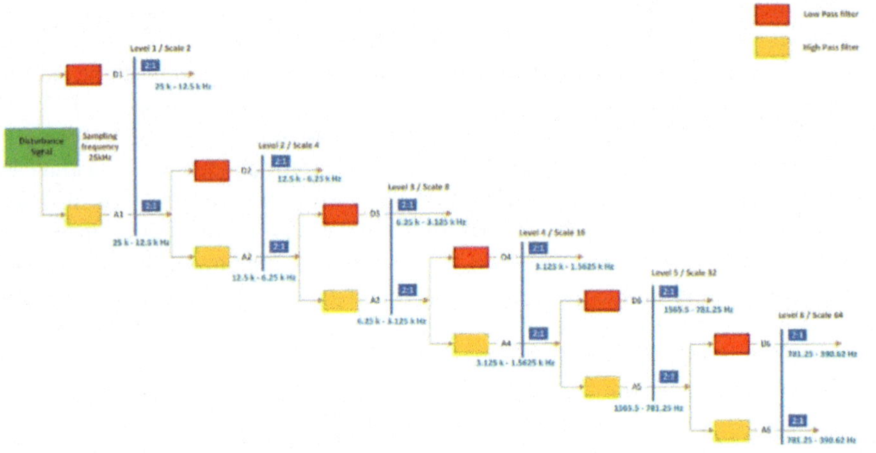


**Fig. 2** Six-level wavelet decomposition tree

**Table 1** DWT decomposition level frequency bands at each decomposition level

| Decomposition Level | Frequency bands (Hz) |
|---|---|
| 1 | 25 k–12.5 k |
| 2 | 12.5 k–6.25 k |
| 3 | 6.25 k–3.125 k |
| 4 | 3.125 k–1.5625 k |
| 5 | 1565.5–781.25 k |
| 6 | 781.25–390.62 |

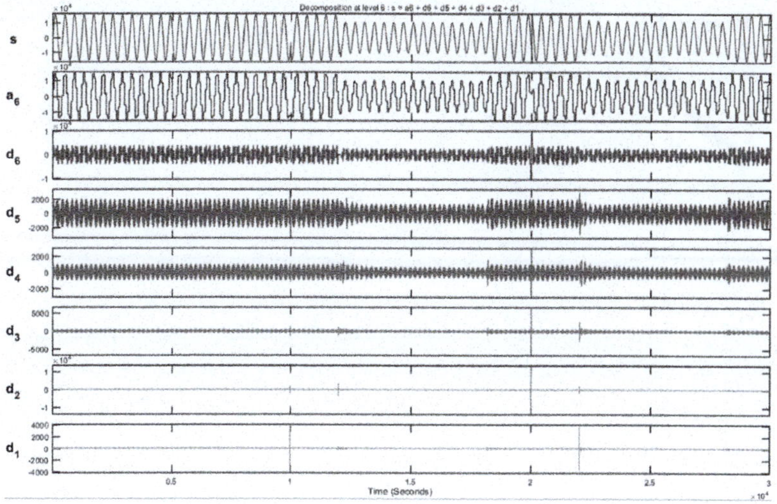


**Fig. 3** Voltage signal DWT analysis

forms of every element, also the approximation and recording of distinctive signals, are precisely defined.

Leveraging regulatory approaches, a Phase-Locked Loop (PLL), provides an output signal with the same phase as the input signal. Figure 4 shows an example of a block diagram. The input and output frequencies ought to be constant to maintain the phases synchronized. $V_{\text{abc}}$, $V_d$, and $V_q$ are calculated employing the input voltage. $V_d$ and $V_q$ are voltage levels in the *d*- and *q*-axes at the Point of Common Coupling (PCC). Figure 4 depicts the suggested PLL design.

In contrast to the diagram depicted in Fig. 2, the fault signal $e(t)$ corresponds to the waveform variation, whilst the Proportional-Integral (PI) regulator and Integrator (I) correspond to the loop filter and Voltage Controlled Oscillator (VCO), respectively. Clarke's transformation is used to convert the 3 phase voltages ($V_a$, $V_b$, and $V_c$) that are delivered into PLL to quantities. Figure 5 is a schematic of a PQ-PLL block. The values are then converted into a frame of reference. An error signal is generated based on the closeness of the values to the reference frame. A PI regulator receives an error signal and regulates the error to zero.

## 3 Deep Learning

The term "deep" in DL indicates to the use of several of the network's layers. DL neural networks try to emulate the human brain by combining various inputs, weights, and biases. DNN based on supervised learning has a significant benefit. These Elements collaborate to correctly detect, categorize, and characterize objects inside the Data. Deep learning's capacity to tackle complex situations that require

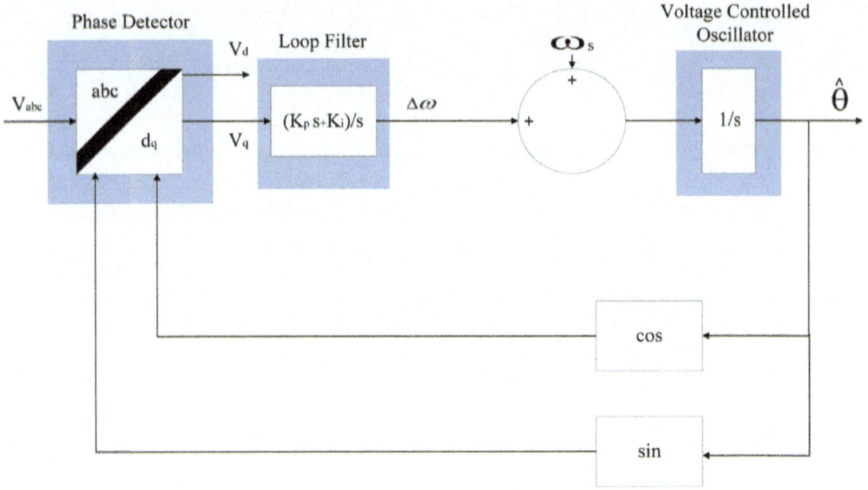


**Fig. 4** PLL block diagram

**Fig. 5** PQ-PLL block diagram

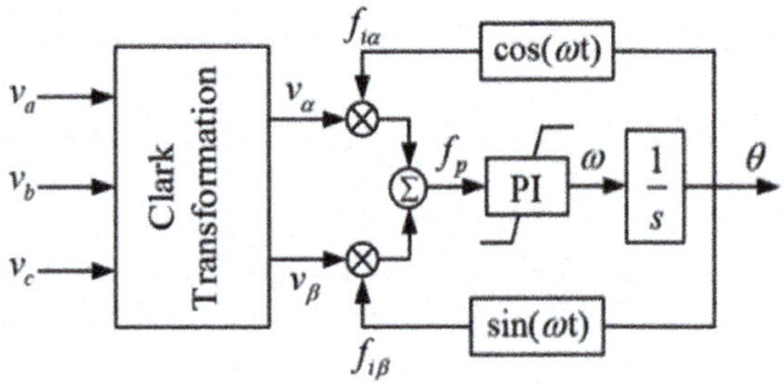


the finding of hidden patterns in data and/or a complete understanding of intricate connections between numerous interdependent variables is one of its key advantages.

DNNs are employed in the identification of distribution system faults due to their efficiency in analyzing massive volumes of data and extracting significant trends and characteristics that may not be immediately apparent to humans. The complicated correlations between numerous parameters and fault states in the distribution network can be learned by DNNs utilizing labeled datasets containing previous fault data. These networks are capable of being utilized to categorize newer data and identify errors in actual time once they have been trained. DNNs can also analyze and incorporate vast volumes of heterogeneous data, including data from sensors, safety relays, and defect indications, to give a deeper comprehension of the functioning of the system.

DNNs are effective for fault detection in distribution networks due to their ability to analyze large amounts of data and extract meaningful patterns. However, there are potential disadvantages to consider. DNNs can be computationally expensive, require large amounts of high-quality labeled data for training, and can be difficult to interpret. Additionally, DNNs can be susceptible to overfitting and lack robustness in response to changes in the distribution network. Careful design, training, and implementation are necessary to ensure their effectiveness.

Dynamic learning algorithms are built to traverse various complexity of neural networks, each of which is basically a collection of decision-making networks that have been pre-trained to perform a specific task. Deep learning algorithms are divided into the following categories:

### *3.1 Generative Adversarial Networks (GANs)*

GANs are employed to build new and innovative data instances that fit the training examples. GANs are usually composed of two components: a generator that learns to create false data and a discriminator that regulates by analyzing this false data. GANs growth in acceptance throughout time as they are regularly employed to enhance astronomical pictures and imitate lensing the gravitational dark matter. GANs operate in simulation by creating and comprehending both false and real data. During the training to grasp such data, the generator generates various types of falsified information, to which the discriminator swiftly adapts and responds as inaccurate facts. The identified findings are then sent to GANs for updating.

### *3.2 Convolutional Neural Network (CNN)*

CNN is made up of numerous layers that are utilized for image analysis and object recognition. CNNs are widely used in satellite image recognition, series forecasting, medical image processing, and anomaly detection. CNNs analyze input data by feeding it via multiple levels and extract information in order to perform convolutional computations. The Convolutional Layer is made up of Rectified Linear Units (ReLU), which are used to correct the feature map. The Pooling layer is used to correct these feature maps before they are sent to subsequent feed.

### *3.3 Recurrent Neural Network (RNN)*

RNN is made up of certain directional interconnections, which establish a loop that allows the input out from Long Short-Term Memory Network to be used as input in the current mode of RNNs. Such inputs are strongly embedded as inputs, and the memorizing ability of LSTMs allows these inputs to be taken for a length of time in the internal storage. RNNs are thus dependent on the inputs maintained by LSTMs and operate under the synchronization phenomena of LSTMs.

### 3.4 Long Short-Term Memory Network (LSTM)

LSTM is RNN that is configured to adapt and learn to dependencies over time. It can retain and remember prior material for a longer amount of time, and this is its standard habit. Because LSTMs are meant to retain information across time, they are commonly utilized in time series forecasts since they can restrict memory or past inputs. LSTM operates in a series of actions. For starters, individuals don't recall unimportant facts gained in the prior condition. They then selectively update specific cell-state variables and eventually create elements of the cell-state as output.

### 3.5 Deep Belief Networks (DBNs)

DBNs are called generative models as they contain several levels of latent and stochastic variables. Because the latent construct has binary values, it is referred to as a hidden layer. Greedy algorithms power DBN. The layer-to-layer technique, which uses a top-down approach to produce values, represents the most frequent way DBN functions. DBNs employ a step-by-step Gibbs sampling technique on the top concealed two-layer. Then, by using ancestral sample technique, these phases generate a sample out from visible units. DBN learns from the values included in the latent value from each layer using the bottom-up pass technique.

In this student, LSTM is considered for the development of algorithm. The SIMULINK waveforms were converted to a MATLAB file for feature extraction. The signal was analyzed using DWT to provide the coefficients for fault detection.

## 4 System Modeling

### 4.1 IEEE 33 Bus System

The developed deep learning model is tested on IEEE 33 radial distribution system. The system's base MVA rating and base voltage are 100MVA and 25 kV, respectively. Figure 6 depicts a IEEE 33 bus system.

To analyze the accuracy of the developed deep learning model, a fault is created. Figure 7 depicts the three-phase voltage waveform during an LG fault occurrence (between time 0.4 s and 0.6 s).

This LG fault signal is passed through high-pass and low-pass filters employing DWT. Figure 8 portrays the LG fault voltage signal of a multi-resolution DWT. A six-stage decomposition is created using the Db6 wavelet. The reconstructed variants of every single element, also the approximation and recordings of the individual signal, are evidently displayed at each frequency spectrum.

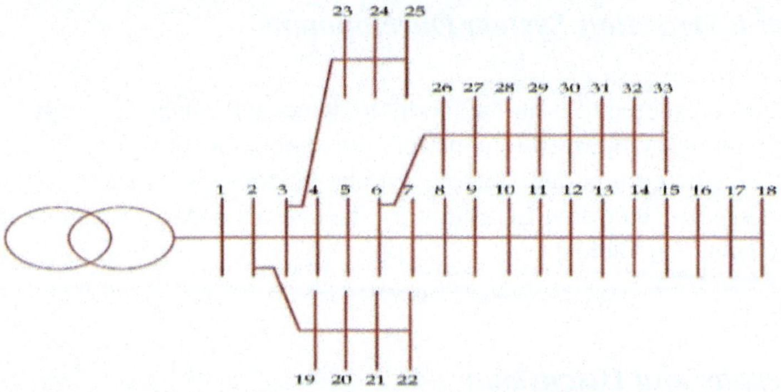


**Fig. 6** IEEE 33 bus system

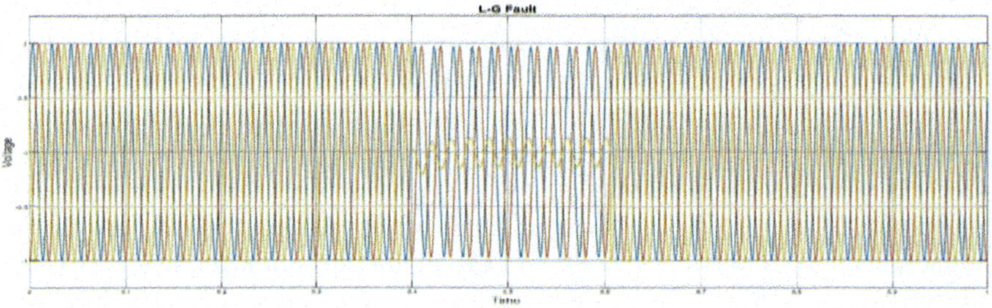


**Fig. 7** LG fault voltage waveform

**Fig. 8** DWT analysis of LG fault voltage signal

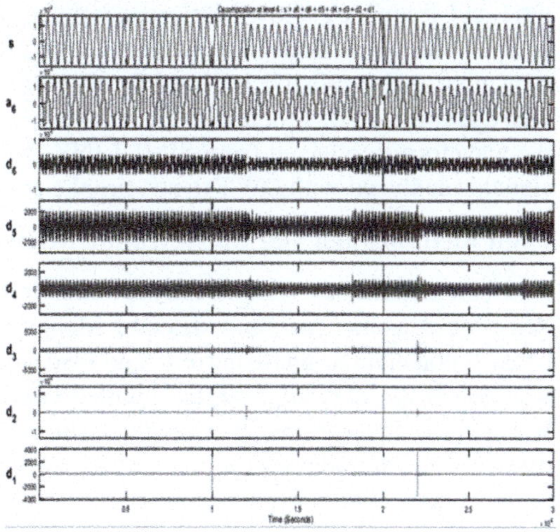

## 4.2 Fault Detection System Development

A dataset is considered to train the model to detect and categorize faults. The IEEE 33 bus system was employed to construct this data collection. The module accepts RMS voltage and current data, feature extraction from DWT at db6 frequency, and load angle determined from PLL as inputs. The data needed to train the deep learning model are shown in Table 2.

## 4.3 Results and Discussion

**Deep Learning Model Development**

A model for detecting the fault and classifying it on distribution lines is developed so that protection schemes can operate quickly and reliably. The simulation code is written in MATLAB, and it generates the fundamental component of the transient voltage and current in the time and frequency domains at the same time. For analysis, one cycle of waveform is abstracted, covering pre- and post-fault information. For data preprocessing, DWT is considered. Different operating and fault conditions on the lines, such as LG fault, LL fault, and LLG fault, are simulated using MATLAB software.

The use of wavelet transforms to estimate the location of a defect on a distribution line has been studied. Wavelets' capacity to divide a signal into frequency bands in time and frequency allows for precise defect identification. The most appropriate wavelet family has been identified for use in estimating the location of a defect on a transmission line. SIMULINK MATLAB SOFTWARE is used to simulate a LG fault on a transmission line for a 735 kV, 300 km transmission line. For feature extraction, the waveforms collected from SIMULINK were converted to a MATLAB file. The signal is analyzed with DWT to acquire the coefficients needed to estimate the fault.

**Fault Identification and Classification Methodology and Algorithm**

To locate a fault, bus systems are separated into four zones and equipped with fault detection sensors. The fault detector identifies the fault by examining the three-phase voltage, three-phase current, and load angle. The observed voltage and current waveforms are derived, and the detailed coefficients are computed using discrete wavelet transformations. If a defect occurs, the suggested deep learning method inspects the parameters zone-by-zone to determine the kind of fault. The operation of the created model is represented in Fig. 9 using flowchart, whereas Fig. 10 depicts the DNN algorithm flowchart.

**Deep Learning Targets for Fault Identification Classification**

The output targets consist of five outputs: LG, LL, LLL, LLG, and LLLG as power system short circuit faults.

**Table 2** DWT input samples

| $V_A$ | $V_B$ | $V_C$ | $I_A$ | $I_B$ | $I_C$ | $V_{\text{co-eff}}$ | $I_{\text{co-eff}}$ | Theta |
|---|---|---|---|---|---|---|---|---|
| 11,560.47 | 29,192.64 | 17,632.17 | 61.16247 | 2.635205 | 88.30267 | −747.831 | 1.988694 | −1.65E−04 |
| 4585.183 | 21,440.69 | 16,855.5 | 97.93996 | 18.24392 | 97.69747 | −1093.11 | 1.818315 | 3.20E−04 |
| 12,337.13 | 22,217.35 | 9880.215 | 100.3327 | 9.637287 | 42.91855 | −842.733 | 1.607185 | −2.48E−04 |
| 1.461998 | 23,023.39 | 23,024.85 | 46.83802 | 46.92917 | 93.76756 | −586.189 | 2.010722 | −0.0011 |
| 2259.38 | 0.140183 | 2259.52 | 101.8693 | 50.80036 | 51.06941 | −1053.46 | −1.34191 | −4.38E−04 |
| 20,763.87 | 20,765.47 | 1.60218 | 4.140239 | 8.101786 | 3.962335 | 0.021369 | 2.347871 | −8.43E−04 |
| 1.461998 | 0.140183 | 1.60218 | 0.028919 | 0.120065 | 0.148985 | 0.021369 | 0.000392 | 0.0459 |
| 1.461998 | 20,824.84 | 20,826.3 | 30.24047 | 30.33161 | 92.53995 | −492.469 | 1.873403 | 0.0066 |

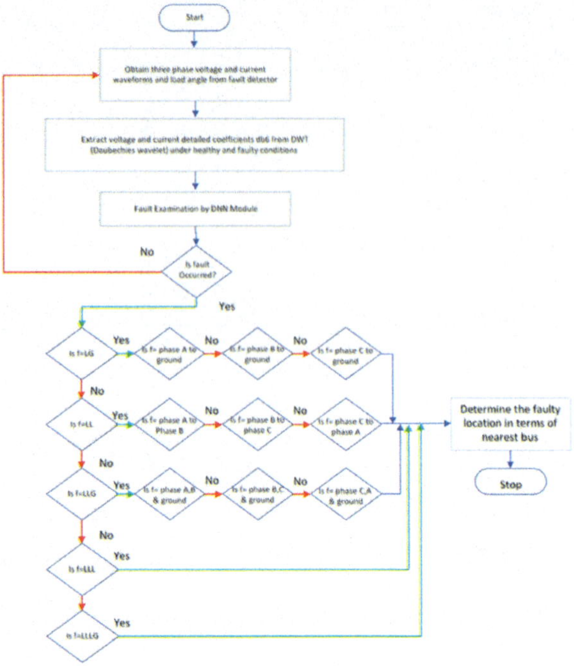


**Fig. 9** Flowchart of fault identification and classification using DNN

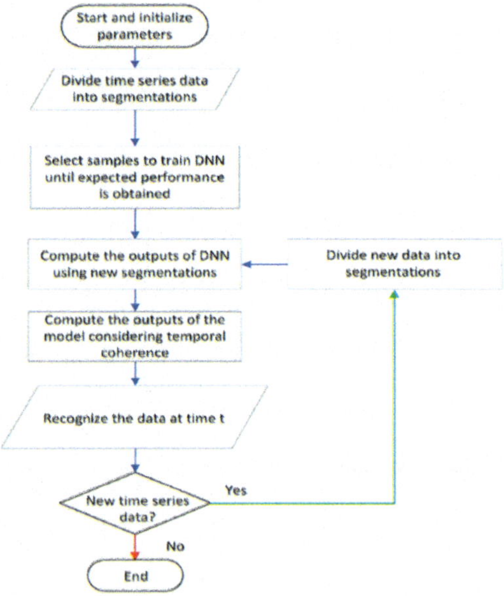


**Fig. 10** DNN-Based Fault Detection Algorithm Flowchart

### Fault Detection and Classification Algorithm

LSTM algorithm is built and parameters related to the model development are provided accordingly. The developed LSTM algorithm code using MATLAB is shown in Fig. 11. To develop the model, a training data set is created by creating various faults and saving the data in.csv format. This dataset is being loaded to train the model, and the model's effectiveness is being evaluated by evaluating it. The testing dataset is different from the training dataset. The training and testing code written in MATLAB is illustrated in Fig. 12.

Making use of the formula. The data that have been imported are put to the test. We can forecast the new data fault type using this method based on existing data. The outputs are depicted in Fig. 13 and Fig. 14.

Minimal Root Mean Square Error (RSME) value is achieved by increasing the iterations. As shown in Fig. 13, at 1000th iteration, RMSE value is 0.41. As illustrated in Fig. 14, the developed model is tested with different fault conditions. The model is able to accurately identify the fault and it also classified the type of fault that occurred. Let's consider LG fault condition, in order to have LG fault data, a LG fault is created in the Simulink environment. When LG fault occurs, one phase will come in contact with the ground resulting in high current and low voltage, and other two phases are

```
inputSize=9;

numResponses=1;

numHiddenUnits=100;

layers=[ sequenceInputLayer(inputSize)

lstmLayer(numHiddenUnits)

fullyConnectedLayer(numResponses)
regressionLayer ];

opts=trainingOptions('adam',...
    'MaxEpochs',1000,...
    'GradientThreshold',0.01,...
    'InitialLearnRate',0.0001);

net=trainNetwork(XTrain,YTrain,layers,opts);
```

**Fig. 11** Defining and creating LSTM algorithm

```
%% Training

XTrain=xlsread('classData.csv',1,'A1:J7862')'
;
YTrain=xlsread('classData.csv',1,'L1:L7862')';


%%Testing

XTest=xlsread('classData.csv',1,'A1:J7862')';

YTest=xlsread('classData.csv',1,'L1:L7862')';
```

**Fig. 12** Training and testing in developed deep learning model

Training on single CPU.

| Epoch | Iteration | Time Elapsed (hh:mm:ss) | Mini-batch RMSE | Mini-batch Loss | Base Learning Rate |
|---|---|---|---|---|---|
| 1 | 1 | 00:00:04 | 4.95 | 12.3 | 1.0000e-04 |
| 50 | 50 | 00:02:38 | 4.13 | 8.5 | 1.0000e-04 |
| 100 | 100 | 00:05:00 | 3.27 | 5.3 | 1.0000e-04 |
| 150 | 150 | 00:07:37 | 2.47 | 3.1 | 1.0000e-04 |
| 200 | 200 | 00:10:14 | 1.77 | 1.6 | 1.0000e-04 |
| 250 | 250 | 00:12:49 | 1.20 | 0.7 | 1.0000e-04 |
| 300 | 300 | 00:15:20 | 0.81 | 0.3 | 1.0000e-04 |
| 350 | 350 | 00:17:58 | 0.63 | 0.2 | 1.0000e-04 |
| 400 | 400 | 00:20:39 | 0.54 | 0.1 | 1.0000e-04 |
| 450 | 450 | 00:23:13 | 0.50 | 0.1 | 1.0000e-04 |
| 500 | 500 | 00:25:46 | 0.48 | 0.1 | 1.0000e-04 |
| 550 | 550 | 00:28:23 | 0.47 | 0.1 | 1.0000e-04 |
| 600 | 600 | 00:31:02 | 0.46 | 0.1 | 1.0000e-04 |
| 650 | 650 | 00:33:46 | 0.46 | 0.1 | 1.0000e-04 |
| 700 | 700 | 00:36:33 | 0.44 | 9.9e-02 | 1.0000e-04 |
| 750 | 750 | 00:39:11 | 0.44 | 9.7e-02 | 1.0000e-04 |
| 800 | 800 | 00:41:46 | 0.43 | 9.4e-02 | 1.0000e-04 |
| 850 | 850 | 00:44:12 | 0.43 | 9.2e-02 | 1.0000e-04 |
| 900 | 900 | 00:46:41 | 0.42 | 8.9e-02 | 1.0000e-04 |
| 950 | 950 | 00:49:19 | 0.42 | 8.6e-02 | 1.0000e-04 |
| 1000 | 1000 | 00:52:00 | 0.41 | 8.3e-02 | 1.0000e-04 |

**Fig. 13** Output tested results

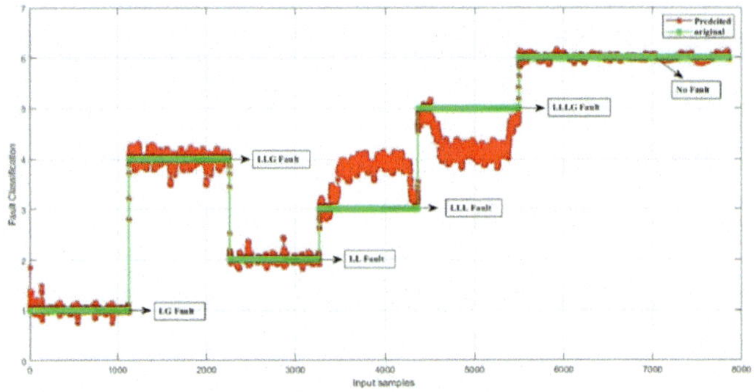


**Fig. 14** Fault identification and classification visual representation

open-circuited resulting in high voltage and low current, and these data of all three phases (Voltage and Current), $V_{\text{co-eff}}$, $I_{\text{co-eff}}$, and theta are captured in order to train and test the proposed model. More the data, the better the model. Similarly, different fault conditions like LLG, LL, LLL, and LLLG faults are created, and their values are used to train and test the model for fault identification and classification. The predicted fault and original fault are shown in Fig. 14. When there is no fault, the model identified that there is no fault, and this is possible because model is trained and tested with data when there is no fault. In a 33-bus system, possible samples that can be considered are 8000. Using these 8000 samples, the model is trained with 80% of the data, and remaining 20% data are used to test the performance of the model.

## 5 Future Scope

A promising field of research with many potential directions for advancement is the application of deep neural networks for fault detection. By enhancing the neural network design, utilizing more training data, and creating more complex fault detection algorithms, the accuracy of identifying defects can be increased. To find issues in electrical distribution networks as they happen, continuous monitoring systems can be created. To produce smarter and adaptive electrical distribution systems, DNNs have capable of being integrated with additional innovative technologies, such as smart grids and Internet of Things (IoT) devices. DNN can be integrated with service management systems to make sure that defects are found and fixed quickly, eliminating downtime and lowering repair costs.

## 6 Conclusion

This study dives deeper into the use of DL for fault recognition and characterization in a distribution network. For fault identification and classification, the created model used regularized values of instantaneous current and voltage as inputs. A waveform-built fault categorization methodology for electrical distribution networks was provided in this study. The vast majority of the wave data processing exercise was managed efficiently by disseminating the computational stress appropriately. DWT is used to extract features from recorded wave data. In compared to other traditional relaying techniques, the fault classifier and detector findings show that deep learning is effective at recognizing and detecting fault type in a reasonable timeframe.

## References

1. Xin Y et al (2018) Machine learning and deep learning methods for cybersecurity. IEEE Access 6:35365–35381. https://doi.org/10.1109/ACCESS.2018.2836950
2. Das S, Santoso S, Navalpakkam Ananthan S (2022). Fault location on transmission and distribution lines: principles and applications. https://doi.org/10.1002/9781119121480

3. Du X, Cai Y, Wang S, Zhang L (2016) Overview of deep learning. In: 2016 31st youth academic annual conference of Chinese association of automation (YAC), Wuhan, 2016, pp. 159–164
4. Chen K, Hu J, He J (2018) Detection and classification of transmission line faults based on unsupervised feature learning and convolutional sparse autoencoder. IEEE Trans Smart Grid 9(3):1748–1758. https://doi.org/10.1109/TSG.2016.2598881
5. Wang Y, Liu M, Bao Z (2016) Deep learning neural network for power system fault diagnosis. In: 2016 35th Chinese Control Conference (CCC), Chengdu, pp 6678–6683
6. Deng L, et al. (2013) Recent advances in deep learning for speech research at Microsoft. In: 2013 IEEE international conference on acoustics, speech and signal processing, pp 8604–8608, doi: https://doi.org/10.1109/ICASSP.2013.6639345
7. Siami-Namini S, Tavakoli N, Namin AS (2019) The performance of LSTM and BiLSTM in forecasting time series. IEEE Int Conf Big Data (Big Data) 2019:3285–3292. https://doi.org/10.1109/BigData47090.2019.9005997
8. Brownlee J (2019) How to tune LSTM hyperparameters with keras for time series forecasting. Machine Learning Mastery
9. Adam K, Smagulova K, James AP (2018) Memristive LSTM network hardware architecture for time-series predictive modeling problems. IEEE Asia Pac Conf Circuits Syst (APCCAS) 2018:459–462. https://doi.org/10.1109/APCCAS.2018.8605649
10. Huang S-H, Le T-H (2021) Chapter 5 – training neural network. In: Huang S-C, Le T-H (eds.) Principles and labs for deep learning. Academic Press, pp. 117–146, ISBN 9780323901987, https://doi.org/10.1016/B978-0-323-90198-7.00007-0
11. Ungrad H, Winkler W, Wiszniewski A (1995) Protection techniques in electrical energy systems (1st ed.). CRC Press. https://doi.org/10.1201/9781003067504
12. Shen Y, Chen Y, Zhang J, Sang Z, Zhou Q (2019) Self-healing evaluation of smart distribution network based on uncertainty theory. IEEE Access 7:140022–140029. https://doi.org/10.1109/ACCESS.2019.2939537
13. Sharma P, Saini D, Saxena A (2016) Fault detection and classification in transmission line using wavelet transform and ANN. Bull Electr Eng Inform 5. https://doi.org/10.11591/eei.v5i3.537
14. Mnyanghwalo D, Kundaeli H, Kalinga E, Hamisi N, Lam J (Reviewing editor) (2020) Deep learning approaches for fault detection and classifications in the electrical secondary distribution network: Methods comparison and recurrent neural network accuracy comparison. Cogent Eng 7:1, DOI: https://doi.org/10.1080/23311916.2020.1857500
15. Yu JJQ, Hou Y, Lam AYS, Li VOK (2019) Intelligent fault detection scheme for microgrids with wavelet-based deep neural networks. IEEE Trans Smart Grid 10(2):1694–1703. https://doi.org/10.1109/TSG.2017.2776310
16. Baskar D, Selvam P (2020) Machine learning framework for power system fault detection and classification. Int J Sci Technol Res 9:2
17. Rudin F, Li G, Wang K (2012) An algorithm for power system fault analysis based on convolutional deep learning neural networks
18. Ponukumati BK, Sinha P, Maharana MK, Kumar AVP, Karthik A (2022) An intelligent fault detection and classification scheme for distribution lines using machine learning. Eng Technol Appl Sci Res 12(4):8972–8977. https://doi.org/10.48084/etasr.5107
19. Ponukumati BK, Sinha P, Maharana MK, Jenab C, Pavan Kumar AV, Akkenaguntla K (2021) Pattern recognition technique based fault detection in multi-microgrid. In: 2021 IEEE 2nd international conference on applied electromagnetics, signal processing, & communication (AESPC), 2021, pp. 1–6, doi: https://doi.org/10.1109/AESPC52704.2021.9708541
20. Sinha BKPP, Maharana MK, Jena C, Pavan Kumar AV, Akkenaguntla K (2021) Power system fault detection using image processing and pattern recognition. In: 2021 IEEE 2nd international conference on applied electromagnetics, signal processing, & communication (AESPC), 2021, pp. 1–5, doi: https://doi.org/10.1109/AESPC52704.2021.9708475

21. Sapountzoglou N, Lago J, De Schutter B, Raison B (2020) A generalizable and sensor-independent deep learning method for fault detection and location in low-voltage distribution grids. Appl Energy 276:115299. https://doi.org/10.1016/j.apenergy.2020.115299
22. Balamurali Krishna P, Sinha P (2018) Detection of power system harmonics using NBPSO based optimally placed harmonic measurement analyser units. In: 2018 second international conference on computing methodologies and communication (ICCMC), Erode, India, 2018, pp. 369–373, doi: https://doi.org/10.1109/ICCMC.2018.8488114

# Solar Radiation Prediction Using Regression Methods

**Saurabh Tikariha and Yash Pal**

**Abstract** Solar irradiation forecasting is becoming an important technique in load demand management due to recent developments in modern society and the economy. Accurate prediction of produced solar power can aid in estimating the size of the system, load measurement of the system, and calculating return on investment (ROI). For estimating solar irradiance at any geographical location, various methods can be employed such as averaging solar irradiance. For predicting irradiance, different regression techniques have been applied. This paper performed a comparison between the regression methods such as Multiple Linear Regression (MLR), Random Forest Regression (RFR), and Gradient Boosting Regression (GBR) methods and examines the efficient method for forecasting solar irradiance.

**Keywords** Multiple linear Regression (MLR) · Random Forest Regression (RFR) · Gradient Boosting Regression (GBR) · Solar irradiation

## 1 Introduction

According to the Central Electricity Authority of India (CEA), the total installed capacity of power generation is 407.8GW, and 29.43% of energy comes from renewable energy sources (RES such as Wind, Solar, Biomass, and small Hydro). As of September 2022, solar power generation is 60.81GW, i.e., 37% of total RES. Forecasting the output power of the solar system is required for the good operation of the power grid.

When demand is at its highest in the power-generating sector, the renewable energy production of renewable energy takes precedence and also lowers power

S. Tikariha (✉) · Y. Pal
National Institute of Technology, Kurukshetra, Haryana, India
e-mail: saurabhtikariha30@gmail.com

Y. Pal
e-mail: yashpal@nitkkr.ac.in



O. H. Gupta et al. (eds.), *Soft Computing Applications in Modern Power and Energy Systems*, Lecture Notes in Electrical Engineering 1107,
https://doi.org/10.1007/978-981-99-8007-9_23

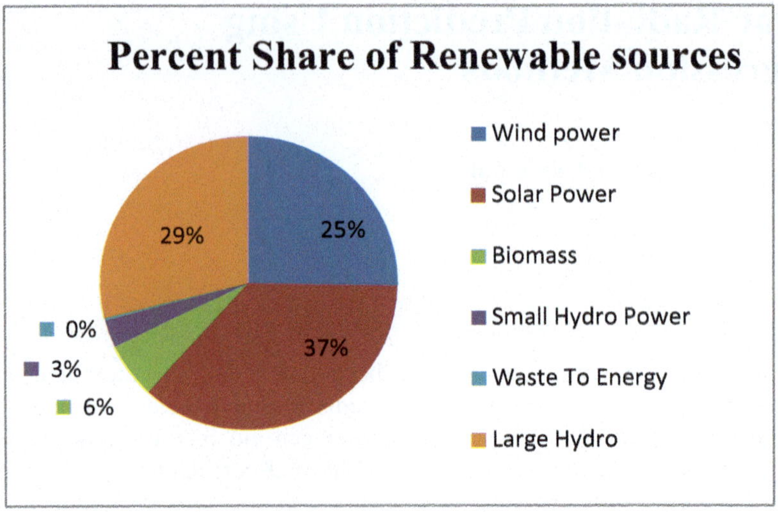


**Fig. 1** Percent share of renewable sources

quality problems. This makes the power supply more reliable. Solar energy is particularly essential compared to other renewable energy sources because it requires less maintenance and provides continuous electricity during bright days. But because of several technical challenges and the high cost of monitoring methods, solar energy has significant complications with solar data projection. For adequate input and output, the availability of solar radiation is primarily dependent on geographical shapes. The radiation forecasting data are utilized to estimate the generation capability of a PV plant at a particular location, such that demand management can be done based on generation capability (Fig. 1).

The time span for forecasting in the current need (short-term forecasting) is 1–30 days. Intermediate-range (mid-term forecasting) is 1–12 months and lastly, long-range (long-term forecasting) is more than a year [1]. Forecasting is planning and controlling an operation that requires an estimate of the demand for the product or the service that an organization expects to provide in the future. Independent traditional methods for predicting solar irradiation and power can be somewhat accurate in predicting solar energy output, their effectiveness may be constrained by a variety of factors, including seasonal fluctuations, weather changes, and cloud cover. As a result, to increase the precision and dependability of solar energy projections, hybrid forecasting systems that may incorporate various data sources and modeling methodologies are needed.

Numerous researches on solar irradiation forecasting have been published. In Choi and Hur [2], the Bagging model is used to predict solar energy output where an ensemble model as a base such as random forest, XGBoost, and LGBM. In Srivastava et al. [3], solar radiation forecasting for the period of 1 to 6-day-ahead using M5, Classification and Regression tree (CART) and Random forest models. Outcomes

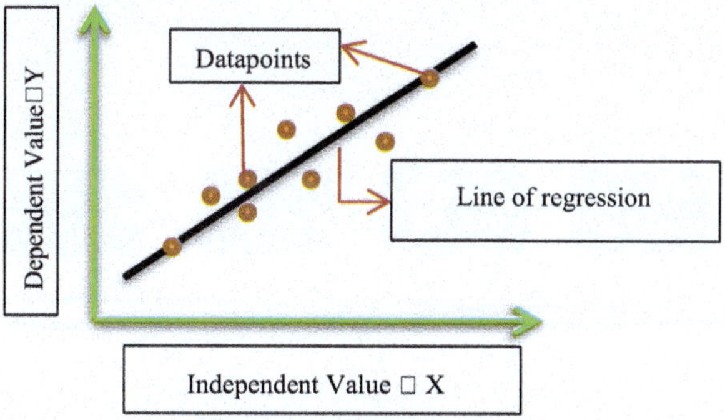


**Fig. 2** Graph of multilinear regression

from all four models were determined and the random forest model produced better outcomes than the CART model. In Bacher et al. [4], in order to get the best prediction results, this study defines weather forecasting and models such as recurrent neural networks, artificial neural networks, and dynamic neural networks. Performance is evaluated using root mean squared error, correlation, and mean absolute error. In Naik et al. [5], three ML regression variants, including random forest, gradient boosting, and adaptive boosting for training, have been used in this research. The network can be tested for prediction short-term forecasts of irradiation. It is determined after prolonged testing and training that RFR produces superior results to the others. In Lukhyswara et al. [6], the solar irradiation forecasting and modeling using a time series model approach to training dataset is done. Multiple linear regression, hybrid fast Fourier transform-autoregressive (FFTAR), and SVR are the techniques used for predicting solar irradiation. Every approach considered is subjected to parameter optimization in an effort to identify the model with the highest level of forecasting accuracy.

In this paper, different regression algorithms are compared for a particular dataset and a comparison has been made based on their R-squared, RMSE, MSE, and MAE scores, the best suitable algorithm is found, which can be used to predict solar radiation forecasting, which can give optimum result (Fig. 2).

## 2 Regression Methods

Finding the relation between a dependent and an independent variable is the main concern of regression analysis. It aids in the prediction of continuous variables and uses a mapping function to convert input variables (X) into continuous output variables (Y), such as the prediction of solar irradiation [7, 8], to produce better forecasting results shown in Fig. 3.

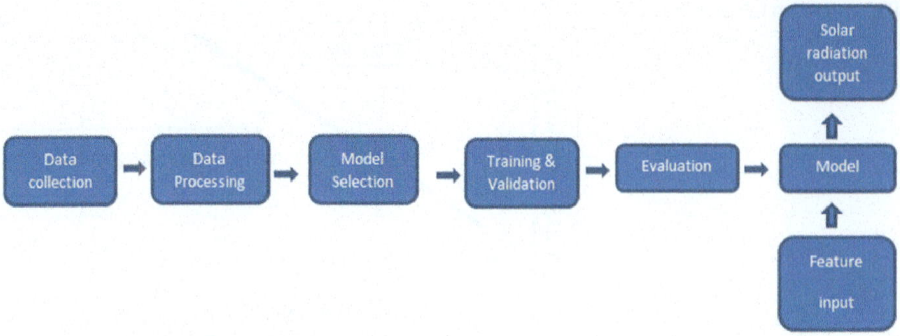


**Fig. 3** Proposed methodology for prediction

## 2.1 Multiple Linear Regression

Simple linear regression is a straightforward relationship. A regression can successfully represent the link between two variables. The multiple linear regression model makes use of more than one independent variable and is a statistical approach used to examine a dataset where several independent variables have an impact on the dependent variable. This is frequently true when the forecasting has a more intricate relationship. This method, which is used for solar forecasting, incorporates a variety of independent variables, including temperature, wind speed, wind direction, and humidity. To evaluate how well the model fits the regression line the R-squared value is determined by the variance of the dependent variable, which is calculated as the divergence from the line. Further, the detailed methodology can be obtained from Hirose et al. [9].

$$Y = \beta_0 + \beta_1 X_{i1} + \beta_2 X_{i2} + \beta_3 X_{i3} \cdots + \beta_k X_{ik} \tag{1}$$

The mathematical modeling of multiple linear regression is represented in Eq. 1. Where $X_{i1}, X_{i2}, X_{i3} \ldots\ldots X_{ik}$ are predictor variables and their corresponding coefficients $\beta_0, \beta_1, \beta_2, \ldots \beta_k$.

## 2.2 Random Forest Regression

This method, RFR, is an ensemble-learning technique that may be applied to both regression and classification. But with RFR, both classification and regression trees are established using distinct trees. In each decision tree, this input is routed via a number of classification and regression decision nodes. Two alternate routes branch off from each node, each of which leads to a different node. An output value is tied to the designated inputs after the method reaches an end node. The performance of a single decision tree can be improved by training many regression trees with

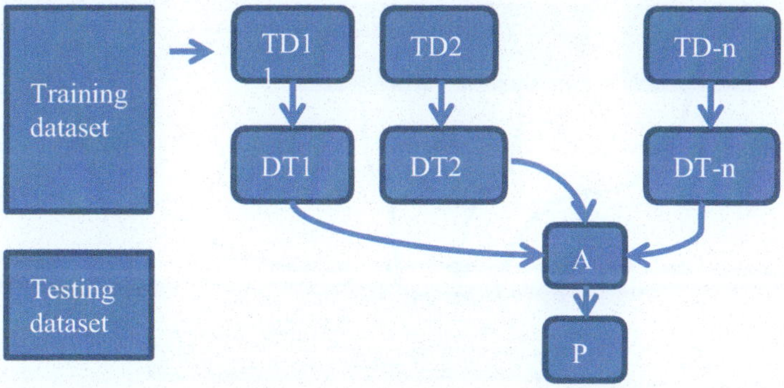


**Fig. 4** Methodology of the RFR method

various topologies and then averaging their predictions. RFR involves randomly choosing features at each node for an additional variation of decision tree models. Although the structure and feature choices of individual decision trees may cause bias, an overall RFR-averaged decision tree considerably lowers error biasing in the outcome prediction.

The RFR approach provides a completely distinct estimate validity of model interpretation inside recognized machine learning-to-know methodologies. The fundamentals of implementing a decision tree approach such as RFR are because it incorporates random sampling and advanced precedence of techniques in the ensemble methodology, the RFR technique provides more generality and valid estimations. A detailed study of random forests is done in Srivastava et al. [10] and Sasirekha et al. [11].

The block diagram of random forest regression is shown in Fig. 4. Here, the training dataset is split into n datasets (TD1-TDn) in order to obtain n decision trees (DT1-DTn). A is referred to as the average value and P is termed as prediction.

## 2.3 *Gradient Boosting Regression*

Gradient Boosting Machine is a high-powered ensemble training approach that is often used in order to solve regression and classification problems. It amalgamates a group of weak forecasting models like decision trees in order to generate a final forecasting model that can make predictions very accurately and precisely. The Gradient Boosting Machine algorithm utilizes gradient-boosted trees in a case where the decision trees perform very low. This algorithm minimizes the loss function of the model with the help of the introduction of gradient descent. This gradient descent is optimization process, which is of first order and its task is to find out the local minima of a function that is differentiable. One of the numerous implementations of Gradient Boosting Machine is regression and multiclass classification as it can minimize any

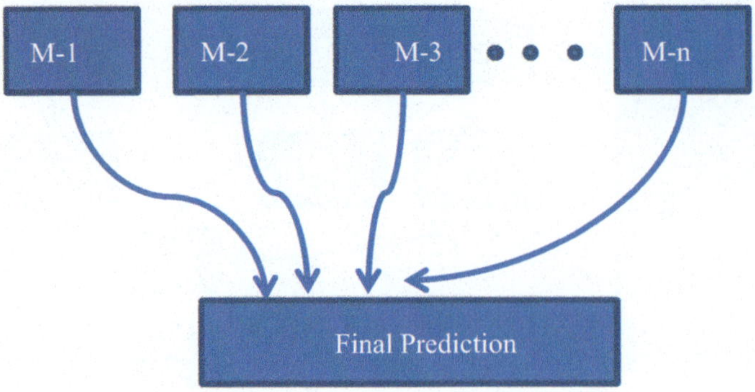


**Fig. 5** Block diagram of GBR method

type of loss function according to the requirements. In a case in order to strengthen the weak learners are trained on the remaining residual mistakes of the good learner, and the sample distribution is not affected by the GBM. An alternative way is to provide misclassified observations with extra weight with the help of training the residuals of the model. The contribution of each weak learner for the final forecasted datapoint is constructed on the gradient optimization technique which later on minimizes the gross error of the good learner. The flexibility, accuracy, and robustness of noisy data are some of the features that provide GBM an edge over the other machine learning algorithms. A detailed study of GBR is done in Aksoy and Genc [12] and Singh and Singh [13].

The block diagram of GBR is shown in Fig. 5. Here, the training dataset is split into n datasets and uses a different method (M−1, M−2, M−3,...M−n) like Decision Tree, Random Forest, etc. The final prediction consists of the average of all the above-mentioned methods.

## 3 Methodology

Preparing Dataset: The dataset used here is the HI-SEAS weather station of four months that is obtained from Kaggle. It consists of 10 columns. The given dataset has nine features as input like Date-Time, Temperature, Pressure, Humidity, Wind direction, Wind Speed, Sunrise-Sunset Time, and the output is solar radiation. In this paper, different machine learning algorithms like Multiple Linear Regression (MLR), Random Forest Regression (RFR), and Gradient Boosting Regression (GBR) have been used (Table 1).

The dataset contains solar radiation and other feature inputs from September 29, 2016 to December 01, 2016. The solar radiation data are plotted as shown in Fig. 6. These data have a total of 32686 data points.

**Table 1** Feature and labels used in the dataset

| Dataset variables | Units |
|---|---|
| Solar radiation | Watts/m$^2$ |
| Temperature | °F |
| Wind speed | m/s |
| Wind direction | Miles/hr |
| Humidity | Percentage |
| Pressure | Mm of Hg |

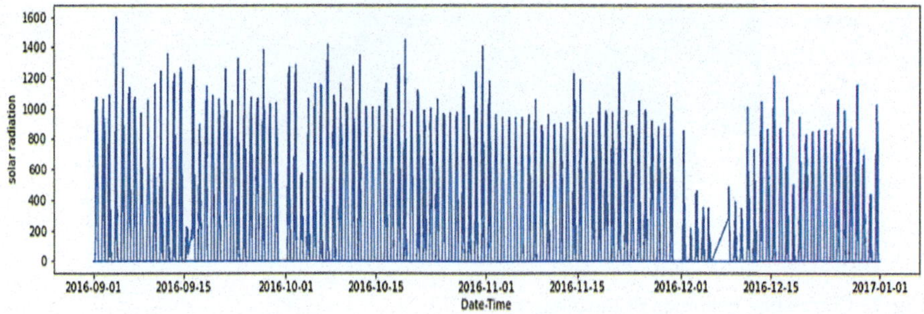


**Fig. 6** Solar radiation data from September 29, 2016–December 01, 2016

Creating an accurate prediction model using machine learning techniques is based on the dataset, which is more correlated to the output. The dataset correlation is shown in Fig. 7.

The histogram displays changes in humidity, temperature, pressure, and radiation depending on the day of the week and the season. The figure shows that solar irradiance reaches its maximum between 11 a.m. and 1 p.m. when the temperature is at its peak. Additionally, the humidity is higher at night and in the rainy season. Also, pressure varies seasonally; in the summer, it is low, and in the winter, it is high as shown in Fig. 8.

## *3.1 Step of Prediction of Solar Irradiation*

1. The dataset is categorized into training datasets and testing datasets. The training and testing datasets contain 70 and 30% of the input dataset respectively.
2. The training dataset contains wind speed, wind direction, temperature, humidity, barometric pressure, sunrise, and sunset time. The input variables and output variable (solar radiation) data are the responsible variables.
3. The Unix Time feature in the dataset is not relevant as it can be obtained using the Date and Time and because of the above-mentioned reason this feature is dropped. The dependency of the radiation is majorly on the time of the day

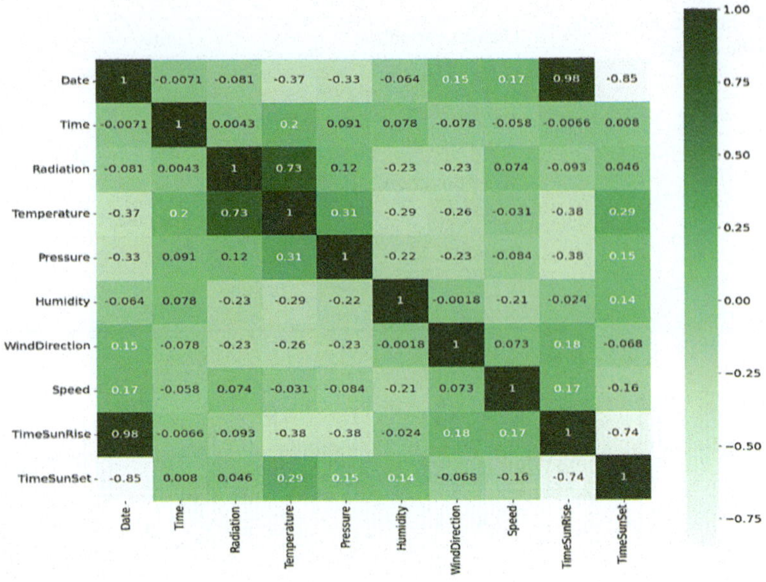


**Fig. 7** Correlation heat-map of the dataset

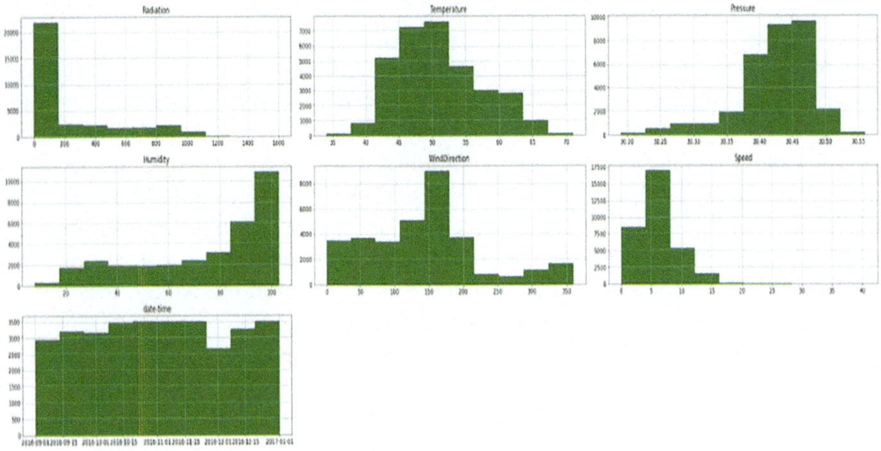

**Fig. 8** Histogram of the dataset

which is shown in the scatter plot. In the early morning, the radiation is very low, which gradually increases and attains a peak around noon which again starts decreasing as the day passes by. The correlation of radiation with temperature is very high as shown by the correlation matrix.

4. The correlation values of radiation with the time of sunrise and sunset are very low as shown in the correlation matrix. Therefore, these features are dropped.

**Table 2** Performance comparison of different regression approaches

| METHOD | R-squared | MAE | MSE | RMSE |
|---|---|---|---|---|
| MLR | 0.6230 | 148.250 | 38419.51 | 196.00 |
| RFR | 0.8784 | 39.5474 | 12384.42 | 111.28 |
| GBR | 0.9376 | 31.3427 | 6357.31 | 79.73 |

5. All the models are trained using the dataset. Different models are developed using methods such as Multiple Linear Regression, Random Forest, and Gradient Boosting Regression. The predicted data are evaluated with the test data by determining the MAE, MSE, RMSE, and R-squared for each observed regression model. The model having the lowest MAE, MSE, RMSE, and R-squared close to 1 is selected, indicating that it is the most accurate forecasting model. The results are shown in Table. 2.

## *3.2 Evaluate Matrix*

1. **Mean absolute error (MAE)**. The difference between the predicted value and the base value of solar radiation is used to calculate the absolute error value. Furthermore, averaging the error values for each sample in the data collection is done.

$$\mathrm{MAE} = \frac{1}{M}\sum \left|y_i - \hat{y}_p\right| \tag{2}$$

2. **Mean square error (MSE)**. This is determined by averaging the square of the difference between the based and predicted values of solar radiation.

$$\mathrm{MSE} = \frac{1}{M}\sum \left(y_i - \hat{y}_p\right)^2 \tag{3}$$

3. **Root mean square error (RMSE)**. It simply takes the square root of MSE.

$$\mathrm{RMSE} = \sqrt{\frac{1}{M}\sum \left(y_i - \hat{y}_p\right)^2} \tag{4}$$

   $M$ = total sample size of the dataset.

   $y_i$ = base value of the solar radiation value.

   $y_p$ = predicted value of the solar radiation value.
4. **R-squared**. R-square shows how well the model fits the line and is determined by the variance of the dependent variable, which is calculated as the deviation

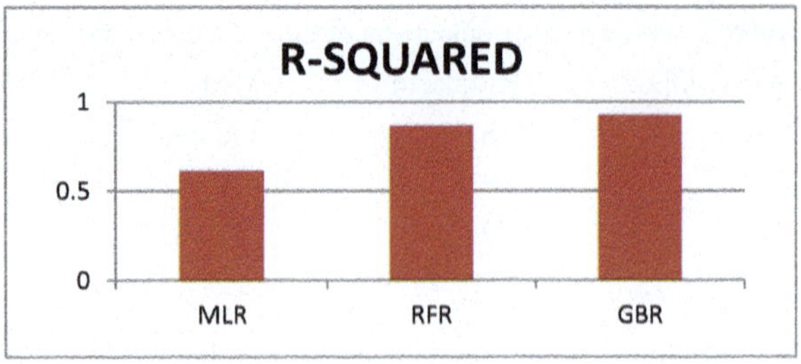


**Fig. 9** Graphical representation of R-squared of MLR, RFR, GBR

from the regression line.

$$R^2 = \left(1 - \frac{SSR}{SST}\right) \tag{5}$$

$SSR$ = sum of the square of residual (error).

$SST$ = sum of the square of the total data size.

## 4 Result

A detailed comparison of the above-mentioned algorithms is done with the results from Singh and Singh [14] and Nayak et al. [15]. While comparing the metrics scores of the above-mentioned models with the metrics scores from Singh and Singh [14], it is observed that all the metrics scores of the above-mentioned models are better than that of Singh and Singh [14]. Furthermore, comparing the models elaborated in this paper with the models developed in Nayak et al. [15], it is observed that the R-squared score of the Gradient Boost Regression models of this paper is 0.93, whereas, the R-squared score of the Gradient Boost Regression models used in Nayak et al. [15] is 0.86. From this, it can be commented that the Gradient Boost Regression model has high accuracy as compared to that of Nayak et al. [15]. The dataset used in Nayak et al. [15] is the same which is used in this paper (Fig. 9).

The appropriate parameters are set in the models as mentioned above and different metrics value of each model is shown in Table 2. The output shown in Fig. 10 is obtained from the GBR model.

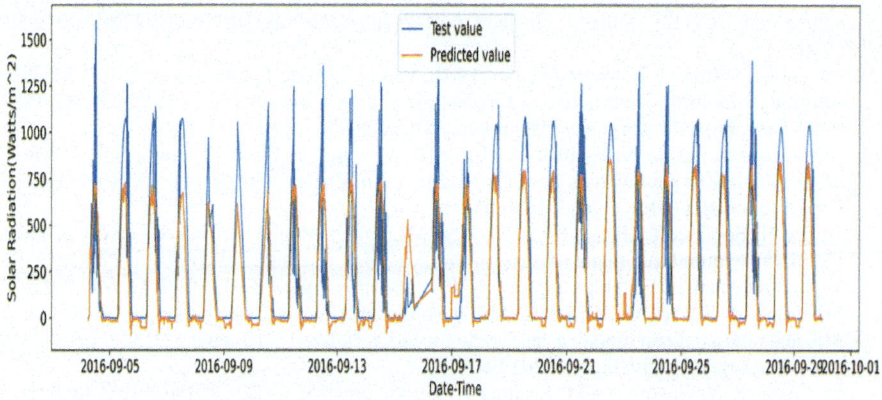


**Fig. 10** Comparison between the predicted value and test value of the GBR method

# 5 Conclusion

The analysis of solar radiation considers primarily temperature, humidity, and pressure as labels. The MLR approach appears to be the least effective of the ones used in this paper, with an R-squared value of 0.67 and MAE is 148.250. It is clear that approaches having higher R-squared value such as GBR and RFR perform better than the rest. Analyzing the developed models furthermore, it is evident that the GBR model outperformed the rest. Concluding this paper, it is perceivable that regression methods like GBR and RFR can be utilized to predict solar radiation.

# References

1. Patel MD, Shah S, Sinha S, Gupta RK, Bharti SK, Singh VP (2021) Solar irradiation forecasting-comparative analysis of various methods. In: 2021 IEEE international conference on technology, research, and innovation for betterment of society (TRIBES), Raipur, India, pp 1–9. https://doi.org/10.1109/TRIBES52498.2021.9751626
2. Choi S, Hur J (2020) An ensemble learner-based bagging model using past output data for photovoltaic forecasting. Energies 13:1438. https://doi.org/10.3390/en13061438
3. Srivastava R, Tiwari AN, Giri V (2019) Solar radiation forecasting using MARS, CART, M5, and random forest model: a case study for India. Heliyon 5:e02692. https://doi.org/10.1016/j.heliyon.2019.e02692
4. Bacher P, Madsen H, Nielsen HH (2011) Online short-term solar power forecasting. https://doi.org/10.1016/j.solener.2009.05.016
5. Naik SS, Priyanka P, Behera S (2022) Comparison of machine learning methods for short-term solar irradiation forecast. In: 2022 2nd Odisha international conference on electrical power engineering, communication and computing technology (ODICON), Bhubaneswar, India, pp 1–5. https://doi.org/10.1109/ODICON54453.2022.10009941
6. Lukhyswara P, Putranto LM, Ariananda DD (2019) Solar irradiation forecasting uses time series analysis. In: 2019 11th International conference on information technology and electrical

engineering (ICITEE), Pattaya, Thailand, pp 1–6. https://doi.org/10.1109/ICITEED.2019.8929990
7. Voyant C, Notton G, Kalogirou S, Nivet ML, Paoli C, Motte F, Fouilloy A (2017) Machine learning methods for solar radiation forecasting: a review. Renew Energy 105:569–582. ISSN 0960-1481. https://doi.org/10.1016/j.renene.2016.12.095
8. Antonanzas J, Osorio N, Escobar R, Urraca R, Martinez-de-Pison FJ, Antonanzas-Torres F (2016) Review of photovoltaic power forecasting. Sol Energy 136:78–111. ISSN 0038-092X, https://doi.org/10.1016/j.solener.2016.06.069
9. Hirose H, Soejima Y, Hirose K (2012) NNRMLR: a combined method of nearest neighbor regression and multiple linear regression. In: 2012 IIAI international conference on advanced applied informatics, Fukuoka, Japan, pp 351–356. https://doi.org/10.1109/IIAI-AAI.2012.76
10. Srivastava R, Tiwari AN, Giri VK (2019) Solar radiation forecasting using MARS, CART, M5, and random forest model: a case study for India. Heliyon 5(10):e02692. ISSN 2405-8440. https://doi.org/10.1016/j.heliyon.2019.e02692
11. Sasirekha P, Navinkumar TM, Swapna P, Vinothkumar M (2022) Comparative analysis of prediction on solar radiation in energy generation system using random forest and decision tree. In: 2022 International conference on sustainable computing and data communication systems (ICSCDS), Erode, India, pp 899–903. https://doi.org/10.1109/ICSCDS53736.2022.9760819
12. Aksoy N, Genc I (2023) Predictive models development using gradient boosting based methods for solar power plants. J Comput Sci 67:101958. ISSN 1877-7503. https://doi.org/10.1016/j.jocs.2023.101958
13. Persson C, Bacher P, Shiga T, Madsen H (2017) Multi-site solar power forecasting using gradient boosted regression trees. Sol Energy 150:423–436. ISSN 0038-092X. https://doi.org/10.1016/j.solener.2017.04.066
14. Singh Y, Singh A (2022) Forecasting solar radiation by the machine learning algorithm and their different techniques. Int J Res Appl Sci Eng Technol 10. ISSN 2321-9653. https://doi.org/10.22214/ijraset.2022.47345
15. Nayak P, Dash A, Chintawar S (2022) Multi-level statistical model for forecasting solar radiation. In: 2022 International conference on innovative trends in information technology (ICITIIT), Kottayam, India, pp 1–6. https://doi.org/10.1109/ICITIIT54346.2022.9744207

# Forecasting of Solar Power Generation Using Hybrid Empirical Mode Decomposition and Adaptive Neuro-Fuzzy Inference System

**Prashant Singh, Durgesh Kumar, Navneet Kumar Singh, Niraj Kumar Choudhary, and Asheesh Kumar Singh**

**Abstract** Excellent learning capabilities of Artificial Neural Network (ANN) made it widely explored and practiced to predict future points of a time series. Similarly, Fuzzy Logic (FL) has proved its capability in dealing with non-linear and complex models. The combination of these is not widely experimented. This paper not only aims to construct hybrid forecasting models using integration of Adaptive Neuro-Fuzzy Inference System (ANFIS) and Empirical Mode Decomposition (EMD) to forecast solar power generation but also discovers the effects of two different types of fuzzy Membership Functions (MFs), i.e., G-bell and Trapezoidal membership functions on forecasting capability of models. All proposed models were evaluated using two performance measures. Based on the forecasting results obtained, it is observed that the hybrid EMDANFIS model is outperforming when compared with the original ANN and ANFIS models, individually. The empirical results indicate that the models with trapezoidal MF (level-3) and G-bell shaped MF (level-1) are better models, providing the highest level of accuracy.

**Keywords** Photovoltaic (PV) · Fuzzy logic (FL) · Adaptive Neuro-Fuzzy Inference System (ANFIS) · Empirical Mode Decomposition (EMD)

---

P. Singh (✉) · D. Kumar · N. K. Singh · N. K. Choudhary · A. K. Singh
Motilal Nehru National Institute of Technology, Allahabad (Prayagraj), UP, India
e-mail: prashantmits@gmail.com

N. K. Singh
e-mail: navneet@mnnit.ac.in

N. K. Choudhary
e-mail: niraj@mnnit.ac.in

A. K. Singh
e-mail: asheesh@mnnit.ac.in



O. H. Gupta et al. (eds.), *Soft Computing Applications in Modern Power and Energy Systems*, Lecture Notes in Electrical Engineering 1107,
https://doi.org/10.1007/978-981-99-8007-9_24

# 1 Introduction

Electricity has become a very important part of our lives today. With the growth of population, demand for all the resources is increasing. Conventional sources of power generation are limited and take a long time to regenerate hence there is a need to know the amount of electricity consumption and generation unit. It also becomes important to switch to the renewable sources of energy. Solar power has great potential to sustain this demand as compared to other renewable sources of energy.

Some of the techniques used for short-term load forecasts are time-series analysis, multiple linear regression, and advanced techniques like ANFIS, Fuzzy Logic, and its hybrid. Most of these approaches required selection of proper system parameters which is not an easy task. The proposed approach is data dependent, and it utilizes the information from the historical dataset.

Various time-series forecasting methods have been developed in the past out of which ANN models have been widely used due to their excellent learning capabilities. Fuzzy logic is often integrated with ANN to overcome problems of ANN's like overlearning.

Modeling of system implementing ANFIS and fuzzy inference system [1] found applications in several areas and turned out to be effective method in minimizing the forecasted output error. The spontaneous and uncertain nature of the parameters in solar power generation emphasizes to forecast the load for grid integrity and reliability. Heng et al. [2] focus to build an Adaptive Residual Compensation algorithm to forecast the power with high accuracy using the residual errors. Forecasting crude oil and electricity prices has become a major field of research. Salim [3] describes how the Variational Mode Decomposition outperforms Empirical Mode Decomposition (EMD)-based forecasting system. Neeraj et al. [4] suggested a hybrid EMD—Att—LSTM model for building a sustainable power system. The data from Australian Energy Market Operator was used to explore the properties to evaluate the components and reflect what trend and seasonality it contains. The multicarrier system [5] is built using ANFIS model to forecast and the energy flow is optimized using genetics algorithm taking the dynamics of the system into account. The paper [6] suggested a hybrid load forecasting model that combines a neural network, weighted least square state estimation, and an ANFIS in order to make the best use of the power generation resources that are currently available. This model overcomes several existing forecasting models. Wangwei et al. [7] proposed a novel hybrid model combining empirical mode decomposition (EMD), convolutional neural network (CNN), and LSTM is put forth to forecast stock prices. In Ismail [8] the decomposition of input data using IMF and creating appropriate ANFIS models outperformed the old models and came up as new approach to the forecasting of load. Further, the hybrid model of EMD-ANFIS was employed in the flood-prone area to depict water levels and provide insights to help in flood control [9].

This work aims to develop hybrid forecasting models by integrating empirical mode decomposition with adaptive neuro-fuzzy inference system for forecasting solar power generation.

Some past works have also used EMD for decomposing the input values, but in this research work, solar power generation data have been decomposed in order to train the ANFIS models. The most effective result obtained using hybrid EMDANFIS model by tuning the hyper-parameter i.e., the number of neurons, membership function and types of MF, and level of decomposition (EMDANFIS). The whole process of this work is divided into three parts, which are as follows:

- To develop ANN, ANFIS, and EMDANFIS-based models.
- To evaluate the forecasting performance of these models by analyzing various performance parameters for the forecast of solar power generation.
- To evaluate the effect of type of membership functions, and level of decomposition on the performance of the proposed hybrid EMDANFIS model.

The rest parts of this paper have been organized as: Sect. 2 covers the theoretical aspects like methodology for forecasting models and data pre-processing, and Sect. 3 introduces the proposed framework of the hybrid forecasting models, such as EMDANFIS, and finally, Sects. 4 and 5 explain the results, discussions, and conclusions with scope for future research, respectively.

## 2 Data Description/Case Study

The training and testing data used in this report have been taken from European Network of Transmission System Operators for Electricity (ENTSO-E), Greece, Southeast Europe [10]. Dataset has hourly records from January 01, 2018 to December 31, 2019 with 24 entries per day. A total of 17520 entries as raw data are available from this dataset. Useful data fields obtained from this dataset are:

- Solar power generated
- Temperature
- Direct horizontal radiation
- Diffuse horizontal radiation
- Date and Hour of the day

### *2.1 Data Cleaning*

In process of data cleaning, firstly, the night-time records were removed from the data series, for solar power estimation, night-time data always have power as well as radiation values of zero and are therefore useless for training the model. Data entries between 6:00 pm to 4:00 am are removed, and 10220 entries are obtained from raw data for further use. In next step if any record has value that is not a number (NaN) and that row of record will also be removed to prevent errors in model training phase.

### 2.2 Data Normalization

Data normalization is done to give equal weightage to all values of input irrespective of their magnitude. It makes the values uniform and dimensionless, Specially, if an algorithm uses Euclidean distance, a variable with a large value can dominate distance measures over the other values.

In this method, all values are scaled down between 0 and 1. These new values are given by:

$$\overline{K} = \frac{k - k_{\min}}{k_{\max} - k_{\min}} \tag{1}$$

where $k$ is the input value, $k_{\min}$ and $k_{\max}$ refer to the minimum and maximum of the input value, respectively.

### 2.3 Feature Selection

In this step, important input variables are determined, which should be considered while training the model. This helps in reducing number of input variables, which in turn helps to decrease complexity and training time of model. The inputs that affect the output more strongly are selected and kept for further steps. As past values of metrological inputs must be used, these inputs (ambient temperature, diffuse radiation, direct radiation) are shifted up to 24 h to generate predictor variables, and the variables with best correlation are selected from each input.

By observing the correlation plot from Table 1, variables T(t-24), $D_f$(t-1), $D_r$(t-24), and P(t-24) are chosen as past values for temperature, diffuse horizontal radiation and direct horizontal radiation, and solar power, respectively, for further steps.

## 3 Framework of Hybrid EMDANFIS Forecasting Model

### 3.1 Empirical Mode Decomposition (EMD)

A signal can be broken down into physically significant components using this data adaptive multi-resolution technique known as empirical mode decomposition (Fig. 1). By dissecting the signal into components with various levels of precision, it

**Table 1** Observed correlation plot between data fields

| Variable | P(t) | T(t-1) | T(t-24) | $D_f$(t-1) | $D_f$(t-24) | $D_f$(t-48) | $D_r$(t-1) | $D_r$(t-24) | P(t-24) | P(t-48) |
|---|---|---|---|---|---|---|---|---|---|---|
| P(t) | 1.000 | 0.431 | 0.470 | 0.393 | 0.551 | 0.605 | 0.867 | 0.846 | 0.911 | 0.876 |
| T(t-1) | 0.431 | 1.000 | 0.974 | 0.282 | 0.215 | 0.229 | 0.609 | 0.532 | 0.435 | 0.430 |
| T(t-24) | 0.470 | 0.974 | 1.000 | 0.310 | 0.249 | 0.257 | 0.597 | 0.587 | 0.492 | 0.492 |
| $D_f$(t-1) | 0.393 | 0.282 | 0.310 | 1.000 | 0.702 | 0.626 | 0.244 | 0.328 | 0.469 | 0.512 |
| $D_f$(t-24) | 0.551 | 0.215 | 0.249 | 0.702 | 1.000 | 0.781 | 0.395 | 0.245 | 0.471 | 0.553 |
| $D_f$(t-48) | 0.605 | 0.229 | 0.257 | 0.626 | 0.781 | 1.000 | 0.461 | 0.403 | 0.551 | 0.471 |
| $D_r$(t-1) | 0.867 | 0.609 | 0.597 | 0.244 | 0.395 | 0.461 | 1.000 | 0.810 | 0.794 | 0.756 |
| $D_r$(t-24) | 0.846 | 0.532 | 0.587 | 0.328 | 0.245 | 0.403 | 0.810 | 1.000 | 0.924 | 0.844 |
| P(t-24) | 0.911 | 0.435 | 0.492 | 0.469 | 0.471 | 0.551 | 0.794 | 0.924 | 1.000 | 0.911 |
| P(t-48) | 0.876 | 0.430 | 0.492 | 0.512 | 0.553 | 0.471 | 0.756 | 0.844 | 0.911 | 1.000 |

can be utilized to evaluate non-stationary and non-linear signals. It is the oscillatory portion of the signal that results from the signal's Hilbert-Huang transformation.

Intrinsic mode function (IMF) has two important characteristics:

1. The difference in the total number of maxima and minima is at most by 1.
2. The mean of the waveform of IMF is 0.

**Steps for IMF Extraction:**

Step 1: Find the all-local Minima and Maxima points in waveform.

Step 2: Create an envelope of minima and maxima using the array of minima and maxima. To create an envelope of minima and maxima, use a cubic spline for interpolating the values of the minima and maxima.

Step 3: From the envelope containing the minima and maxima, take the center value.

Step 4: Now reduce the genuine signal's value by the middle value of the envelope. On both sides of the wave, there are certain mistakes that resemble a sudden leap. This resulted from a mistake in the cubic spline's interpolation. At the start and finish of the signal, the cubic spline interpolation deteriorates significantly. A boundary condition as the solution must be specified. Before creating the cubic spline, envelope of minima and maxima at the start and end of the signal must be determined.

Step 5: Verify that the signal that was extracted is an IMF.

Because the mean is so close to zero, it may be rounded to zero and ignored. An IMF is received since the number of Minima's and Maxima's likewise meets the criterion.

Step 6: Use this IMF to lower the initial test signal.

One IMF has been extracted till this step. EMD method must finish when the residual signal is monotonic, constant, or only has one extremum (just has 1 minima or

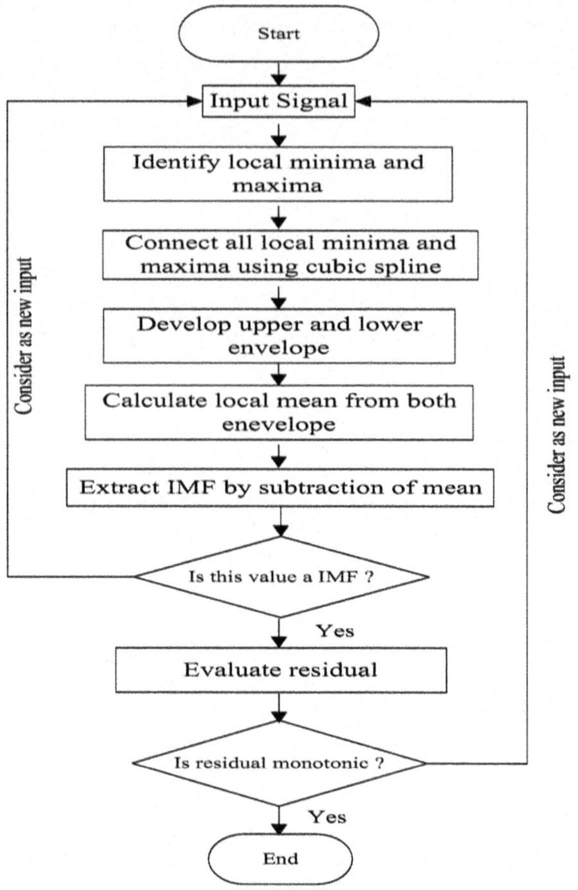


**Fig. 1** EMD algorithm

maxima). Residuum is the term that is used to describe our signal after it has been decreased by the IMF.

By following the above algorithm, the original time series $y(t)$ can be decomposed into multiple IMF signals and a residual and can be expressed as follows:

$$y(t) = \sum_{i=1}^{m} h_i(t) + r_m(t) \tag{2}$$

## 3.2 *Adaptive Neuro-Fuzzy Inference System (ANFIS)*

An Adaptive Neuro-Fuzzy Inference System (ANFIS) is a hybrid of an artificial neural network and a fuzzy inference system of the Takagi–Sugeno type [11]. It

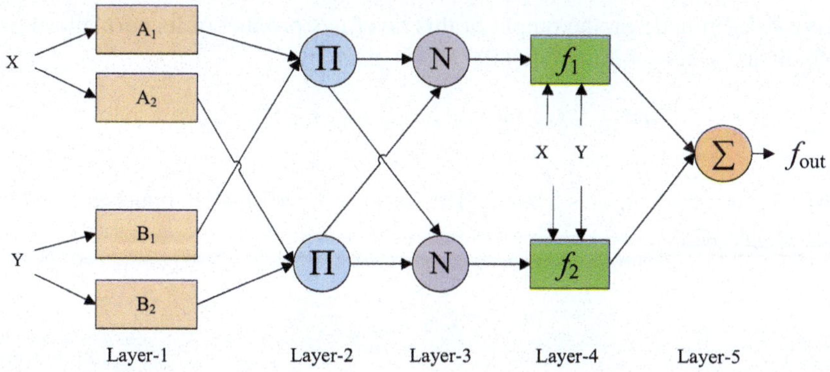


**Fig. 2** Basic architecture of ANFIS

combines neural networks and fuzzy logic rules, allowing it to capture the benefits of both methods [1]. Several decisions are needed to be taken to model the fuzzy systems, like rule bases, proper membership functions, aggregation, and implication operators. Choosing decision parameters manually so that the required fuzzy system achieves the desired performance becomes a difficult task. The more complex the application, the more difficult it is to determine appropriate design parameters. On the other hand, the ANN has the problem of overlearning and local minima. Hence, these individual models are incapable to give the best possible outcomes.

The basic architecture of ANFIS is shown in Fig. 2. The ANFIS models have mainly five layers of nodes [11]:

**Layer 1**: Each node of this layer is indicated by a square, which shows that parameters in this layer are adjustable. This layer gives a membership value between 0 and 1 after receiving input. The membership values produced are distributed in all number of linguistic variables for a given Fuzzy variable. It produces the output as

$$O_i^1 = \mu_{Ai}(X) \quad \text{for } i = 1, 2, 3 \tag{3}$$

**Layer 2**: These layer nodes are labeled with Pie. It computes the firing strength of each rule, which is the minimum of corresponding values of linguistic variables. Values are given by

$$O_i^2 = \mu_{Ai}(X) * \mu_{Bi}(X) \quad \text{for } i = 1, 2, 3 \tag{4}$$

**Layer 3**: This layer computes the normalized strength of all fired rules and use the following formula

$$O_i^3 = w_i' = \frac{w_i}{(w_1 + w_2 + w_3 \cdots\cdots + w_n)} \quad \text{for } i = 1, 2, 3 \tag{5}$$

**Layer 4**: Each node on the output of this layer ia s product of its normalized firing strength and a first-order polynomial and is given by

$$O_i^4 = w_i' \times (p_i x + q_i y + r_i) \quad \text{for } i = 1, 2, 3 \tag{6}$$

**Layer 5**: The final output is computed in this layer by adding all inputs received from the previous layer.

$$O_i^5 = f_{out} = \sum_{i=0}^{n} w_i' f_i \tag{7}$$

### 3.3 *Implementation of Hybrid EMDANFIS Network*

In this work, solar power in training dataset is decomposed into multiple IMFs. Then separate ANFIS models are initialized corresponding to each IMF. An additional parameter, number of IMFs to be extracted, is introduced in this model. Number of input membership functions is kept fixed in case of ANFIS models. Two ANFIS models were developed, one with Generalized bell MF and other with Trapezoidal MF. The training and testing process of the proposed hybrid EMDANFIS model is shown in Figs. 3 and 4. For hybrid EMDANFIS, two models were developed, with level of decompositions was chosen between 1 and 4, and the best was considered for each model.

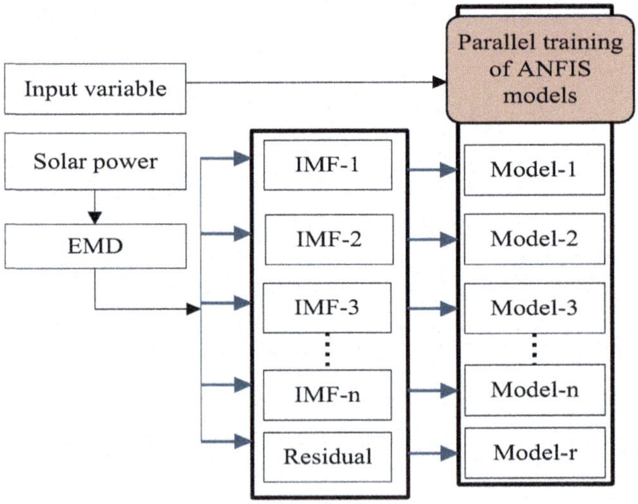


**Fig. 3** Training process of hybrid EMDANFIS model

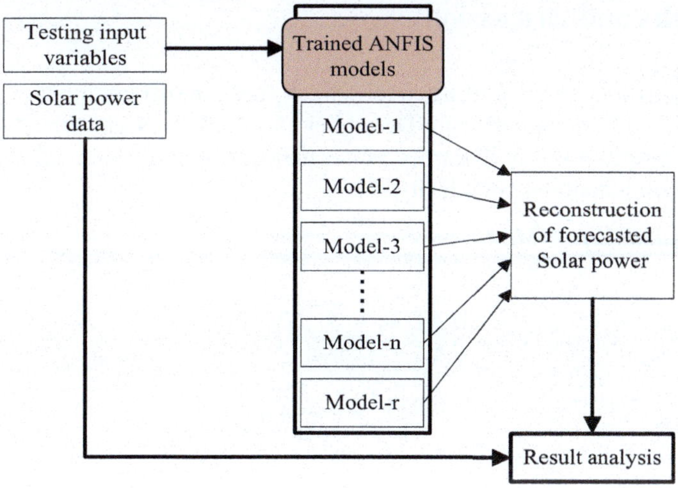


**Fig. 4** Testing process of hybrid EMDANFIS model

Once the models have been trained for these IMFs, a separated solar output power is forecasted by each model for the testing data. Finally, the forecast output is reconstructed by adding outputs of these models and this output is compared with the actual value of power for test data.

Steps involved in Hybrid model forecasting are:

- Choose type of membership function for initial fuzzy inference system (FIS) generation and number of IMfs to be extracted during Empirical decomposition.
- Decompose output of training dataset into IMFs and residuals.
- Total n + 1 ANFIS models are initialized using grid partitioning in MATLAB, where n is number of IMFs extracted.
- Use decomposed output as training target, combined with metrological inputs to train each ANFIS model parallelly.
- Once training is finished, Inputs of testing set are given to each of the trained ANFIS model, and individual forecast outputs of these models are summed to reconstruct the final forecast output.
- Forecasted output is compared with solar power output of testing dataset and models are analyzed in terms of RMSE, MAE, nominal RMSE, and nominal MAE.
- Repeat the above steps after changing type of membership function and level of decomposition during EMD until better results are obtained.

## 4 Results and Discussion

The solar power output is forecasted for 1 week ahead in two sets, one for duration of August 25–31, 2019 and other for December 25–31, 2019. Night-time data between 06:00 pm to 4:00 am were discarded as it is unnecessary for training the models and increases computational complexity.

*Mean Absolute Error (MAE)*:

$$MAE = \frac{1}{n}\sum_{\upsilon=1}^{n} |P_\upsilon - A_\upsilon| \tag{8}$$

$$nMAE = \frac{MAE}{\max - \min} \tag{9}$$

*Root Mean Square Error (RMSE)*:

$$RMSE = \sqrt{\frac{1}{n}\sum_{\upsilon=1}^{n} (P_\upsilon - A_\upsilon)^2} \tag{10}$$

$$nRMSE = \frac{RMSE}{\max - \min} \tag{11}$$

where $n$ is the testing dataset's size, max & min are data value of actual solar power, $P_\upsilon$ is the predicted value of solar power and $A_\upsilon$ is the actual solar power. Numerical results with the hybrid EMDANFIS model are shown in Figs. 5a–f, respectively, for the mentioned dates in August and December month, indicating better performance of hybrid models. Similarly, from Table 2, it is observed that EA-1 model with Generalized bell MF and level 1 of Empirical decomposition performs better in terms of average MAE while in case of RMSE, EA-2 model with Trapezoidal MF and level 3 of Empirical decomposition performs better. Other models such as ANFIS and ANN give less accurate forecast than these two hybrid models.

## 5 Conclusions

In this research, forecasting performance of two conventional time-series forecasting models, i.e., ANNs and ANFIS, was developed and analyzed. The forecasting results obtained from these conventional models are compared with the results obtained from proposed hybrid models, i.e., integrated empirical mode decomposition and adaptive neuro-fuzzy inference system model. It is observed that the forecasting accuracy of the proposed hybrid models (EMD-based ANFIS) is better than conventional ANN,

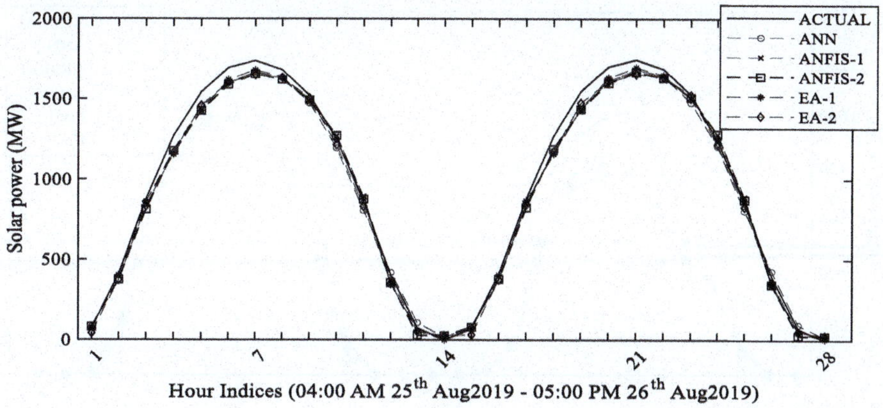


(a)

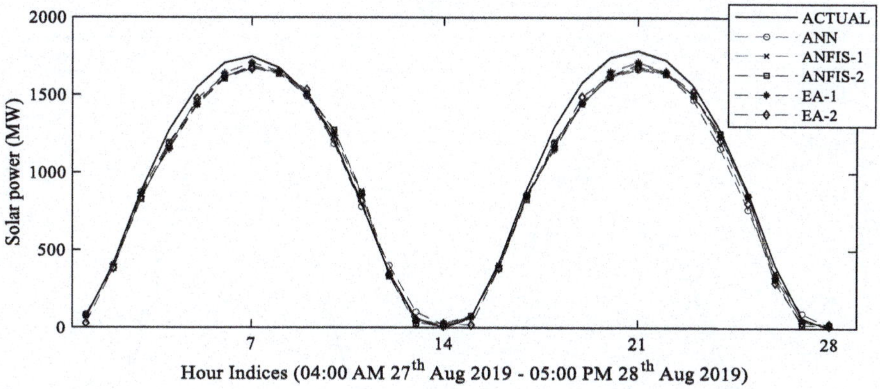


(b)

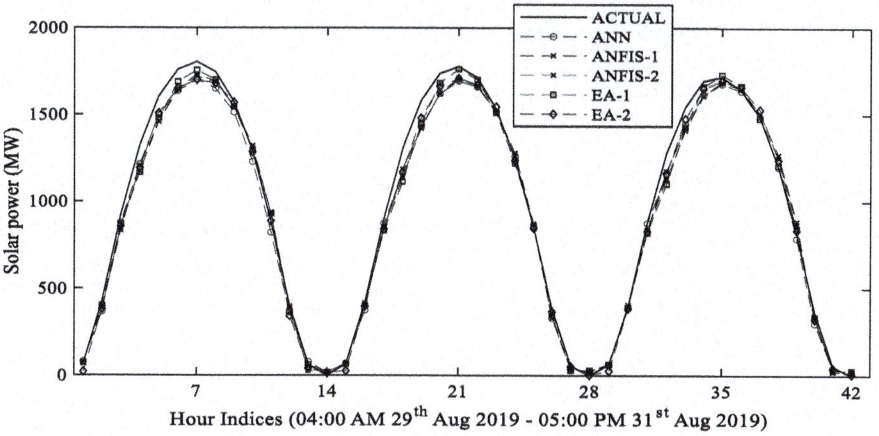


(c)

**Fig. 5** Actual and predicted solar power for **a** two tailing days, August 25–26, **b** two tailing days, August 27–28, **c** three tailing days, August 29–31, **d** two tailing days, December 25–26, **e** two tailing days, December 27–28, and **f** three tailing days, December 29–31

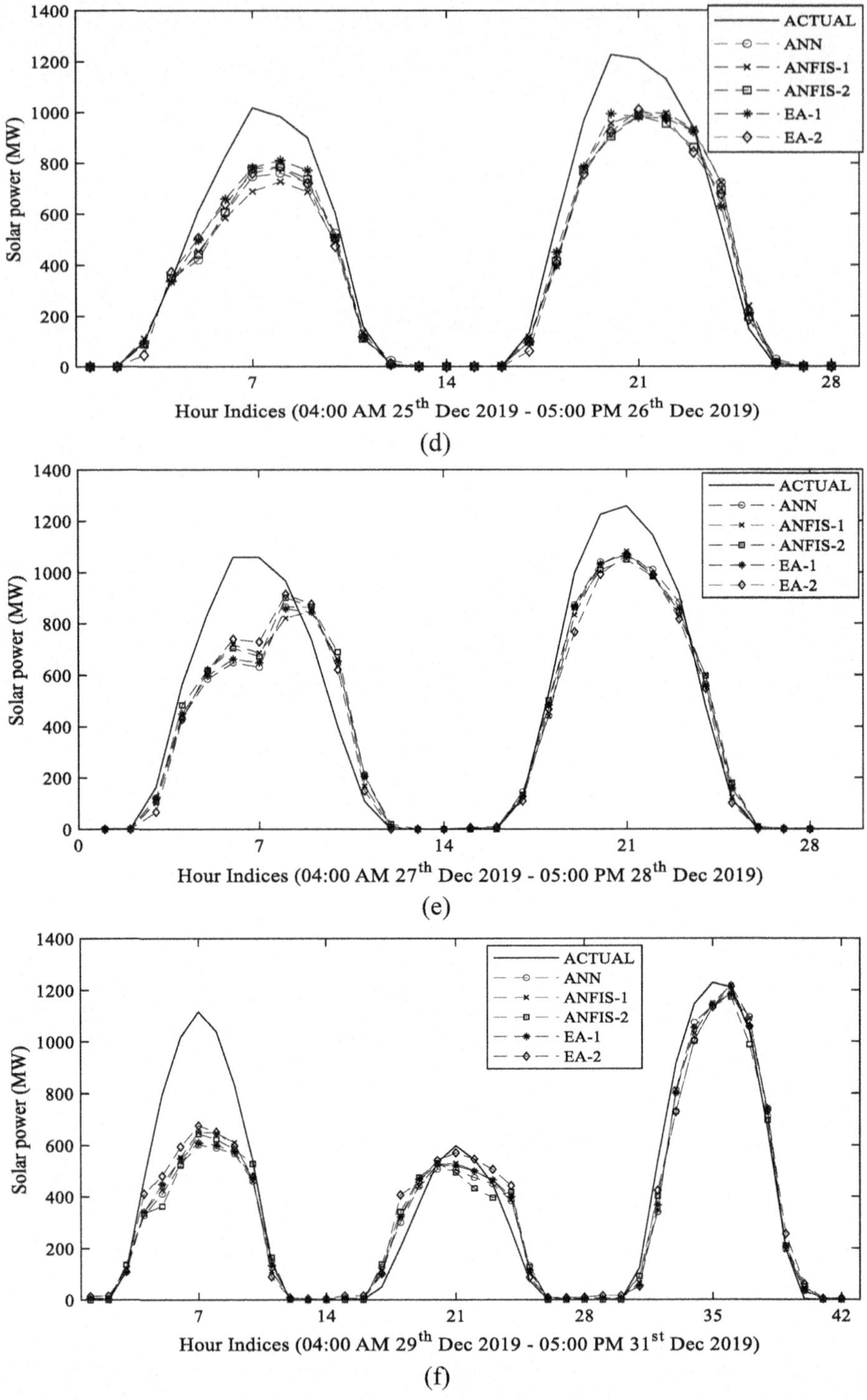


**Fig. 5** (continued)

**Table 2** Forecasting efficiency of different forecasting models

Results of forecasting models

| Date | ErMs | ANN | ANFIS-1 | ANFIS-2 | EA-1 | EA-2 |
|---|---|---|---|---|---|---|
| 25–31 August | MAE | 52.77099 | 48.96406 | 54.84331 | **43.97443** | 49.12973 |
| | RMSE | 65.59518 | 65.90712 | 66.4653 | 61.6221 | **60.19533** |
| | nMAE | 0.029268 | 0.027157 | 0.030418 | **0.02439** | 0.027249 |
| | nRMSE | 0.036381 | 0.036554 | 0.036864 | 0.034178 | **0.033386** |
| 25–31 December | MAE | 92.91436 | 92.21596 | 89.33233 | **85.58546** | 89.44086 |
| | RMSE | 149.8606 | 152.6904 | 142.2396 | 141.5565 | **138.5497** |
| | nMAE | 0.073859 | 0.073304 | 0.071011 | **0.068033** | 0.071098 |
| | nRMSE | 0.119126 | 0.121375 | 0.113068 | 0.112525 | **0.110135** |
| Mean | nRMSE | 0.077754 | 0.078965 | 0.074966 | 0.073351 | **0.071761** |
| | nMAE | 0.051564 | 0.05023 | 0.050715 | **0.046211** | 0.049173 |

[*] ErMs = Error Measures, EA = EMDANFIS

ANFIS-1, and ANFIS-2 models, when evaluated over performance measures like RMSE and MAE.

**Acknowledgements** The authors are interested in expressing their gratitude to the European Network of Transmission System Operators for Electricity (ENTSO-E), Greece, for providing the data on solar photovoltaic power.

## References

1. Jang JSR (1993) ANFIS: adaptive network based fuzzy inference system. IEEE Trans Syst Man Cybern 23(3):665–685
2. Su H-Y, Liu T-Y, Hong H-H (2020) Adaptive residual compensation ensemble models for improving solar energy generation forecasting. IEEE Trans Sustain Energy 11(2):1103–1105
3. Lahmiri S (2017) Comparing variational and empirical mode decomposition in forecasting day-ahead energy prices. IEEE Syst J 11(3):1907–1910
4. Mathew NJ, Behera RK (2022) EMD-Att-LSTM: a data-driven strategy combined with deep learning for short-term load forecasting. J Mod Power Syst Clean Energy 10(5):1229–1240
5. Kampouropoulos K, Andrade F, Sala E, Espinosa AG, Romeral L (2018) Multiobjective optimization of multi-carrier energy system using a combination of ANFIS and genetic algorithms. IEEE Trans Smart Grid 9(3):2276–2283
6. Ali M, Adnan M, Tariq M, Poor HV (2021) Load forecasting through estimated parametrized based fuzzy inference system in smart grids. IEEE Trans Fuzzy Syst 29(1):156–165
7. Shu W, Gao Q (2020) Forecasting stock price based on frequency components by EMD and neural networks. IEEE Access 8:206388–206395
8. Ismail Z (2013) A new approach to peak load forecasting based on EMD and ANFIS. Indian J Sci Technol 6
9. Loh EC, Ismail SB, Khamis A, Empirical mode decomposition couple with artificial neural network for water level prediction. Civ Eng Arch 7:19–32. https://doi.org/10.13189/cea.2019

10. https://doi.org/10.25832/time_series/2020-10-06
11. El-Hasnony IM, Barakat SI, Mostafa RR (2020) Optimized ANFIS model using hybrid meta-heuristic algorithms for Parkinson's disease prediction in IoT environment. IEEE Access 8:119252–119270

# GPSR Compliance

*The European Union's (EU) General Product Safety Regulation (GPSR) is a set of rules that requires consumer products to be safe and our obligations to ensure this.*

*If you have any concerns about our products, you can contact us on ProductSafety@springernature.com*

In case Publisher is established outside the EU, the EU authorized representative is:

Springer Nature Customer Service Center GmbH
Europaplatz 3
69115 Heidelberg, Germany

The manufacturer's authorised representative in the EU is Springer Nature Customer Service Centre GmbH, Europaplatz 3, 69115 Heidelberg, Germany. If you have any concerns regarding our products, please contact ProductSafety@springernature.com

Printed and bound by CPI Group (UK) Ltd, Croydon, CR0 4YY
01/10/2026
02225860-0011